SINGLE-MOLECULE STUDIES OF NUCLEIC ACIDS AND THEIR PROTEINS

Single-Molecule Studies of Nucleic Acids and Their Proteins

David Bensimon
LPS-ENS, PSL, CNRS, France and Department of Chemistry and Biochemistry, UCLA, USA

Vincent Croquette and
Jean-François Allemand
LPS-ENS, PSL, CNRS, France

Xavier Michalet
Department of Chemistry and Biochemistry, UCLA, USA

Terence Strick
IBENS, PSL, CNRS, France

OXFORD
UNIVERSITY PRESS

OXFORD
UNIVERSITY PRESS

Great Clarendon Street, Oxford, OX2 6DP,
United Kingdom

Oxford University Press is a department of the University of Oxford.
It furthers the University's objective of excellence in research, scholarship,
and education by publishing worldwide. Oxford is a registered trade mark of
Oxford University Press in the UK and in certain other countries

First Edition published in 2019
Impression: 1

Published in the United States of America by Oxford University Press
198 Madison Avenue, New York, NY 10016, United States of America

British Library Cataloguing in Publication Data
Data available

Library of Congress Control Number: 2017962121

ISBN 978–0–19–853092–3

DOI: 10.1093/oso/9780198530923.001.0001

Printed and bound by
CPI Group (UK) Ltd, Croydon, CR0 4YY

This work is dedicated to our spouses for their patience and support.

Preface

Biophysics and biochemistry are undergoing a revolution with the development of new tools and techniques to manipulate and observe single molecules and their interactions. In this monograph, our purpose is not to present a comprehensive review of this vast and rapidly expanding field, but rather to introduce these novel techniques and exemplify their use in the study of nucleic acids and their interactions with proteins. In other words, our goal is to provide an introduction to the main techniques which are used to study single molecules, both via their manipulation and visualization, to allow the reader to understand the concepts behind these methods.

We discuss in detail the mechanical properties of nucleic acids, where a comprehensive understanding was achieved thanks to the use of single-molecule manipulation approaches. This agreement between theoretical models and observations has set the stage for the single-molecule study of the interaction between nucleic acids (DNA or RNA) and their associated proteins.

While we will not review this quickly expanding area, we will exemplify the use of single-molecule manipulation and visualization methods in the study of some specific DNA motors: DNA and RNA polymerases, helicases, and topoisomerases. Our purpose is to show that these techniques can open up new vistas on the understanding of the mechanism of molecular motors and be used to address a variety of questions pertaining to their rate, processivity, step size, efficiency, stoichiometry, etc.

We acknowledge that in the choice of the examples presented here, we have been biased by our own investigations and apologize to our colleagues for perhaps not giving proper exposure to their work. We believe that a comprehensive review would have been impossible at that stage.

Acknowledgements

The work presented here would have been impossible without the help of our many friends, students, post-docs, and colleagues, and in particular: L. Jullien, S. Weiss, F. Ding, M.N. Dessinges, T. Lionnet, C. Gosse, G. Lia, E. Praly, F. Mosconi, A. Meglio, E. Cavatore, M. Peliti, B. Maier, O. Saleh, K. Neuman, N. Dekker, M. Manosas, R. Lavery, T. Owen-Hughes, L. Finzi, M. Duguet, N. Cozzarelli, N. Crisona, R. Ebright, A. Volgodskii, S. Benkovic, F.X. Barre, and E. LeCam. We thank C. Bagshaw for a critical reading of this manuscript and making many suggestions. We are also grateful for the support of our institutions: CNRS, PSL, ENS, Univ. Paris 6 and 7, UCLA, and our granting agencies: ANR, PUF, INSERM, EU (via FET-Open, ERC, Marie-Curie, and EURYI grants), HFSP, NSF and NIH.

Contents

1

Introduction to DNA

1.1 Introduction

When, in 1953, Watson and Crick suggested that deoxyribonucleic acid (DNA) adopted a double-helical structure [1], the molecule was already known to be the support of genetic heredity [2, 3]. However, the discovery that DNA consisted of two complementary polymers (two strands) winding around each other significantly changed the way that we thought about cellular processes, such as DNA replication [4]. It provided a much needed molecular and structural basis for DNA replication and transcription. It became clearer that the study of molecular interactions was a necessary step in understanding their cellular function. In the thirty years following the discovery of the double helix, numerous techniques have emerged to advance this study, which by now constitute the bulk of "Molecular Biology". These techniques allow for the transforming, synthesizing, and sequencing of DNA molecules as well as for studying and quantifying the interactions between biomolecules (such as proteins and DNA). A landmark in this genomic era was set when the human genome sequencing programme reached its goal less than fifty years after Watson and Crick's discovery.

The decade beginning around 1990 has witnessed the emergence of a wealth of new techniques and tools devoted to the physical study of single molecules. Methods as diverse as optical, acoustic, and magnetic tweezers, microfibres, and atomic force microscopy are used in many labs to manipulate (move, stretch, or twist) single biomolecules (DNA, proteins, carbohydrates, etc.). In parallel, optical methods based on fluorescence (by resonant energy transfer or directly with evanescent waves, two photons, or confocal configurations) have also been developed to study biochemical processes at the single-molecule level. Many groups are combining both aspects to monitor or affect the activity and visualize the displacement of a single molecule (myosin, RNA polymerase, etc.) *in vitro* and *in vivo*.

From a physicist's point of view, such experiments have provided new data against which to test models of polymer elasticity, with or without specific types of interactions, such as electrostatic, self-avoidance, or base-pairing. Single-molecule studies have also provided quantitative constraints on more complex polymer problems, such as protein folding.

Single-Molecule Studies of Nucleic Acids and Their Proteins. David Bensimon, Vincent Croquette, Jean-François Allemand, Xavier Michalet, Terence Strick 2019. © Oxford University Press 2019. Published in 2019 by Oxford University Press.
DOI: 10.1093/oso/9780198530923.001.0001

The mechanical properties of nucleic acids and proteins play a fundamental role in their biological function. For instance, the double-helical structure of DNA stores the genetic content of the molecule within the core of the double helix, thus preventing proteins easy access to the genetic code. It is, however, clear that the initial steps of many fundamental biological processes (such as DNA replication and the transcription of DNA into messenger RNA) depend on the unwinding, or melting, of regulatory DNA sequences. Once the DNA has been opened locally by a combination of mechanical and enzymatic effects, the proteins can then read and copy the genetic code. As we shall see in the following, the use of single-molecule approaches has made it possible to quantitatively reproduce and analyse such effects, leading to a much better understanding of the role these structural transitions play *in vivo*.

1.2 Single-molecule versus bulk studies

This book is devoted to a review of the interactions between some proteins and DNA at the *single-molecule level.* Since DNA/protein interactions have been extensively studied for the past fifty years using ensemble (bulk) enzymology methods, one may legitimately wonder what can be learned from single-molecule investigations. The short answer is individual behaviour. In ensemble assays, the measured observable is an average over a population of enzymes that may be extremely heterogenous. Some enzymes may be inactive, others might not be bound to their substrate, while those that are bound will usually be in different stages of their catalytic cycle, such as a pause state. As a consequence, the enzymatic activity deduced from these studies often significantly underestimates the "real" activity of a single enzyme.

Moreover, since enzymes are stochastic machines (their passage from one state to the next is a probabilistic process), an ensemble of enzymes cannot be synchronized over many catalytic cycles. This fact sets a fundamental limit to a detailed study of enzymatic mechanisms at the ensemble level. However, these details are in principle accessible when investigating the activity of a single enzyme. For example, the step size of a translocating enzyme can be directly observed in single-molecule but not ensemble studies. Single-molecule assays allow the study of subpopulations in an ensemble of enzymes. In fact, by providing the whole probability distribution of enzymatic activities, single-molecule assays provide in principle all the information that can be obtained on these molecular machines.

The great difficulty of single-molecule assays is putting these considerations into practice. Practical issues (temporal and spatial resolution, Brownian fluctuations, noise, photobleaching of dyes, etc.) set other limits on these studies. Moreover, the degree of parallelism of single-molecule assays is, at present, very limited (molecules are usually studied one at a time), so the sampled population is quite small (at best, a few hundred molecules). As a result, the statistical uncertainties on the measured observables are quite large ($> 5\%$), and certainly much larger than what can be measured in ensemble studies. It is, for example, very tedious to extract basic enzymology parameters (such as the

Michaelis–Menten parameters of a first order enzymatic reaction) from single-molecule assays. Similarly, ATP consumption rates are much easier to measure in bulk studies than in single-molecule assays.

One must, therefore, consider single-molecule assays as a useful addition to the toolbox of enzymologists. They can provide details on the enzymatic mechanism which are impossible to get using standard approaches, but yield much less precise average kinetic parameters than the latter (although they may be more accurate if the bulk study is dominated by a large subpopulation of inactive enzymes).

1.3 DNA structures

Most biomolecules (DNA, RNA, proteins) are linear polymers—i.e., they consist of a linear chain made of repeating structural subunits. For DNA and RNA, the monomeric subunits are nucleic acids of four different types (adenine, guanine, cytosine, and thymine (uracil for RNA)). For proteins, the building blocks are the twenty different amino acids. The structure of the biomolecule is an essential determinant of its function. This is particularly true for proteins for which the prediction from the sequence of the structure (and probable function) of the molecule has emerged as a major goal of the post-genomic area. This is also true for RNA whose 3D structure plays a crucial catalytic role in many reactions, such as the synthesis of proteins by the ribosomal complex.

In contrast with these examples in which the structure of the biomolecule is apparently uniquely determined by the sequence of its subunits, the structure of DNA is a double helix independent of its sequence. This structure has profound implications on all genetic processes. For transcription to occur, the two strands of DNA have to be locally unwound by an RNA polymerase, a process that might sensitively depend upon the torque acting on the molecule. During DNA replication, the two strands have to be unzipped by a helicase to make way for the enzymatic complex polymerizing two new DNA strands on the templates provided by the unwound single strands. This DNA unwinding generates a large torque ahead of the complex which is released by specialized enzymes, the topoisomerases. These enzymes are also responsible for disentangling DNA prior to cell division. Understanding how these enzymes and others work, therefore, requires an understanding of the mechanical properties of DNA.

Before studying the elastic behaviour of DNA in greater detail, it is important to recall its physico-chemical properties. DNA can be described as a molecule having a hierarchical series of structures. The primary structure describes the chemical/genetic sequence of the molecule. The secondary structure describes the way that the two constituent strands of the molecule wrap around each other to form the double helix. The tertiary structure represents the trajectory of the double helix's axis: the molecule might, for example, writhe about itself or wrap around globular histone proteins to form the chromatin fibres which are the main constituents of chromosomes. These higher-level structures are less well known, and would correspond to the series of packaging steps by which DNA is organized into a chromosome.

1.3.1 Primary structure of DNA

In its most common B-form, DNA consists of two anti-parallel strands wrapping around each other with a right-handed helical symmetry. Each strand is made of a polymer chain of deoxyribose rings linked by a phosphodiester bond between their 3′-OH moiety and the phosphate group on the 5′ carbon of their neighbour (see Fig. 1.1a). The distance along a strand between nearby phosphate groups is about 7 Å. One of four possible different chemical groups (bases), adenine (A), guanine (G), cytosine (C), and thymine (T) is linked to the 1′ carbon of the sugar ring. These groups consist of planar and hydrophobic aromatic cycles, resulting in a strong hydrophobic attraction between neighbouring bases. In an analogy of the piling of plates, this attraction is referred to as the "stacking" interaction.

1.3.1.1 Secondary structure

The presence of primary and secondary amines on the aromatic cycles of the bases enables guanine (G) and cytosine (C) to form three hydrogen bonds and adenine (A) and thymine (T) to share two hydrogen bonds (see Fig. 1.1a). These bases are complementary: the adenine (resp. guanine) group on one strand is paired with a thymine (resp. cytosine) group on the adjacent (complementary) strand. Hydrogen bonding between A (resp. G) and T (resp. C) yields a stable right-handed helix of complementary and anti-parallel strands (see Fig. 1.1b), which, from an information point of view, are to each other as a film is to its negative. These pairing and stacking interactions contribute to the exceptionally high bending rigidity of double-stranded DNA as compared to artificial polymers, such as polyethylene (see section 4.1).

The global structure (handedness, helical pitch, effective diameter, etc.) adopted by the double helix depends on a number of environmental parameters, such as solvent, ionic conditions, and temperature [6]. The canonical form of DNA in solution is named B-DNA and exists in the right-handed double helix just described. B-DNA has a diameter of about 24 Å, vertical spacing between bases of roughly 3.4 A, and helical pitch of 10.4 base pairs (bp) per turn. The various interactions which stabilize the B-form of DNA also give it local rigidity: its persistence length ξ (i.e., the distance over which the molecule is more or less straight in spite of thermal agitation that tends to shake and bend it, see section 4.1) is about $\xi = 50$ nm or 160 bp. This is much larger than the persistence length of most synthetic polymers, such as polyethylene, polystyrene, PVC, etc., which have a $\xi \sim 1$ nm; dsDNA is also much stiffer than other natural polymers, such as RNA or single-stranded DNA (ssDNA). An RNA or ssDNA molecule can adopt complex shapes and change direction over distances of a few bases, forming, for example, "hairpin" structures with a double-stranded stem ending with a single-stranded loop.

Let us mention a few non-canonical forms which may appear punctually within B-DNA:

- If the bases on the two opposite strands are not complementary, a "bulge" of unpaired bases may appear.

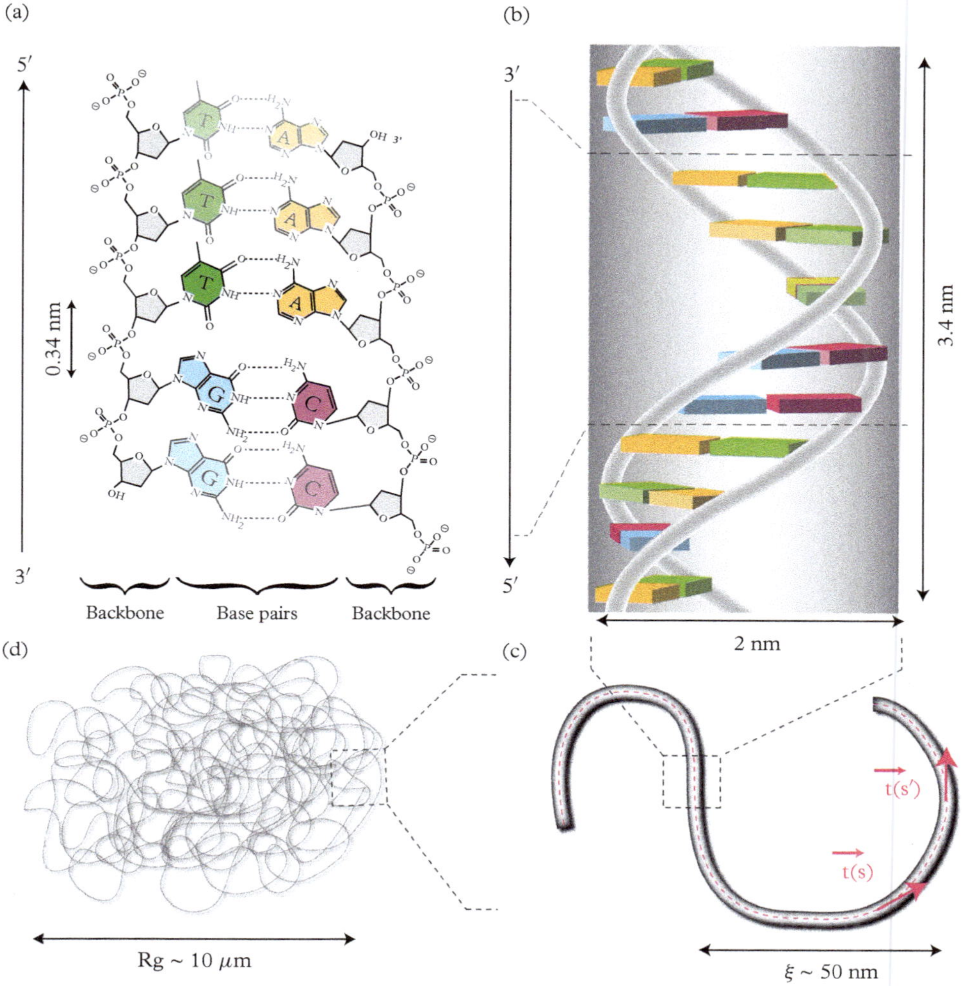

Fig. 1.1 Primary, secondary, and tertiary structure of the DNA molecule. *(a) Each of the two strands of the DNA molecule is made of a ribose–phosphate backbone. Chemical groups (adenine (A), guanine (G), cytosine (C), and thymine (T)) are attached on the ribose sugar. The distance between base pairs is 3.4 Angstrom for the standard B-form of DNA. Note that the two strands are anti-parallel: orientation can be defined using the anchoring point of the phosphate group (carbon 5′) on the sugar. (b) Secondary structure of DNA: each strand wraps around each other in a non-symmetric helical fashion. The pitch of the helix is 10.5 base pairs for the standard B-DNA, which means that the twist angle is* $\approx 36°$. *(c) At a larger scale, DNA behaves as a worm-like chain whose orientation decorrelates over a typical lengthscale of* $\xi = 50$ *nm. (d) Long DNA molecules (*$L \gg \xi$*) look like random coils with a typical gyration radius [5]* $R_g = \sqrt{2\xi L} \approx 10\ \mu m$ *in the case of an E.coli chromosome (four million base pairs).*

- In the presence of a palindromic sequence, the DNA may locally adopt a "cruciform" shape.
- A denaturation bubble is formed if the base pairs facing each other on the two strands of the double helix are complementary but dissociated.

X-ray diffraction data collected on DNA crystallized in various conditions of humidity or salinity show that the double helix is structurally polymorphic. For example, if the molecule undergoes crystallization in low humidity and in the presence of low amounts of cations, a structure known as A-DNA is observed. It is more compact than the B-form (the spacing between bases is 2.6 Å [6]), but has the same handedness. There also exists a left-handed double helix with a pitch of twelve base pairs per turn [6], known as Z-DNA. This DNA structure not only requires strong ionic conditions to be stabilized, but also a sequence of alternating purines (i.e., adenine or guanine) and pyrimidines (i.e., thymine or cytosine).

The fact that the bases in B-DNA are buried inside the double helix confers a very low reactivity to the molecule and drastically reduces the interactions of the polymer with itself (in contrast with single-stranded DNA or RNA which can form complex tertiary structures, such as hairpins). This feature is essential to preserve the genetic code from being damaged or modified easily, but it presents a challenge to the protein machinery which has to regulate, edit, copy, and transcribe the molecule. This machinery must access the bases and, therefore, open the DNA molecule.

As originally mentioned by Watson and Crick [1] and proved by Meselson and Stahl [7], the existence of two copies of the genetic information allows for a simple DNA replication mechanism (the so-called "semi-conservative" replication model of DNA): each original strand is used as a template for the synthesis of its complementary strand. Another essential feature of the two strands' complementarity is to allow DNA mismatches and mispairing to be corrected using the original strand as a template. This ensures the extremely high fidelity of DNA replication (the error rate is of the order of one base per 10^9).

The DNA replication mechanism raises a major question: if each strand is used as a template for a new molecule, the mother DNA has to be unwound to be duplicated but then how are the daughter strands separated "without everything getting tangled" [4]? This problem was solved by the discovery of enzymes, known as topoisomerases, which are capable of unknotting and untangling the DNA molecule. These topoisomerases control both the torsional stress in DNA and its overall topology.

1.3.2 Tertiary structures in DNA

The primary and secondary structures of DNA confer upon the molecule a large bending stiffness, so that the typical lengthscale, denoted by the persistence length ξ, over which thermal fluctuations are able to bend the axis of the molecule under physiological conditions, is about 50 nm (as compared to 1 nm for usual polymers). In other words, if $\vec{t}(s)$ and $\vec{t}(s')$ are two vectors tangent to the DNA's axis at positions s and s', ξ is the typical decorrelation length (see Fig. 1.1c):

$$\langle \vec{t}(s) \cdot \vec{t}(s') \rangle = e^{-|s-s'|/\xi}$$

DNA can thus be modelled as a flexible rope whose random coil geometry constitutes *de facto* the tertiary dynamical structure of DNA. Given the crystallographic length L of the DNA, one can estimate the typical radius of gyration R_g of the molecule using the model of a "semi-flexible" polymer in solution [5], see section 4.1:

$$R_g = \sqrt{2\xi L}$$

For the *E.coli* chromosome, L $\approx$ 1.5 mm so $R_g \approx 10\ \mu$m (to be compared with the typical size of the bacteria $\approx 1\ \mu$m), see Figs 1.1d and 1.2.

Therefore, to fit into the cell, DNA has to be packed. The way this packaging is achieved implies major changes in the DNA topology. In bacteria, the space occupied by the circular chromosomal DNA is minimized by maintaining the molecule in a highly supercoiled state, exhibiting compact braided structures known as "plectonemes", see Fig. 1.2, reminiscent of the interwindings observed in a tangled phone cord.

In eukaryotes (organisms possessing a nucleus), DNA is wrapped about protein octamer complexes, known as histones, with about 145 bp of DNA wrapped around each histone to form a nucleosome core. Such nucleosomes occur every 200 bp along the DNA to form the bead on a string structure of chromatine. Chromatine itself is further compactified into supercoiled structures of higher order, eventually resulting in the packaging of a total DNA about 1 metre long (in humans) into a cell nucleus a few microns in diameter, see Fig. 1.3. One of the major challenges of biology is to understand how this extremely compacted molecule is regulated, copied, transcribed, repaired, and disentangled prior to cell division.

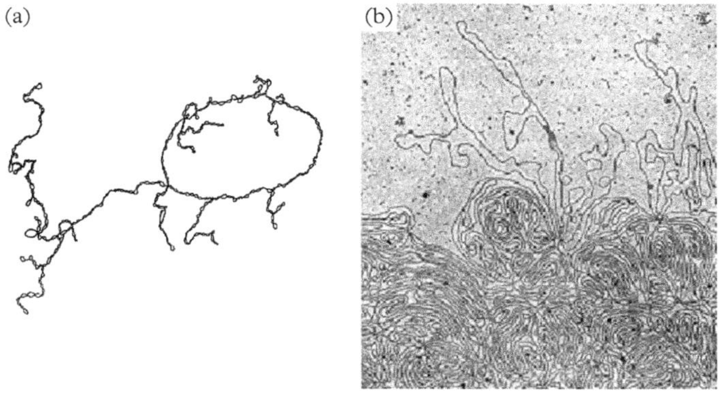

Fig. 1.2 *E.coli* and human chromosomal DNA. *(a) Sketch of a supercoiled E.coli chromosome: its total length is about 1.5 mm. (b) Electron microscopy picture showing a portion of a typical human chromosome after depletion of its proteins [8]. Though its total length is a few centimetres, thermal fluctuations tend to entangle the molecule so that the typical length of a DNA coil is a few tens of microns.*

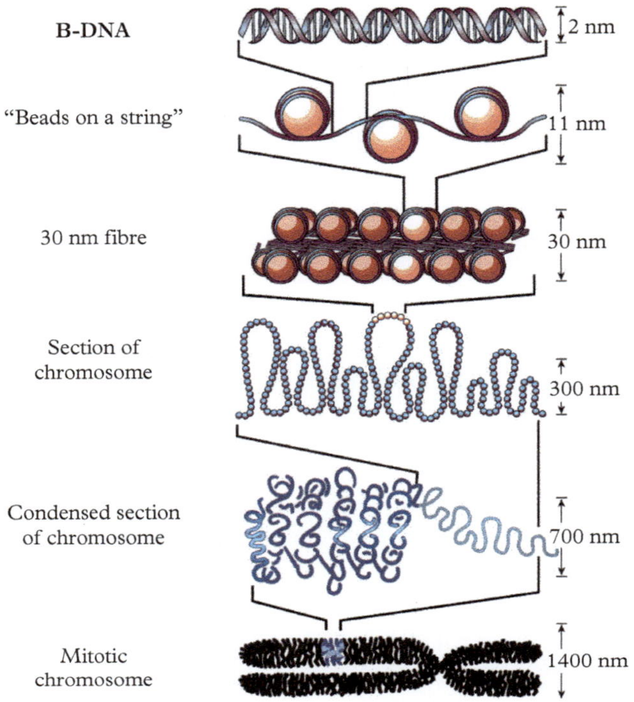

Fig. 1.3 Schematics of the various stages of DNA compaction in the chromosomes of eukaryotes. *The DNA is wrapped twice around a positively charged protein, known as a histone, to form the basic "bead on a string" structure of chromatine. This structure is further coiled into the so-called 30 nm chromatine fibre, which undergoes various poorly understood levels of supercoiling and looping to form the compact metaphase chromosome structure shown at the bottom. Adapted from [9].*

1.3.3 RNA structures

Whereas double-stranded DNA is a rather simple linear polymer chain, which has a dynamical tertiary structure and is well described by models of polymer elasticity discussed in later sections, RNA (and to some extent single-stranded DNA) is capable of forming complex stable 3D structures. These structures are stabilized by base-pairing interactions along the RNA chain and are essential for the function of some RNA molecules (such as transfer RNA (tRNA) and ribosomal RNA (rRNA), ribozymes, riboswitches, etc.). Indeed, it is by now clear that the cellular role of RNA is not just to serve as a disposable copy of information encoded in DNA (the principal role of messenger RNA (mRNA), but also to act as a major regulatory and catalytic unit in the cell-a role long thought to be played only by proteins encoded in the mRNA. In fact, the 5′-untranslated region (5′-UTR) at the end of mRNA often plays a regulatory role in transcription of DNA into mRNA or translation of mRNA into protein catalysed by the ribosome. The structure adopted by this portion of RNA (known as a riboswitch) can be altered by the binding of small metabolites, thus regulating (up or down) the mRNA translation into a protein involved in the metabolic pathway.

Ribosomal RNA (rRNA) is also an essential player in the ribosome [10, 11], not only by ensuring proper positioning of the tRNAs but also by catalysing elongation of the peptidic chain. The role of proteins in the ribosome complex is mostly structural—to serve as a scaffold for the catalytic unit formed by the rRNA.

Finally, small RNA molecules (siRNA, miRNA, etc.) play a major regulatory role in gene expression by mechanisms which are still not fully understood. This dual role of RNA, both as a support of genetic information and a catalytic molecule, has led to the suggestion that life may have originated in a purely RNA world.

From a theoretical point of view, the determination of RNA secondary and tertiary structures is somewhat simpler than the equivalent problem for proteins. First, there are fewer nucleotide types (essentially four) than amino acids (twenty). Second, their interaction is much better understood. Besides the usual self-avoidance and electrostatic repulsion, they can form Watson–Crick base pairs (and other known types of pairings, such as Hoogsteen base pairs) which can be stabilized by stacking interactions. Due to these interactions, the secondary structure of RNAs consists in a nested configuration of hairpin loops, see Fig. 1.4. This simple picture is complicated by the possibility of interactions between nucleotides on different hairpins (so called pseudo-knots).

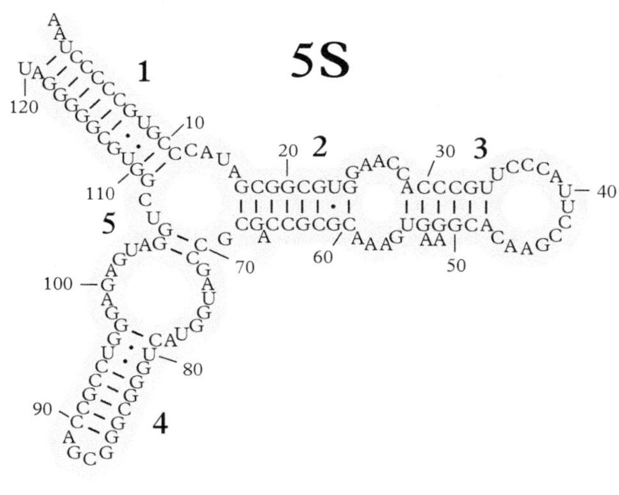

Fig. 1.4 Secondary structure of a small RNA *(the 5S-rRNA associated with the 5S subunit of the ribosome). Notice the configuration of nested hairpins characteristic of RNA secondary structures: the paired sections, labelled (2,3) and (4,5), are part of secondary hairpins in the overall 5S-rRNA hairpin which is composed of paired sections (1–5).*

Nonetheless, the prediction and design of RNA structures is better understood [12] than in their protein equivalents.

As we shall see later, by allowing one to probe the structure of single RNA molecules, single-molecule techniques are ideally positioned to unravel the mechanisms of RNA folding and structural switching.

2

Manipulating DNA

The first step in any DNA manipulation experiment is to anchor appropriately labelled DNA to pre-treated surfaces (glass slides, beads, micropipettes, etc.). Several methods have been developed to achieve specific DNA binding to surfaces. They have found useful applications, from gene mapping, sequencing, and analysis [13] to the development of very sensitive immunological assays (immuno-PCR) [14]. Most of these applications achieve the required binding specificity via biochemical reactions between a labelled DNA molecule, as described in section 2.1, and an appropriately treated surface. For example, the extremity of the molecule can be functionalized with biotin (or digoxigenin) which can interact specifically with streptavidin bound to a surface (or an antibody to digoxigenin) [15]. Similarly, surfaces coated with oligonucleotides can be used to hybridize to the complementary extremity of DNA molecules. Finally, there exists a large range of chemical methods to anchor (with various degrees of specificity) the extremities of DNA to surfaces bearing reactive groups (e.g., primary or secondary amine, carboxyl, or thiol moieties).

2.1 Tailoring and labelling DNA

Compared to the chemical synthesis of artificial polymers (whose length and sequence are poorly controlled), the biochemical synthesis of DNA is a marvel of accuracy and control. Due to the importance of DNA as the support of genetic information, nature has evolved a whole panoply of enzymes which act on DNA and perform a number of "editing" functions: copy (DNA polymerases), cut (restriction enzymes), paste (DNA ligases), and correct spelling (mismatch-repair system). Although the "letters" typically used are A, G, C, and T, other "letters" are also found in nature: e.g., uracil (U) and thymine (T) in messenger RNA, or methylated cytosine (C) and adenine (A) in DNA. Artificial "letters" have also been introduced by researchers for specific purposes: e.g., labelling the four bases with different fluorescent groups for sequencing [16] or with certain chemical moieties (biotin, digoxigenin (DIG)) for specific binding to certain biomolecules (streptavidin or an antibody against digoxigenin (anti-DIG)).

The availability of these powerful molecular editing tools gives scientists unprecedented control over the synthesis of DNA, which allows them to tailor DNA sequences

Single-Molecule Studies of Nucleic Acids and Their Proteins. David Bensimon, Vincent Croquette, Jean-François Allemand, Xavier Michalet, Terence Strick 2019. © Oxford University Press 2019. Published in 2019 by Oxford University Press.
DOI: 10.1093/oso/9780198530923.001.0001

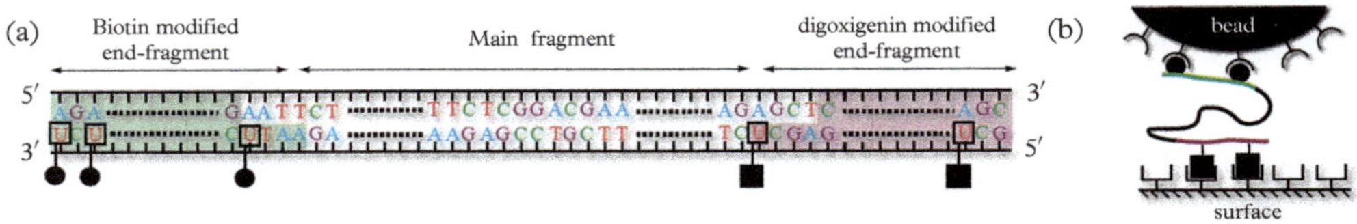

Fig. 2.1 DNA construct used for micromanipulation. *(a) The DNA molecule used for micromanipulation experiments [17, 18] is composed of three parts: the main part (a few kilobases long) is found along with two oligonucleotides (a few hundred base pairs long) which are synthesized by the polymerase chain reaction (PCR) using biotin- or digoxigenin-labelled uracil nucleotides (uracil, like thymine, is complementary to adenine). These two fragments prepared separately from the DNA main part are cut using appropriate restriction enzymes and ligated using a ligase enzyme. (b) The labels introduced in the small fragments allow for binding on appropriate surfaces (coated with streptavidin to bind biotin or anti-DIG to bind digoxigenin-labelled DNA).*

for their particular purposes. For instance, the polymerase chain reaction (PCR) allows for the generation of faithful copies of a given DNA sequence—the copy number increases exponentially with the number of cycles. The use of restriction enzymes and ligases permits a controlled modification of the underlying DNA sequence and is widely used by molecular biologists, e.g. to insert or delete genes. Finally, the availability of off-the-shelf, well-characterized DNAs from various sources (plasmids or phages) lets one use a number of standard sequences without having to write one's own from scratch.

To micromanipulate DNA molecules, one uses this panoply of enzymes to prepare a DNA molecule (typically between 10^3 and 10^4 bp long) ligated at its extremities with properly PCR-labelled, small oligonucleotides (see Fig. 2.1): one oligonucleotide multiply tagged with biotin so that it sticks to a streptavidin-coated surface and another multiply labelled with DIG to stick to an anti-DIG-coated bead or surface.

A single DNA molecule can, thus, be designed to anchor a small streptavidin-coated bead to an anti-DIG coated surface. Pulling and rotating the small bead, using one of the techniques described in Chapter 3, allows for stretching and twisting of the DNA molecule.

2.2 Molecular combing

An interesting alternative to specific binding to antibodies or streptavidin, which does not require any modification or end-labelling of the molecule, relies on the specific adsorption of DNA by its ends on hydrophobic surfaces at a pH $\sim$ 5.5 [19]. On many different hydrophobic materials (teflon, polystyrene, graphite, silanized glass, etc.), DNA was observed to adhere strongly and non-specifically at a low pH, and weakly or not at all at a high pH. In between, there exists a narrow pH range (pH $= 5.5 \pm 0.2$) where DNA binds to the surface by its extremities only.

Once DNA is anchored to a surface by its end(s), a very easy way to stretch it is to drain the solution (e.g., by pulling the surface out or letting the solution evaporate), see

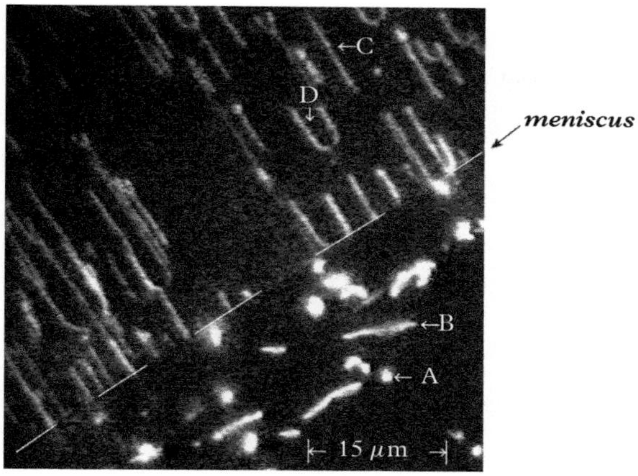

Fig. 2.2 Molecular combing of λ−DNA stained with YOYO-1. *The molecules in solution (bottom) are bound at one (A) or both (B) ends. The meniscus (shown via an arrow and dashed white lines) extends across the image from the lower left to upper right. The extended molecules left behind the meniscus are visible as straight segments (C) if bound at one end or loops (D) if bound at both ends. Adapted from [20].*

Fig. 2.2. In this process, known as molecular combing, the anchored DNA molecules are aligned on the surface by the receding meniscus, just like algea on the shore by the receding tide. It has been shown that the force applied on the molecule by the receding air/water interface (estimated to be ∼160 pN) is large enough to stretch it but not large enough to break its bond(s) with the surface [20, 21]. Because this force is acting locally (at the interface), DNA is stretched uniformly [19]: ∼1 μm for every 2,000 bases (b). Fluorescent *in situ* hybridization (FISH) on combed molecules (bacterial (BAC) or yeast (YAC) artificial chromosomes or even whole genomes) yields accurate (∼1 kb) genomic maps [22, 23]. Thus, the order and orientation of, and distance between genes and the presence of genomic rearrangements (e.g. deletions, see Fig. 2.3), can be determined.

Observation of the combing of DNA molecules grafted at both ends to a surface is instructive, see Fig. 2.4. First, as the meniscus moves past the anchoring points of the molecule, it stretches its two anchored segments (legs) *perpendicular* to the interface. The portion of the molecule in solution decreases until it spans the distance between the two legs. It is then stretched *parallel* to the contact line, its length diminishing as the meniscus recedes to form a loop. Then, as the tension in the molecule increases, the loop breaks. The length of the DNA molecule at the breaking point l_b can be measured and compared with the unstretched length $l_{b,0}$, see Fig. 2.4. In this experiment, $l_b/l_{b,0} = 2.14 \pm 0.2$ [21]: in other words, DNA can be extended to more than twice its length before breaking. As we shall see in Chapter 5, this large deformation of DNA (it is comparable to the ratio of the interphosphate distance to the distance between base pairs) must be accommodated by a profound rearrangement of the Watson–Crick double helix.

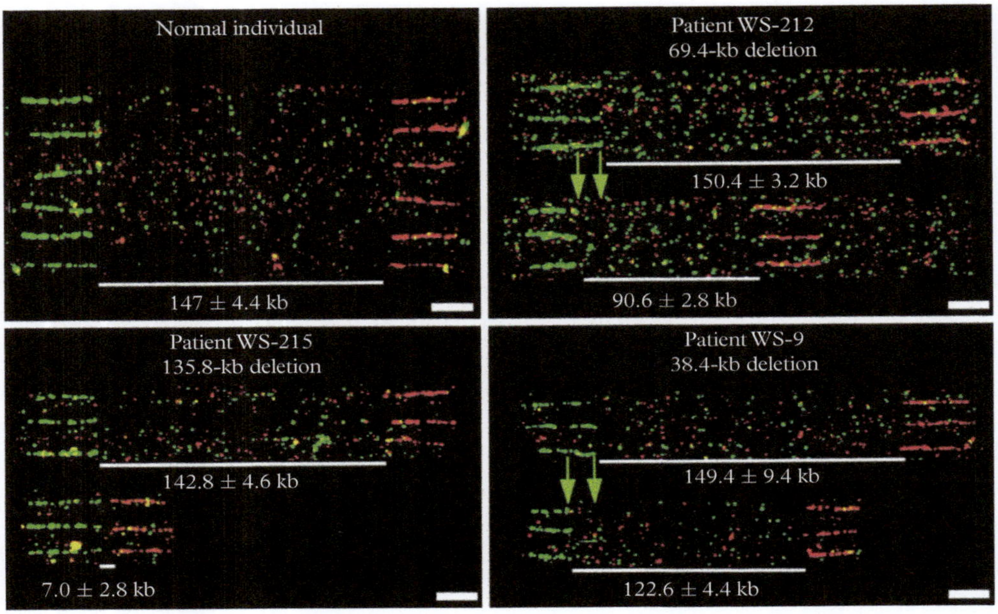

Fig. 2.3 Molecular combing of genomic DNA *and hybridization of probes complementary to the ends of the TSC2 (tuberous sclerosis 2) gene. Both alleles in a normal individual present the correct gene size, whereas in affected individuals one of the alleles exhibits a major deletion (many kb) of a gene fragment. The possibility of detecting large genetic rearrangements in a single allele is a unique feature of molecular combing. Adapted from [23].*

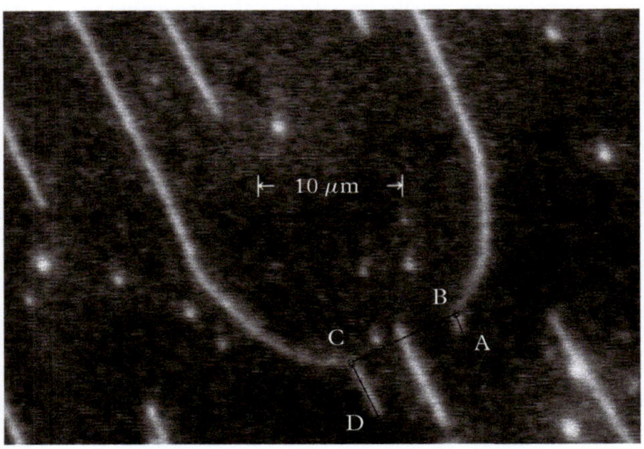

Fig. 2.4 Fragment of one of the authors' DNA combed on a hydrophobic surface. *The DNA bound at both ends forms a typical broken loop. From measurements on such figures, we deduce the extension of DNA at the breaking point: the stretched molecule length is* $l_b = \overline{BC}$*, its unstretched length* $l_{b,0}$ *is deduced by dividing the length* $\overline{AB} + \overline{DC}$ *by the extension factor observed for a straight molecule combed on a hydrophobic surface (e.g., 1 micron for 2,000 base pairs, with a crystallographic length of about 0.67* μm*).*

2.3 Forces at the molecular scale

Manipulating single molecules requires an understanding of the types of forces encountered at the molecular level. The mimimal force on a microscopic object (bacterium, bead, cell, etc.) is set by the force exerted on it via its random collisions with the molecules of the environment, the so-called Langevin force f_L. It sets the lower limit to force measurements and is a result of the Brownian fluctuations (of energy $k_B T = 4\ 10^{-21}$ J = 0.6 kcal/mol at room temperature) of the object of size d (sensor, cell, membrane) anchored by the molecule. For a $d = 2\ \mu$m diameter bead or cell in water (viscosity $\eta = 8.9\ 10^{-4}$ Pa·s), the mean Langevin force (averaged over a time $\tau = 1$ s) is $\langle f_L \rangle = \sqrt{12\pi k_B T \eta d/\tau} \sim 10$ fN (notice that the faster the measurement (the smaller τ), the larger $\langle f_L \rangle$). This can be compared to the typical weight of a cell: ~10 fN, i.e., every second a cell experiences a thermal knock equal to its weight!

Just above these forces lie the entropic forces, which result from a reduction of the number of possible configurations of the system consisting of the molecule (e.g., protein, DNA) and its solvent (water, ions). As an example, a free DNA molecule in solution adopts a random coil configuration that maximizes its configurational entropy [5]. Upon stretching, the molecular entropy is reduced so that, at full extension, there is but one configuration left: a straight polymer linking both ends. To reach this configuration, work against entropy has to be done, a force has to be applied. The entropic forces are rather weak. Since the typical energies involved are of order $k_B T$ and the typical lengths are of the order of a nanometre, entropic forces are of order $k_B T/\text{nm} = 4$ pN ($4\ 10^{-12} N$). These are typically the forces exerted by molecular motors, such as myosin on actin [24], or the force necessary to stretch a DNA molecule to its contour length [15] or unzip the two strands of the molecule [25, 26].

Noncovalent (e.g., ligand/receptor) bonding forces are much stronger. They usually involve modifications of the molecular structure on a nanometre scale: breaking and rearrangement of *many* van der Waals, hydrogen, or ionic bonds, and stretching of covalent bonds. The energies involved are typical bond energies, of the order of one electron-volt (1 eV = 1.6 10^{-19} J = 36 kcal/mol). The elastic forces are, thus, of order eV/nm = 160 pN. These are typically the forces necessary to break receptor/ligand bonds [27, 28, 29, 30, 31] or deform the internal structure of a molecule [32, 33, 34].

Finally, the strongest forces encountered at the molecular scale are those required to break a covalent bond, of the order of 1 eV/Å ~1600 pN.

2.4 Micromanipulation tools

There are, by now, many tools to manipulate single molecules: atomic force cantilevers [27], microfibres [35, 32], optical [36] or magnetic tweezers [37] and traps [17, 33, 38], hydrodynamic drag [15], acoustic traps, and biomembrane force probes (BFP) [39]. In all these techniques, a DNA molecule (but also a protein or some other polymer) is first anchored to a surface at one end and a force sensor at the other. The force sensor is usually a trapped micron-sized bead or cantilever, the displacement of which is used to measure the force (see Fig. 2.5).

2.4.1 Microfibres

The simplest tool to pull on a polymer (DNA, RNA, protein, polysaccharides, etc.) consists of a microfibre or the cantilever of an Atomic Force Microscope (AFM) to which one end of the polymer is attached, the other end being attached to a movable surface (a microscope coverslip, a bead, or a stiffer fibre). As this surface is displaced, tension builds up in the polymer which pulls on the microfibre. The displacement δx of the microfibre yields the stretching force from Hooke's law: $F = k_{fibre}\delta x$, where k_{fibre} is the cantilever stiffness. This stiffness can be calibrated against the stiffness of known fibres or simply deduced from the amplitude of the microfibre fluctuations $\langle \delta x^2 \rangle$ at zero force: $k_{fibre}\langle \delta x^2 \rangle = k_B T$.

2.4.2 The biomembrane force probe

The biomembrane force probe (BFP) is a technique pioneered by E. Evans [39]. It consists in using the deformation of a 10–20 μm vesicle under tension as a force sensor. The tension in the vesicle is controlled by a suction pipette which sets the hydrostatic pressure difference across its membrane, see Fig. 2.5. The advantage of this technique is that the stiffness of the force sensor (i.e., the tension) can be set at will, allowing for the measurement of a very large range of forces (from 10^{-15} to 10^{-9} N). This method has been used to monitor the force required to break bonds between pairs of receptor/ligand, antigen/antibody, etc. It has not been used to study the elastic properties of nucleic acids or monitor their interactions with various proteins, mostly for practical reasons (i.e., ensuring the binding of a single DNA molecule between two beads: one held in a pipette and the other bound to the membrane).

2.4.3 Hydrodynamic drag

Once DNA is bound at one end to a surface and at the other to a small bead (as is usually the case in single-molecule manipulations), the easiest way to pull on the molecule is to use the hydrodynamic drag of a flow past the bead to pull on the molecule [15, 42]. The advantages of this technique are its simplicity (flush the buffer at constant rate) and possible high throughput (applying a force and observing many thousands of beads in the same field of view [43]). The disadvantage is that the bead fluctuations along the stretched molecule are coupled to the transverse ones (by geometry and the gradient in the fluid velocity near the surface). Moreover, the force on the molecule is not well controlled as the drag on the system depends on both the bead size, its distance from the surface (which sets the fluid velocity), and the molecule's extension. As a result, the

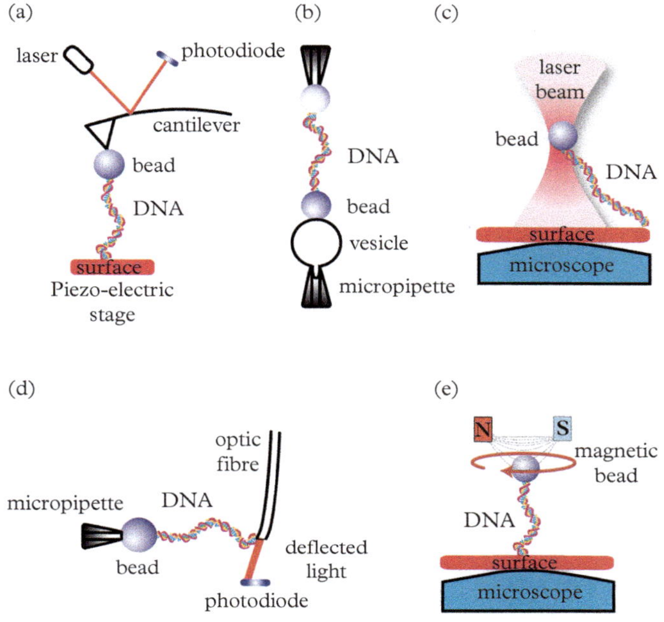

Fig. 2.5 Examples of force transducers used in various experiments. *(a) An AFM cantilever is often used as a force transducer during intermolecular force measurements [27], protein unfolding experiments [40], or DNA stretching. Its deflection upon pulling is detected by the displacement of a laser beam reflected from the cantilever. (b) Biomembrane force probe, where a vesicle deformation is used as a force sensor. Its stiffness is adjusted through the pressure in the micropipette. (c) In optical tweezers, an object with a refractive index higher than that of water is attracted at the focalization point of a laser. The position in the trap is observed with a microscope and, together with the trap stiffness, is used to assess the trapping force. (d) In some of the experiments involving DNA pulling [32, 41] and unzipping [25], the force transducer is an optical fibre and its deflection is detected optically (by measuring the displacement of either the fibre directly on a microscope stage or a light beam emitted from its pulled end). (e) Principle of magnetic traps: a magnetic bead is subjected to the gradient of a magnetic field.*

precision of extension measurements with this approach is usually poor (> 100 to 1000 bps). It has nonetheless been used to study DNA replication [42, 44].

2.4.4 Optical tweezers and traps

As shown by Ashkin [36], laser beams can be used to trap small transparent beads. To understand how a force is exerted by a light beam passing through a transparent bead, whose size is larger than the wavelength justifying the use of the laws of geometrical optics, consider the configuration shown in Fig. 2.6. A ray propagating in a medium of refraction index n_m is incident with an angle θ on a bead of radius r and index of refraction $n_b > n_m$. By Snell–Descartes's law: $n_m \sin\theta = n_b \sin\phi$, where $\phi < \theta$ is the angle of refraction of the ray in the bead.

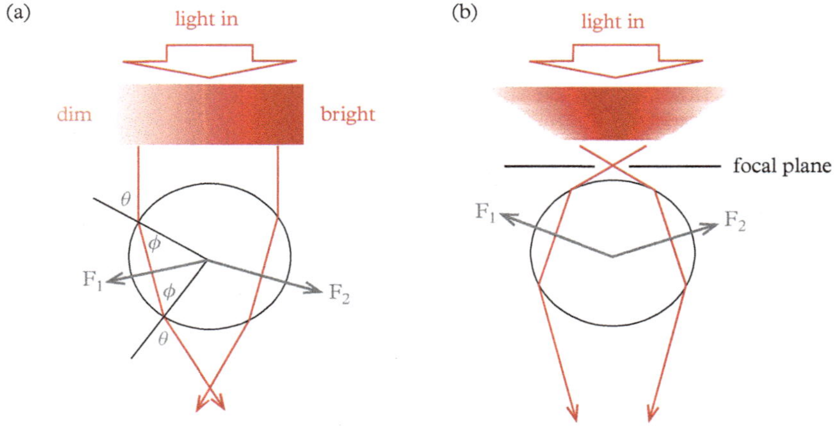

Fig. 2.6 Light ray incident on a bead with angle θ and refracted with angle ϕ. *(a) The change in direction, i.e., momentum, of the beam due to this refraction exerts a net force ($\vec{F}_2$—$\vec{F}_1$) on the bead which pulls the bead sideways to regions of high beam intensity. (b) For a strongly focused laser beam, the change in momentum can result in a longitudinal force, pulling the bead to the focal point of the beam, which may counter the radiation pressure from less strongly focused rays that push the bead away from that focal point.*

For simplicity, we shall assume that the incident ray is fully transmitted, i.e., no absorption ($T \sim 1$; $R \sim 0$). In that case, only the direction of the ray, i.e., its momentum, has been altered: therefore, a force was exerted on the ray. A simple geometric calculation yields:

$$F = \frac{2n_m I_0}{c} \sin(\theta - \phi)$$

where I_0 is the power (energy/time) of the ray and $n_m I_0/c$ is the momentum carried per unit time (for a fully reflected 1mW laser beam ($\theta - \phi = \pi/2$), the force $F = 7$ pN). By the law of action and reaction, a similar force must be exerted by the ray on the bead and directed outwards. If the bead is illuminated by a homogeneous light beam (many rays), the components of the forces on the hemispheres that are transverse to the incident light cancel out and the total force is directed along the light beam. This so-called radiation pressure pushes the bead forward.

Imagine, now, that the bead is illuminated by a non-homogeneous light beam, so that the density of rays is higher on the right hemisphere than on the left one. Then, the transverse component of the force pulling the bead towards the higher light intensity (on the right hemisphere) will be larger than the transverse component on the left hemisphere, see Fig. 2.6a. As a result, the bead (while still being pushed forward) will move transversely to a position in the beam where the intensity gradient is null (and the intensity maximal). In a Gaussian beam, this corresponds to the centre of the beam and the bead is, thus, trapped at its centre, see Fig. 2.6b.

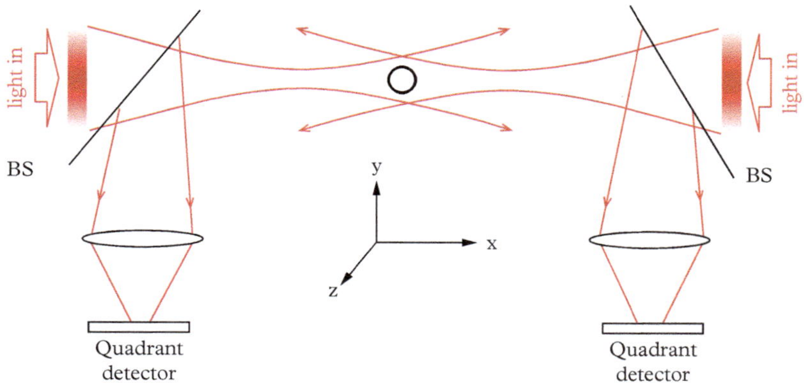

Fig. 2.7 Two counter-propagating Gaussian beams define a stable trap position for a small transparent bead. *The trapping along the optical (x-)axis is ensured by the divergence of the beams. As a result, the radiation pressure decreases downstream from the beam's focal point. The deflections of the refracted beams by beam splitters (BS) on quadrant photodetectors provide a direct measure of the momentum transferred by the beams to the bead and, thus, of the transverse trapping force.*

The bead can also be trapped longitudinally by a single laser beam if the beam is focused with a high numerical aperture objective (typically NA = 1.4), i.e., if the angle of the light cone at the focal point is sufficiently large, see Fig. 2.6b. In this case, the backward force exerted by the more off-centre rays may counterbalance the radiation pressure exerted by the low angle rays and the bead can be trapped close to the focal point of the laser beam. The bead can be easily moved in 3D, by displacing the laser beam (using small movable mirrors), hence the definition of this setup as optical tweezers. The force exerted by optical tweezers on a bead can be deduced from the displacement of the bead δx from its equilibrium position: $F = k_{trap}\delta x$. The optical trap stiffness k_{trap} has to be determined prior to any force measurement—e.g., by pulling on the bead with a known force, such as the hydrodynamic drag F_s of a fluid (of viscosity η) flowing with a velocity v around the bead: $F_s = 6\pi\eta r v$. Alternatively, one may determine k_{trap} by measuring the intensity of Brownian fluctuations $\langle\delta x^2\rangle$ of the trapped bead. By the equipartition theorem, they obey the relation [36, 45, 46]:

$$\frac{k_{trap}\langle\delta x^2\rangle}{2} = \frac{k_B T}{2} \tag{2.1}$$

A conceptually simpler (but practically more complex) trapping configuration uses two counter-propagating, diverging Gaussian beams to trap a bead in a fixed 3D position [47], see Fig. 2.7. As previously described, the bead is trapped in 2D at the centre of the beams and pushed along the optical (x-)axis from both sides by the counter-propagating beams. Since the beams are diverging, there is a stable equilibrium position where the force exerted by one beam is balanced by the force exerted by the other: the bead is trapped in 3D. The force exerted on the bead when pulled perpendicular to the optical axis can be measured directly from momentum transfer [48]. Namely, it can be deduced from the angular displacement θ of the trapping beam resulting from its refraction in the bead: $F = In\theta/c$, where I is the beam intensity, n the index of refraction of the solution, and c the velocity of light [47]. This absolute measurement of the force eliminates the need for a calibration of this optical trap.

While the forces achieved with optical tweezers (at intensities that do not damage the biological sample) are typically smaller than 50 pN, optical traps can reach twice that value. Notice that, by using trapped anisotropic particles, it is possible with optical tweezers to apply a torque on the particle [49, 50].

2.4.5 The magnetic trap

To twist and stretch a DNA molecule and study its interactions with proteins, a magnetic trapping technique [17] has proved particularly convenient. Briefly, it consists of stretching a single DNA molecule bound at one end to a surface and at the other to a magnetic micro-bead (1−4.5 μm in diameter), see Fig. 2.8. Small magnets, whose position and rotation can be controlled, are used to pull on and rotate the micro-bead and, thus, stretch and twist the molecule. This system allows applying and measuring forces ranging from a few femtoNewtons (fN $= 10^{-3}$pN) to nearly 100 pN (see [18]) with a relative accuracy of $\sim 10\%$. In contrast with other techniques, this force measurement is absolute and does not require a calibration of the sensor. It is based on the analysis of the Brownian fluctuations of the tethered bead, whose position can be measured by an analysis of its image on a CCD camera. Identifying the centre of the bead allows for the

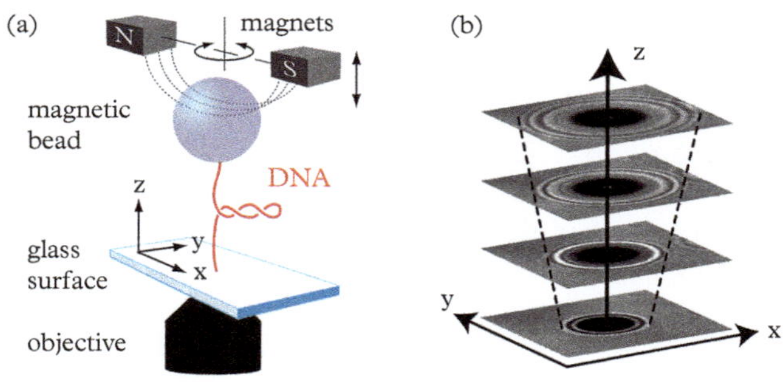

Fig. 2.8 Schematic representation of a magnetic trap. *(a) Small super-paramagnetic beads are bound to the surface of a flow cell by one DNA molecule (or a few). Magnets positioned above the sample exert a force on the beads and, thus, on the tethering molecules. Un-nicked DNA can be twisted by rotating the magnets since the beads follow the rotation like the needle of a compass. The beads are observed on an inverted microscope where they appear decorated by diffraction rings (b) whose size varies with the distance to the focal point. From the real-time analysis of these images, one deduces the beads' 3D position and, thus, their distance from the surface (i.e., the extension of the tethering molecule) and the exerted force (from the beads' transverse fluctuations, see text in section 2.4.5).*

determination of its x,y-coordinates, while characterizing the size of its diffraction rings allows for the determination of its distance to the focal point, i.e., its z-coordinate.

The DNA-bead system is, thus, completely equivalent to a damped pendulum of length $l = \langle z \rangle$ pulled by a magnetic force F (along the vertical axis). Its longitudinal δz^2 $(= \langle z^2 \rangle - \langle z \rangle^2)$ and transverse δx^2 fluctuations are characterized by effective rigidities $k_{||} = \partial_z F$ and $k_{\perp} = F/l$. By the equipartition theorem, they satisfy [45, 46]:

$$\delta z^2 = \frac{k_B T}{k_{||}} = \frac{k_B T}{\partial_z F} \tag{2.2}$$

$$\delta x^2 = \frac{k_B T}{k_{\perp}} = \frac{k_B T l}{F} \tag{2.3}$$

Thus, from the bead's transverse fluctuations $(\delta x^2, \delta y^2)$, one can extract the force pulling on the molecule (the smaller the fluctuations, the larger F) and, from δz^2, one obtains its first derivative, $\partial_z F$. This measurement method is valid with magnetic (but not optical) traps since the variation of the trapping force occurs on a scale (~ 1 mm) much larger than the scale on which the elasticity of the molecule changes (~ 0.1 μm). In other words, the stiffness of a magnetic trap is very small compared to F/l (or $\partial_z F$; the reverse is true for optical tweezers).

A further advantage of the magnetic trap technique is that measurements on DNA at constant force are trivial (just keep the position of the magnets fixed).[1] With cantilevers or optical tweezers, to work at constant force requires an appropriate feedback to ensure that the displacement of the sensor is kept constant. Moreover, magnetic traps allow twisting of the molecule by simply rotating the magnets. However, because its stiffness depends on the force, the magnetic trap technique has a lower spatial resolution (of ~ 10 nm) at low forces (< 1 pN) than the other manipulation methods.

2.4.6 The acoustic trap

A recent addition to the single-molecule manipulation toolbox is the acoustic trap [51, 52, 53]. It consists of creating a standing acoustic wave along the pulling direction (e.g., the optical axis). A polystyrene or glass bead, which has a density and compressibility that differs from the solution, will be attracted to the acoustic wave nodes, i.e., the minima of acoustic pressure, see Fig. 2.9. Since that pressure varies over a lengthscale set by the wavelength of sound in water ($\lambda_s \sim 100$ μm), which is much larger than the typical size of a tethering DNA molecule, the force exerted by the acoustic wave is constant over all the DNA anchored beads in the sample [53]. That force can be tuned by varying the intensity of the sound wave and scales as the volume of the trapped bead (reaching 100 pN for a polystyrene microsphere of diameter 4.5 μm). Acoustic and magnetic traps are,

[1] Because the tension on the molecule (like the pressure on a gas) is an extensive property (i.e., it is independent of the system size), it is often the preferred control variable and means to study the system's response.

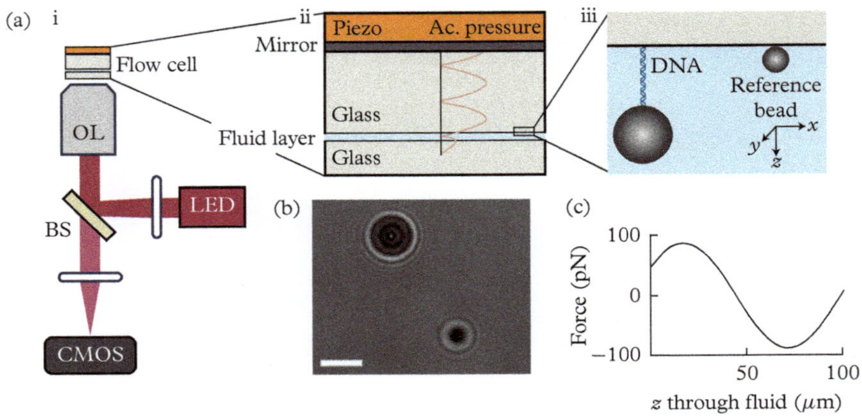

Fig. 2.9 Schematic representation of an acoustic trap positioned on top of an inverted microscope. *(a) Small polystyrene microspheres are anchored by a DNA molecule to the upper surface (iii) of a flow cell defined by a glass plate and a coverslip (ii). An aluminium layer deposited on top of the cell serves as a mirror for illumination. A piezoelectric transducer coupled to the upper glass slide generates an acoustic stationary wave along the z-axis of the flow cell (at a 6.8 MHz resonance frequency). The intensity of the acoustic pressure is shown as the red curve in (ii). Notice the pressure node (minimum) in the solution away from the top surface. The tethered bead (whose position can be determined as in a magnetic trap from its image (b)) is pulled to that node by a force F which has a magnitude that can be computed as a function of the distance from the top surface (c). As for a magnetic trap, the force can also be deduced from the bead's vertical distance and transverse fluctuations. Adapted from Sitters et al. [53].*

thus, similar in many respects: they are both constant-force (i.e., force-clamp) devices where the force is set and the resulting molecular extension is measured. In contrast, optical tweezers are constant extension (extension clamp) devices where the position of the trap (i.e., the molecule's extension) is set and the resulting force is measured. Notice, however, that in contrast with both optical tweezers and magnetic traps, acoustic traps have so far not been shown to allow for rotation of trapped tethered beads (though that might be achievable with asymmetric beads [54]).

2.5 Choosing a single-molecule manipulation tool

As the range of forces at the molecular scale spans several decades from 0.01 pN to 1000 pN, different instruments are necessary to cover the full range of forces. It is interesting to compare the force range and bandwidth of measurements done using cantilevers, glass microneedles, optical/magnetic/acoustic tweezers, or traps. All single-molecule force measurement techniques involve measuring the displacement of a small sensor (bead or AFM cantilever) tethered to a fixed surface by a biomolecule (DNA or protein). Consequently, they are all equivalent to a damped pendulum subjected to continuous thermal shocks, which limits the accuracy of force and displacement measurements.

Without loss of generality, the deviation x of the particle from its equilibrium position is given by:

$$\gamma \frac{dx}{dt} + kx = f_L \qquad (2.4)$$

where k is the stiffness of the manipulating tool (e.g., AFM cantilever or optical trap) and γ is the dissipation (friction) associated with motion of the particle in the solution of viscosity η ($= 0.89\ 10^{-3}$ N·s/m^2 for water). For a microsphere of radius r: $\gamma = 6\pi\eta r$, f_L is the thermal (Brownian) noise due to the shocks of the molecules in the solution on the particle. The dissipation–fluctuation theorem relates the strength of these fluctuations to the dissipation:

$$\langle f_L(t) f_L(t') \rangle = 2k_B T \gamma \delta(t - t') \qquad (2.5)$$

Using the Fourier transform:

$$x(t) = \frac{1}{\sqrt{2\pi}} \int_{-\infty}^{\infty} \tilde{x}(\omega) e^{i\omega t} d\omega \qquad (2.6)$$

Eq. 2.4 then becomes:

$$i\omega\gamma\tilde{x} + k\tilde{x} = \tilde{f}_L \qquad (2.7)$$

where $\langle | \tilde{f}_L|^2 \rangle = k_B T\gamma/\pi$. The particle fluctuation power spectrum becomes, see Fig. 2.10:

$$\langle |\tilde{x}|^2 \rangle = \frac{\langle |\tilde{f}_L|^2 \rangle}{k^2 + \gamma^2\omega^2} = \frac{k_B T}{\pi k \omega_0} \frac{1}{1 + (\omega/\omega_0)^2} \qquad (2.8)$$

with $\omega_0 = k/\gamma$. Using Parseval's relation, one can verify that the average variance of the fluctuations $\langle x^2 \rangle$ satisfy:

$$\frac{k\langle x^2 \rangle}{2} = \frac{k}{2} \int_{-\infty}^{\infty} \langle |\tilde{x}|^2 \rangle d\omega = k_B T/2 \qquad (2.9)$$

which is the result of the equipartition theorem (another way to state the dissipation–fluctuation theorem).

Quite often, to improve the signal/noise ratio (SNR), the signal is averaged over a bandwidth: $\Delta\omega = 2\pi\Delta f$. If $\Delta\omega \ll \omega_0$, then the variance σ_x^2 of the position fluctuations is:

$$\sigma_x^2(\Delta f) \equiv \langle x^2 \rangle_{\Delta f} = \int_{-2\pi\Delta f}^{2\pi\Delta f} \langle |\tilde{x}|^2 \rangle d\omega \sim \frac{4k_B T\gamma\Delta f}{k^2} \qquad (2.10)$$

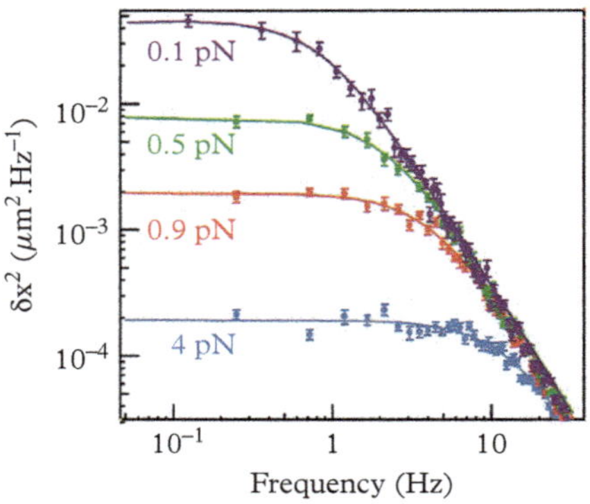

Fig. 2.10 The average power spectrum of the transverse fluctuations $\langle|\tilde{x}|^2\rangle$ *for a bead tethered by a single DNA molecule in a magnetic trap at various forces.*

and the variance σ_F^2 of the fluctuations $\delta F = kx$ in the force measurement is:

$$\sigma_F^2(\Delta f) \equiv \langle\delta F^2\rangle_{\Delta f} = k^2\langle x^2\rangle_{\Delta f} \sim 4k_B T\gamma\Delta f \tag{2.11}$$

Of course, for Eq. 2.10 to be correct: $\Delta f < \omega_0/2\pi$ or

$$k\langle x^2\rangle_{\Delta f} = \sqrt{\langle x^2\rangle_{\Delta f}\langle\delta F^2\rangle_{\Delta f}} < k_B T$$

Note that for a given frequency range Δf (determined by the dynamics of the system studied), the fluctuations in force depend only on the dissipation γ: precise force measurements require low dissipation, i.e., small sensors (small microspheres rather than AFM cantilevers). For a 1 μm bead in water, the force fluctuations over a $\Delta f = 1$ *Hz* bandwidth are: $\sigma_F \sim 20$ fN.

In a force-clamp configuration, the force is fixed and the molecule's extension is measured. The molecular stiffness k_m determines the standard deviation of the longitudinal fluctuations: $\sigma_z = \sigma_F/k_m$. For a 1 μm dsDNA molecule ($\xi = 50$ nm) pulled by a magnetic trap with a force of 1 pN: $k_m \sim 1.4\ 10^{-5}$ N/m and $\sigma_z \sim 1.4$ nm (averaged over 1 s). To reduce the force and displacement noise, the dimensions of the sensor and molecule studied have to be reduced (to reduce the dissipation γ), and high loads (resulting in high molecular stiffness) should be used whenever possible. For magnetic (and acoustic) traps where a bead tethered by a DNA of length l is pulled by small magnets with a force F, the stiffness k of the sensor (the tethered bead) is essentially given by the ratio F/l (for transverse fluctuations) or $\partial_z F$ for longitudinal fluctuations, so that it varies as $1/l$. Reducing the length of the tethering DNA by a factor 2 increases the S/N ratio of position measurements by a factor 4 (it reduces $\langle\sigma_x^2(\Delta f)\rangle$ in Eq. 2.10 by a factor 4).

Glass microneedles have a typical stiffness $k \sim 10^{-5}$ N/m [55, 25]: with a $\sigma_x \sim 10$ nm resolution on the fibre's displacement [25], forces in the range of tens of fN could be measured. However, the large size of the glass fibre imposes a relatively large Langevin force, of the order of 0.5 pN/$\sqrt{Hz}$. Fast measurements done by sampling between 10 and 100 Hz, therefore, limit the force measurement to a resolution σ_F of a few pN. In the case of an AFM cantilever with a typical stiffness of 10^{-3} N/m [56], a spatial resolution σ_x of the order of one Angström should allow for the measurement of forces as low as 1 pN. However, the Langevin force acting on the cantilever is large, of the order of 0.1 pN/$\sqrt{Hz}$ for high-quality cantilevers, and rapid measurements done with these instruments are typically limited to a resolution of 15 pN.

In the case of optical tweezers, a trap with a minimum stiffness of about 10^{-5} N/m is created, but the trap's dimensions are small, of the order of 0.5 μm. This corresponds to forces of the order of 1 pN. To generate higher forces, the laser power needs to be increased. If the power is too low, the trapping potential is too weak and does not function effectively. The spatial resolution σ_x of the bead's position within the trap can be of the order of a few Angströms, implying that forces as small as a few fN should be measurable. Once again, though, the thermal forces acting on the bead hinder the measurement of such low forces, and for a bead with a micron-scale diameter the limiting thermal resolution is of the order of 10 fN/$\sqrt{Hz}$. Fast experiments done with a bandwidth of a few hundred Hz are associated with an error in the force estimate $\sigma_F \sim 0.1$ pN. This precludes the use of optical tweezers at very low forces. Attempts to reduce that error (or noise) in force measurements by reducing the bead size results in a weaker trapping potential. Magnetic tweezers, using roughly the same sensor—a small bead—experience similar thermal limitations. With magnetic tweezers, however, the trapping potential can be reduced until the bead's weight (or buoyancy) becomes apparent. At this point, decreasing the bandwidth allows for precise force measurements.

To summarize, AFM cantilevers with a very high stiffness (typically $0.1 < k < 0.01$ N/m) are good position sensors but poor force measuring devices. Magnetic (and acoustic) traps have a low stiffness set by the tethering molecule (typically $k < 10^{-5}$ N/m) which makes them good sub-pN manipulation tools, albeit with a low precision in position measurement (at sub-pN forces). Optical tweezers with stiffness of O(10^{-5} N/m) are often a good compromise between these extremes: however, to apply a constant force on a molecule, magnetic and acoustic traps, which are natural force-clamps, are often the chosen option. Optical tweezers, glass fibres, and AFM cantilevers are natural extension clamps, though they can be converted to force-clamps with appropriate feedback loops.

2.6 Problems

Calibration of optical tweezers. Consider the optical tweezers formed by a laser focalized on a 1 μm bead in water (viscosity $\eta \sim 10^{-3}$ Pa·s). These tweezers create an effective potential well V in which the bead is trapped:

$$V(x) = -V_p \exp^{\frac{-\delta x^2}{2a^2}} \tag{2.12}$$

where $\delta x = x - x_0$ is the deviation of the bead from the centre of the trap at x_0, $a = 500$ nm is the half-width of the trap, and V_p is its depth. To calibrate the trap, one measures the mean square fluctuations of the bead at rest (in the centre of the trap): $\langle \delta x^2 \rangle =$ 125 nm^2 for a laser power $P = 10$ mW.

1. Find the stiffness k of these optical tweezers. How does it vary with the laser intensity?
2. Estimate the depth of the potential V_p.
3. Compute the maximal force F_{max} that can be exerted with these tweezers. Is it adapted to the study of molecular motors? To which displacement δx does this force correspond?
4. What is the cut-off frequency of the bead's movement? How does it vary with the laser intensity? What is its physical significance?
5. What is the probability that the bead will thermally hop over its potential at zero force and at a force $F = \frac{\sqrt{e}}{2} F_{max}$?
6. What is the order of magnitude of the minimal laser power for such an escape from the well to occur within 100 s?

Acoustic tweezers. Consider a standing acoustic wave $P(z,t) = P_0 \cos kz \sin \omega t$ of peak pressure $P_0 = 1$ MPa $= 10^6$ N/m^2, wavelength $\lambda = 2\pi/k = 1$ mm, and frequency $\omega = c_0 k$, where c_0 is the sound velocity in the medium (for water : $c_0 =$ 1,500 m/s). A bead of radius $a = 1$ μm and density ρ_1 equal to the fluid density $\rho_0 = 10^3$ kg/m^3 is experiencing a force : $\vec{F} = -\vec{\nabla} V$, where the potential:

$$V = \frac{4\pi a^3}{3} \frac{\bar{P^2}}{2\rho_0 c_0^2}(1 - \frac{c_0^2}{c_1^2})$$

Here, c_1 is the sound velocity in the bead (for polystyrene: $c_1 = 2,300$ m/s) and $\bar{P}^2$ is the time average value of the acoustic pressure.

1. What is the trap stiffness for a free bead trapped at a pressure node?
2. What is the maximal force that can be applied on such a bead tethered by a DNA molecule?
3. Compare the sound intensity $I = P_0^2/2\rho_0 c_0$ required to generate that force to the light intensity (in W/cm^2) required to generate a similar force in optical tweezers.

3

Fluorescence Spectroscopy and Microscopy Techniques

This chapter briefly reviews single-molecule fluorescence spectroscopy and microscopy techniques. Single-molecule fluorescence can be used for a number of unique applications, such as detecting, identifying, counting, localizing, tracking, and sizing individual molecules, the latter term referring generically to single-molecule conformational studies that are performed with various spatial and temporal resolutions. Single-molecule fluorescence can also be used to probe material properties or image labelled samples at the molecular-scale level. In all these applications, it can be critical to understand the underlying optical processes at play, including optical properties of the fluorophore in various physico-chemical environments, the excitation source, excitation and collection optics, as well as the detector: add to this list of field of expertise biochemical modification, conjugation and purification techniques, surface treatment, microfluidics, and advanced mathematical analysis methods, and it becomes painfully clear that a mere chapter cannot provide more than a glancing overview of the use of single-molecule fluorescence techniques for studying nucleic acids and their associated enzymes. With this caveat in mind, this chapter is organized as follows. After a brief introduction, section 3.2 introduces the basic principles of fluorescence (light absorption, stimulated and spontaneous emission, non-radiative decay, etc.), followed in section 3.3 by a brief overview of various illumination methods. Section 3.4 presents various detection techniques and discusses the sources of noise in fluorescence spectroscopy. The use of these fluorescent methods is exemplified in the tracking of single fluorescent molecules presented in section 3.5, while the last section briefly outlines the future prospects of single-molecule fluorescence approaches. As this is a well documented but fast-evolving field, the reader is encouraged to consult a growing number of reviews and books for complementary information [57, 58, 59, 60, 61, 62].

Single-Molecule Studies of Nucleic Acids and Their Proteins. David Bensimon, Vincent Croquette, Jean-François Allemand, Xavier Michalet, Terence Strick 2019. Published in 2019 by Oxford University Press.
DOI: 10.1093/oso/9780198530923.001.0001

3.1 Introduction

Optical spectroscopy methods, which are methods that measure the optical signature of an object[1] probed with one or several light beams, are particularly adapted to the study of single molecules because of their non-invasiveness and sensitivity to molecular conformation and environment. The optical signature of a molecule indeed reports on its electronic structure, which depends on the molecule's conformation and electronic interactions with its immediate environment. These methods are, therefore, complementary to the single-molecule manipulations presented in Chapter 2, which provide very little, if any, information on the molecular conformations or environment. Both approaches are increasingly combined.

Single-molecule optical methods developed in the 1990s have reached the sensitivity required to detect [63, 64], image [65], manipulate [66], and follow the spectroscopic evolution of single light-emitting molecules (e.g., fluorophores) on surfaces [67], in solids [68], liquids [69], and fluid membranes [70], as well as inside living cells [71, 72]. Initially performed at cryogenic temperatures, spectroscopic studies of single molecules and their evolution under the influence of different external parameters have allowed precise tests of molecular quantum electrodynamics and microscopic theories of molecules' interaction with their environment [73, 74]. At higher temperatures, the temporal evolution of nanodomains in macroscopic materials such as crystals and glasses has been elucidated using embedded single molecules as reporters of their local environment [64, 75, 76]. In biological and biochemical processes, whose actors are often individual molecules (DNA, RNA, ribozymes, enzymes, motor proteins, etc.), the development of reliable and versatile fluorescent labelling techniques has made most of these events amenable to microscopic observation at the single-molecule level [77], *in vitro* and, increasingly, *in vivo* [78, 79]. The ability to fully understand enzyme behaviour and dynamics at the single-molecule level has revolutionized the way we look at these processes and is poised to become a standard method in biochemical and biophysical studies [80, 81, 82]. In the following section, we provide a mere overview of these developments, which are described more thoroughly in numerous books and reviews [74, 80, 83, 84, 85, 86, 87, 60, 61, 62].

3.1.1 Brief overview of single-molecule spectroscopy

In a typical experiment, incident light of known characteristics (wavelength, polarization, intensity, spatial and temporal profile, etc.) is directed onto the sample and the light transmitted, scattered, or emitted by the sample is collected and analysed using various spectroscopic techniques. The term "spectroscopy" is used here in a loose way and indicates that one or more of the several characteristics of the collected light (not necessarily its wavelength) can be selected and measured. Although most studies use a light source for excitation, some light emission processes do not require any, and can

[1] Properties of emitted or scattered photons, such as wavelength, intensity, polarization, lifetime, etc.

instead be triggered by chemical reactions (chemiluminescence) or enzymatic activity (bioluminescence), among other possible mechanisms. While chemiluminescence and bioluminescence can be studied at the single-molecule level [88], they have not yet been used to study nucleic acids and associated proteins and will, thus, not be discussed further.

In general, spectroscopic techniques developed for macroscopic samples can be adapted and used at the single-molecule level, provided a large enough signal/noise ratio (SNR) can be obtained. In practice, however, this may require modifying optical setups, using more sensitive detectors, employing dyes with better photophysical properties, adapting sample preparation procedures, and resorting to different data analysis methods from those used for ensemble measurements. In particular, most single-molecule measurements are performed in a low photon-count regime where statistical fluctuations (shot noise) are significant. This has several important consequences.

First, since the relative precision of a measurement (e.g., the position of a fluorophore) is inversely proportional to $\sqrt{N}$, where N is the number of collected photons, itself proportional to the measurement duration T, a trade-off needs to be found between measurement precision and temporal resolution. The photon flux ($\sim N/T$) can be increased to some extent using higher excitation power, but is fundamentally limited by the maximum rate of the photophysical process being measured, by photo-induced modifications of the molecule (photobleaching, blinking, etc.) and possible photo-induced damage to live cells and tissues. It is in general necessary to extensively characterize the various physical processes involved in the measurement (i.e., the photophysics of the molecule) to distinguish between effects due to the phenomenon of interest (e.g., changes in conformation of a biomolecule resulting in variation in the fluorescence resonant energy transfer between two dyes) and purely photophysical effects (e.g., blinking of one of the fluorescent dyes). This remark is particularly relevant when dynamics is studied, as photophysical phenomena can have characteristic time scales covering several orders of magnitude and be easily confounding.

This sensitivity of single-molecule optical measurements to a variety of influences and the possibility of photobleaching means that single-molecule optical measurements need to be repeated on many molecules to obtain a statistically significant number of observations from which effects due to photon statistics and photophysics can be deconvolved from raw data. Whereas this may sound like an inconvenience, it is, in general, relatively straightforward to achieve, as will be discussed in more detail in the section describing some of the most common experimental designs. Repeating the measurement on several hundreds or thousands of molecules has additional advantages—in particular, when the molecular properties to be studied are heterogeneous. In the case of equilibrium properties (static over the time scale of the measurement), such measurements will give access to the full distribution of properties, rather than just the average (and possibly the variance) of these properties, as would be obtained in the corresponding ensemble measurement. In the case of out-of-equilibrium properties (such as the dynamic evolution of a molecular reaction), measurements on multiple molecules may reveal the stochastic nature of the dynamics, which would be hidden and averaged out in ensemble measurements.

Note that in some favourable cases, it is possible to observe or perform spectroscopic measurements on individual molecules for long periods of time, but this is, in general, not sufficient to fully characterize the population (static heterogeneity) or the dynamics (dynamic heterogeneity), due to the breaking of ergodicity in some systems with deep energy traps or a large number of metastable states [89, 90, 91]. The fact that large numbers of individual measurements are necessary and possible with optical methods makes these techniques complementary to single-molecule manipulation ones, which are not as easily amenable to parallelization, but instead draw their strength from their ability to repeat a measurement over and over using the same molecule. However, optical methods are not necessarily high-throughput, and much progress in this direction is to be expected in the future.

While this chapter discusses light emission by a single molecule and which information can be extracted from its measurement, it is worth mentioning that recent developments have also shown how to control enzyme activity or molecular conformation by light [92, 93]. These novel "optogenetic" tools, combined with the spectroscopic and single-molecule manipulation techniques described in this book, promise to transform the way single-molecule experiments are performed, and the information that can be extracted from those experiments.

3.1.2 Historical outlook

Single-molecule optical methods have an interesting history, starting with the pioneering studies of J. Perrin on Brownian motion at the beginning of the 20th century and ending in the rapid and spectacular developments of the last decade of that century. The first (though unsuccessful) attempt to detect individual fluorescent molecules can probably be attributed to Perrin himself [94], who described his experiments in the following terms:[2]

> S'il est maintenant difficile de douter de la réalité des molecules, nul ne contestera de l'intérêt qu'il y aurait à percevoir directement ces molécules. Je me suis demandé si quelque phénomène permettrait de rendre visible, sinon les molécules individuellement, du moins les fluctuations de concentration qui doivent exister à chaque instant, par suite de l'agitation moléculaire, au sein d'une solution étendue. Espérant trouver dans la fluorescence un tel phénomène, j'ai observé au microscope des solutions fluorescentes, soit en couche très minces dans des cuves spéciales, soit en lame minces (bulles de savon). Je n'ai pas réussi à percevoir les molécules isolées, ni même les fluctuations, mais chacune de ces deux recherches m'a permis de constater des phénomènes nouveaux.

[2] "As it is now difficult to doubt the reality of molecules, nobody will discuss the interest of being able to see these molecules directly. I have asked myself whether some phenomenon could allow to make visible, if not the individual molecules, at least the concentration fluctuations, which must take place at any time, due to molecular motion inside an extended solution. Hoping to find such a phenomenon in fluorescence, I have observed fluorescent solutions with a microscope, either as very thin slabs in special cuvettes, or as thin layers (soap bubbles). I have not been able to see isolated molecules, not even fluctuations, but both researches have led me to the discovery of new phenomena."

Despite his failure to detect single molecules or their number fluctuations, Perrin had captured one of the key ingredients of single-molecule sensitivity, namely the high dilution necessary to have a chance of observing them individually. All he was missing was a powerful light source and efficient way of rejecting background light. Half a century later, Hirschfeld noted that the total number of photons emitted by a molecule before bleaching does not depend on its quantum yield [95], concluding that a brief and intense illumination flash resulting in a detectable signal/noise ratio should permit detection of molecules labelled with very few fluorophores. He confirmed this prediction by detecting individual antibody molecules labelled with one hundred fluorophores as bursts of fluorescence of the expected duration and intensity when mechanically moved through a region illuminated in total internal reflection mode [95].

Although very little needed to be changed to detect individual fluorescent molecules, it took another decade before this became technically feasible. Actin molecules labelled with fewer than eight fluorophores were first observed with video-microscopy [96]. Shortly after, the first demonstration of true single-molecule detection in solution was presented in a flow-cytometry experiment designed for DNA sequencing [69]. The collimated laser beam and photomultiplier tubes (PMT) of that experiment were later superseded by a confocal excitation and detection geometry and more sensitive single-photon avalanche photodiodes (SPAD)[97], which are now amongst the most commonly used arrangements for single-molecule bursts [98, 99] or Fluorescence Correlation Spectroscopy (FCS) analysis [100, 101, 102] of freely diffusing fluorescent molecules.

The first image of a single molecule at room temperature was obtained shortly after using near-field scanning optical microscopy (NSOM)[65], followed by similar achievements using epifluorescence or Total Internal Reflection Fluorescence (TIRF) microscopy [88] and confocal microscopy [103]. Although imaging resolution was lower than in NSOM, the loss was more than compensated for by experimental simplicity. TIRF (used in the original experiment of Hirschfeld) and epifluorescence imaging have been used to image and study the motion of single fluorescent molecules in model systems such as supported bilayers [70] or polymer films [104] and, eventually, in live cells [72, 71]. More sophisticated illumination schemes have been recently introduced, allowing enhanced localization resolution along one or more directions. Super-resolution imaging, based on the localization of individual molecules with nanometre resolution, is one of the many interesting recent applications of this kind of approach, and is discussed later. A parallel and independent path was explored in the domain of spectroscopic studies of single molecules embedded in crystals at cryogenic temperatures [63, 64]. These studies were initially performed as precise tests of molecular quantum electrodynamics and the microscopic theories of the interactions of molecules with their environment [73, 74]. They were later extended to higher temperatures to monitor the evolution of nanodomains and study the role of defects and impurities in macroscopic materials, such as crystals and glasses [75].

Although a majority of single-molecule optical studies exploit fluorescence and measure one or more of its characteristics (intensity, polarization, lifetime, etc.), other spectroscopic signatures have been measured at the single-molecule level as well, such as Raman (in surface-enhanced Raman spectroscopy, SERS) [105], hyper-Raman [106],

Stark [107], and absorption spectroscopy [108]. In fact, complete control of the electronic quantum mechanical state of individual molecules can now be achieved using optical means [109]. This chapter will focus on single-molecule fluorescence techniques due to their preponderance and versatility, made possible by the relative simplicity of the instrumentation needed and availability of multiple fluorescent labelling techniques, which allow the observation of most molecular actors of biological and biochemical processes (DNA, RNA, ribozymes, enzymes, motor proteins, etc.) at the single-molecule level, *in vitro* and *in vivo* [80, 110, 82, 77].

3.2 Principles of fluorescence

Light–matter interaction (by absorption or scattering) is fundamentally a quantum process, requiring an understanding of the interaction between an electromagnetic (EM) wave (i.e., light of frequency ν or wavelength $\lambda = c/\nu$) and a quantum object (an atom or molecule) [111, 112]. The key ingredients of this interaction were formulated by Einstein [113] in 1917, see Fig. 3.1, who built on Planck's hypothesis that the EM energy came in quanta (known as photons) of energy $h\nu$, where h is Planck's constant: $h = 6.626\ 10^{-34}$ Js $=$ $4.135\ 10^{-15}$ eVs. Thus, the energy of a photon of green light (with $\lambda = 530$ nm) is 2.34 eV ($= 3.75\ 10^{-19}$ J).

The second fundamental assumption adopted by Einstein was that the energy levels in atoms and molecules are also quantized, so that the transition between an initial state $|i\rangle$ of energy E_i and a final state $|f\rangle$ of energy $E_f > E_i$ can only be induced by absorption of a photon of energy: $h\nu_{if} = E_f - E_i > 0$. The transition rate k_{if} from state $|i\rangle$ to state $|f\rangle$ depends only on the local spectral density of photons $\rho(\nu)$ (the number per unit volume of photons of energy $h\nu$) in a frequency bandwidth $d\nu$ around ν_{if})[3]:

$$k_{if} = B_{if}\,\rho \tag{3.1}$$

where B_{if} is known as the Einstein coefficient for absorption (its unit is m^3s^{-2}, for other definitions see [114, 115]). Einstein noticed that the excited state $|f\rangle$ could decay back to state $|i\rangle$ by a process known, ever since, as stimulated emission, which is the time reversal of absorption: the *light-driven emission* of a photon of energy $h\nu_{if}$ from high energy state $|f\rangle$ to lower energy state $|i\rangle$ as a result of the interaction of the atom/molecule with photons of density ρ:

$$k_{fi} = B_{fi}\,\rho \tag{3.2}$$

B_{fi} is known as the Einstein coefficient for stimulated emission. Due to microscopic time reversal symmetry, $B_{fi} = B_{if}$. The laser (acronym for Light Amplification by Stimulated Emission of Radiation) is a direct outcome of Einstein's insight [115].

[3] The unit of ρ is $m^{-3}s$ and the light intensity I (measured in W m^{-2}) is $I = \int I(\nu)d\nu = c\int h\nu\rho(\nu)d\nu$. In all the discussion below the initial and final states are assumed to be non-degenerate.

Einstein finally pointed out that high energy states are inherently unstable and may revert spontaneously (independently of the illumination intensity, like radioactive decay) to a lower energy state by fluorescent emission, i.e., by emission of a photon of appropriate energy. This process is random and occurs within a time τ characterized by an exponential distribution: $P(\tau) = \exp(-\tau/\tau_s)/\tau_s$. From the mean lifetime τ_s of the excited state, one deduces the spontaneous transition rate: $k_s = 1/\tau_s$. For a typical fluorescent molecule $\tau_s \sim 1-10$ ns.

In steady state, the rate of absorption by the atoms in their ground state (at density n_i) must equal the rate of emission (stimulated and spontaneous) of the atoms in their excited state (with density n_f), see Fig. 3.1:

$$B_{if}\rho n_i = B_{fi}\rho n_f + k_s n_f \tag{3.3}$$

from which one derives:

$$\frac{n_f}{n_i} = \frac{\rho}{\rho + k_s/B_{if}} \tag{3.4}$$

The absorption saturates when the photon spectral density $\rho > k_s/B_{if}$, i.e., when the intensity of the illuminating beam is larger than:

$$I_s = h\nu_{if} c\pi \Delta\nu_f / \tau_s B_{if} \tag{3.5}$$

where $\Delta\nu_f$ is the spectral width of the excited state, see Eq. 3.8 below.

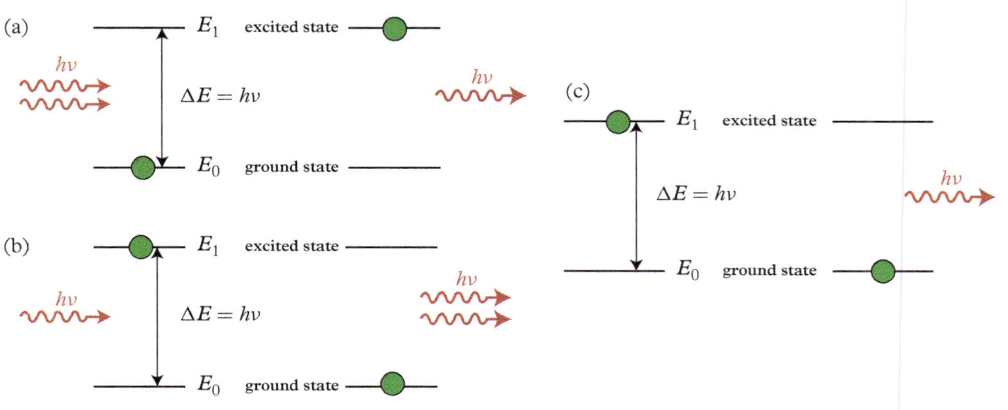

Fig. 3.1 Absorption and stimulated and spontaneous emission. *(a) The absorption of photons of density ρ by an atom in its ground state results in a transition to a higher energy state with rate: $k_{ex} = B_{01}\rho$. (b) The stimulated emission of a photon in presence of radiation by an atom in its excited state. This process is the time reversal of absorption and occurs with rate $k_{st} = B_{10}\rho$ (with $B_{10} = B_{01}$). (c) The spontaneous emission of a photon by an atom in its excited state at a rate k_s which is independent of the illumination intensity.*

Einstein was concerned with a microscopic explanation of Planck's black-body radiation formula. He, thus, considered the case of a thermodynamic equilibrium between the radiation field and absorbing/emitting atoms, for which: $n_f/n_i = e^{-\beta h\nu_{if}}$ ($\beta = 1/k_B T$, where T is the temperature and k_B is the Boltzmann constant). Hence, from Eq. 3.4, Planck's black-body formula is obtained:

$$\rho(\nu_{if}) = \frac{k_s}{B_{if}} \frac{1}{e^{\beta h\nu_{if}} - 1} = \frac{8\pi \nu_{if}^2}{c^3} \frac{1}{e^{\beta h\nu_{if}} - 1} = \frac{8\pi}{c\lambda_{if}^2} \frac{1}{e^{\beta h\nu_{if}} - 1} \tag{3.6}$$

provided the product $B_{if}\tau_s$ satisfies [114, 115]:

$$B_{if}\tau_s = c\lambda_{if}^2/8\pi. \tag{3.7}$$

This equation relates the coefficient of absorption to the spontaneous (or radiative) lifetime τ_s of the excited state. The longer this lifetime, the smaller the absorption coefficient. The radiative lifetime sets a limit on the longest lifetime of an excited state and is observable only at very low temperatures. In practice, for complex molecules at room temperature, due to the existence of multiple excited states (see section 3.2.1) and to non-radiative decay channels (internal conversion, collisions with surrounding molecules, Doppler broadening, intersystem crossing, etc.) the description is more complex [57,60,62]. For simplicity we shall assume here that these non-radiative effects result in a shortening of the lifetime of the excited state τ_f. The decreased lifetime affects (via Heisenberg's uncertainty principle) the density of states around E_f, described by the probability of absorption $g(\nu)$ of a photon of frequency ν close to ν_{if} [116]:

$$g(\nu) = \frac{\Delta\nu_f}{\pi} \frac{1}{(\nu - \nu_{if})^2 + \Delta\nu_f^2} \tag{3.8}$$

with $\Delta\nu_f = 1/4\pi\tau_f$. As a result of this shortening of the excited state lifetime, the intensity I_s of the illuminating beam required to significantly populate the excited state becomes (from Eq. 3.5 and Eq. 3.7):

$$I_s = \frac{2\pi h\nu_{if}}{\lambda_{if}^2 \tau_f} \tag{3.9}$$

For Rhodamine 6G [101] excited at $\lambda_{if} = 514$ nm, with a typical excited state (non-radiative) lifetime $\tau_f \approx 1.6\ 10^{-15}$s:$I_s \approx 500$ kW/cm^2. In practice due to long lived triplet states (see section 3.2.2), the saturation intensity is often smaller by an order of magnitude. While intensities $50 - 100$ kW/cm^2 are easily achieved by focusing a laser beam of a few mW with a microscope objective, too long an illumination is highly detrimental to biological samples. Consequently, these intensities are often only used for in vitro single-molecule studies of diffusing molecules. To limit photo-induced damage to biological tissues, *average* illumination intensities in the visible range are usually kept below 1 W/cm^2. For peak illumination power of a few MW/cm^2 (encountered in beam-scanning confocal microscopes), the illumination time per pixel is kept as short as possible (a few μs; for extended observations of a single fluorophore at a fixed position, stroboscopic illumination is the preferred option, with a total illumination energy below 50 mJ/cm^2 [117, 118]).

The excited state lifetime also sets the absorption cross section of light: σ_{if}. This cross section is defined by considering a light beam of spectral intensity $I(\nu, x)$ propagating in a medium which contains a density n_i of excitable molecules (e.g., fluorophores):

$$\frac{dI(\nu,x)}{dx} = -h\nu B_{if}\rho(\nu,x)g(\nu)n_i = -B_{if}g(\nu)n_iI(\nu,x)/c \equiv -\sigma_{if}n_iI(\nu,x) \tag{3.10}$$

with $\sigma_{if}(\nu) = B_{if}g(\nu)/c$. The larger the absorption cross section, the more efficient the excitation of the fluorophore. From Eqs 3.7 and 3.8, the maximal absorption (at $\nu = \nu_{if}$) is [74]:

$$\sigma_{if}^{max} = \frac{\lambda_{if}^2 \tau_f}{2\pi \tau_s} \tag{3.11}$$

In vacuum and at low temperatures, when the excited state lifetime approaches its limit $\tau_f \to \tau_s$, the absorption cross section is the largest and set by the excitation wavelength [63] (usually in the visible range $\sim 0.5\ \mu$m). However, in more common situations, the excitation lifetime is set by non-radiative effects (e.g., quick collision-induced transition to lower energy states) and can be $10^5 - 10^6$ times shorter than τ_s. The absorption cross section is then closer to atomic dimensions: $\sigma_{if}^{max} \sim 1$ Å$^2 = 10^{-16}$cm^2 [74]. For Rhodamine 6G [101] with $\tau_s = 4\ ns$, $\sigma_{if}^{max} = 1.7\ 10^{-16}$cm^2. As a consequence of absorption, in a solution of fluorophores at concentration n_i particles cm^{-3} (i.e. at molar concentration $c_M = 10^3 n_i/N_A$, where $N_A = 6\ 10^{23}$ particles/mol is Avogadro's constant), the light intensity decreases (by integration of Eq. 3.10), according to the Beer–Lambert law:

$$I(\nu,x) = I(\nu,0)e^{-\sigma_{if}n_i x} = I(\nu,0)10^{-\epsilon c_M x}$$

where $\epsilon = 10^{-3}N_A\sigma_{if}/\log(10)$ is known as the molar extinction coefficient. Its value for good fluorophores is $\epsilon \sim 80000$ M^{-1}cm^{-1} ($\sigma_{if} \sim 3\ 10^{-16}$cm^2).

3.2.1 Transitions between energy levels

Atoms and molecules display quantized energy levels, which are typically of three types: electronic, vibrational, and rotational.

- Electronic energy levels are associated with the various quantum states adopted by the electrons in an atom/molecule. The typical difference between successive electronic energy levels is ~ 1 eV.
- Vibrational energy levels are associated with the possible vibrational modes of the molecule and superpose onto the electronic energy levels. The difference between successive vibrational energy levels is typically ~ 0.1 eV.

- Finally, rotational energy levels are associated with the possible 3D rotational modes of the molecule. The energies associated with these degrees of freedom are usually very small: ~ 1 meV.

These various energies have to be compared to the energy supplied by thermal fluctuations (shocks from the environment) of amplitude: $k_B T \approx 25$ meV. Therefore, at thermal equilibrium, only the lowest electronic and vibrational energy level is occupied, whereas around that ground state $|0,0\rangle$ various rotational energy levels will be populated.

In fluorophores, absorption of a photon of energy $h\nu_{ex}$ causes the molecule to transit to a high electronic and vibrational energy state. From that excited state, it decays rapidly via non-radiative transitions to the lowest vibrational energy level (the emission state), wherefrom it returns to a state in a lower electronic energy level by emission of a photon of energy: $h\nu_{em} < h\nu_{ex}$. In steady state conditions, the rate of absorption k_{if} of light of wavelength $\lambda_{ex} = \lambda_{if}$

$$\begin{aligned} k_{if} = B_{if}\rho(\nu_{if}) &= \frac{c\lambda_{ex}^2}{8\pi\tau_s}\frac{I_{ex}}{ch\nu_{ex}}\frac{1}{\pi\Delta\nu_f} \\ &= \frac{\lambda_{ex}^2\tau_f}{2\pi\tau_s}\frac{I_{ex}}{h\nu_{ex}} = \sigma_{ex}^{max}\frac{I_{ex}}{h\nu_{ex}} \end{aligned} \tag{3.12}$$

is equal to the average rate of emission k_{em} of light of power P_{em}^0:

$$P_{em}^0 = k_{em}h\nu_{em} = \sigma_{ex}^{max}\frac{\nu_{em}}{\nu_{ex}}I_{ex} \approx \sigma_{ex}^{max}I_{ex} \tag{3.13}$$

For a typical fluorophore $\nu_{em} \approx \nu_{ex}$ and Eq. 3.13 simply reflects energy conservation: when non-radiative losses are small, the power emitted is roughly equal to the power absorbed. Notice that k_{em} is bounded by the lifetime of the emission state: $k_{em} \leq k_s = 1/\tau_s$. As the excitation intensity is increased, the emitted power increases until it saturates at an intensity $P_{em}^{max} = h\nu_{em}/\tau_s$, typically of order 10^{-10}W = 100 pW. The excitation intensity required to achieve that saturation is $I_s = h\nu_{ex}/\tau_s\sigma_{ex}^{max}$, see Eq. 3.9, which as already mentioned is an over-estimate since intersystem crossing to triplet states (section 3.2.2) may reduce that estimate by an order of magnitude.

In a typical confocal microscopy experiment (see section 3.3) a laser beam of wavelength $\lambda_{ex} = 0.5$ μm and incident power $P_{ex} = 0.1$ mW is focused by an objective with a high numerical aperture (NA = 1.4) into a diffraction-limited spot of diameter $d = 1.22\lambda_{ex}/\text{NA}$. The excitation intensity at the focal point is thus: $I_{ex} \approx 67$ kW/cm^2. The emitted power for a fluorophore located at the focus and absorbing with a typical cross section $\sigma_{ex}^{max} \sim 10^{-16}$ cm^2 is thus: $P_{em}^0 \approx 6.7$ pW. This value is orders of magnitude smaller than the excitation power and, therefore, requires careful filtering of the excitation light to reduce the background light intensity to a negligible level. One source of background light is due to the elastic (Rayleigh) scattering of the incident beam (photons at frequency ν_{ex}, which can be filtered out).

Another inevitable source of background noise in fluorescence experiments is due to the inelastic (Raman) scattering of the excitation beam by the solvent (e.g., water) molecules, which results in background radiation of photons at frequency ν_{em} that cannot, therefore, be filtered out. Even though the Raman absorption cross sections ($\sigma_{Raman} \sim 10^{-28}$ to 10^{-30} cm^2 [119–121]) are much lower than the absorption cross section of a fluorophore, the number of molecules (water or impurities) involved is much larger than that of the fluorophores of interest. The resulting background fluorescence, proportional to the excitation/detection volume, could potentially overwhelm the fluorescence signal. For example, in the confocal situation previously discussed, the typical excitation volume is of the order of 1 fl (1 μm^3) and contains about $N_{H_2O} = 3.3\ 10^{10}$ water molecules. With a Raman cross section at 500 nm of $\sigma_{Raman} \sim 10^{-28}$ cm^2 per water molecule, the total effective scattering cross section is $\sigma^{tot}_{Raman} = 3.3\ 10^{-18}$ cm^2, a factor only 30 smaller than a typical fluorophore absorption cross section. If that cross section is smaller than the value quoted above or if the excitation volume is larger than 1 fl (for instance, due to improper filling of the back aperture of the objective lens), the background fluorescence intensity due to Raman scattering could grow and possibly compromise the detection of the single-molecule fluorescence signal. Alternatively, since fluorescence emission tends to be broad, an emission filter can be selected that allows most of the fluorescence to be transmitted while blocking the water Raman peak (red-shifted by an amount $\Delta\lambda = \alpha_R \lambda^2_{ex}$ with $\alpha_R \sim 3500$ cm^{-1}). An excitation wavelength λ_{ex} can also be chosen (within limits of available lasers) that does not necessarily match the fluorophore peak value, but gives a Raman peak clear of the main emission wavelength.

With these results in mind, it is an appropriate time to briefly discuss the Quantum Mechanical analysis supporting them.

When a molecule is exposed to visible light, electrons are excited from state $|i\rangle$ (in general $|i\rangle$ is the ground state $|0,0\rangle$) to higher electronic and vibrational energy states, $|f\rangle$ (e.g., state $|1,n\rangle$), see Fig. 3.2. If H_I is the Hamiltonian describing the interaction between the electron and EM wave, Dirac showed [122] that the transition rate was given by:

$$k_{if} = \frac{2\pi\,|\,\langle i|H_I|f\rangle\,|^2}{\hbar}\rho(E) \tag{3.14}$$

where $\langle i|H_I|f\rangle$ is the transition matrix element between states $|i\rangle$ and $|f\rangle$ and $\rho(E)$ is the density of states at energy $E = E_i + h\nu$ (with $\int dE\rho(E) = 1$). The density of states [116] can be written as a Lorenzian centred around E_f (see Eq. 3.8):

$$\rho(E) = \frac{1}{\pi}\frac{\Delta E_f}{(E - E_f)^2 + \Delta E_f^2} \tag{3.15}$$

where $\Delta E_f \equiv h\Delta\nu_f$ is the uncertainty in the excited state energy, related by the Heisenberg uncertainty principle to its lifetime τ_f: $\Delta E_f \cdot \tau_f = \hbar/2$. Equation 3.14, which sets the rates of atomic and molecular transitions is known as Fermi's golden rule.

For a neutral system, the main interaction with an EM wave (with electric field $\vec{E}$) is usually of a dipole type: $H_I = -\vec{p}\cdot\vec{E}$, where $\vec{p} = \sum_i q_i\vec{r}_i \equiv e\vec{r}$ is the dipole moment of a molecule with charges q_i located at positions $\vec{r}_i$ ($e\vec{r}$ is the average dipole moment) and where neutrality implies $\sum_i q_i = 0$. In this case (if $\vec{E} \parallel \hat{z}$), the maximal transition rate becomes [115]:

$$k_{if}^{max} = (2\pi)^2 \frac{e^2}{4\pi\epsilon_0\hbar c} |z_{if}|^2 \, I\rho(E_f) = 4\pi\alpha |z_{if}|^2 \, I/\Delta E_f \tag{3.16}$$

where $\alpha \equiv e^2/4\pi\epsilon_0\hbar c \approx 1/137$ is known as the fine structure constant and ez_{if} is the molecular dipole element coupling between states $|i\rangle$ and $|f\rangle$ (or transition dipole moment). Comparing with Eq. 3.12 yields:

$$\sigma_{if}^{max} = 16\pi^2\alpha |z_{if}|^2 \, \nu\tau_f \tag{3.17}$$

For fluorophores in solution, the excited state lifetime τ_f is determined by non-radiative effects such as collisions with the environment which bring the excited electron down from vibrational excited states to the lowest vibrational level within a sub-picosecond (ps) time scale, see Fig. 3.2. In such a case, $\nu\tau_f$ is typically of $O(1)$ and the absorption cross section is of $O(10^{-16}\ \mathrm{cm}^2)$, as argued earlier.

For atoms or molecules in vacuum, the only de-excitation pathway is via spontaneous radiative decay and, therefore, $\tau_f = \tau_s$, which by Einstein relations, Eq. 3.11, yields [114]:

$$\tau_s = \frac{\lambda^3}{32\pi^3\alpha c |z_{if}|^2} \tag{3.18}$$

In that case, the absorption cross section is: $\sigma_{if}^{max} = \lambda^2/2\pi$, which is orders of magnitude larger than in solution (as shown in Eq. 3.11). The calculation of z_{if} is a complex problem, which is rarely conducted for common fluorophores (and is too complex for fluorescent proteins). However, this approach has been validated for simple di-atomic molecules (such as CO) and it provides an explanatory framework for the excitation and emission behaviour of more complex fluorophores.

It is customary to schematically represent excitation and emission processes by a Jablonski diagram (see Fig. 3.2) depicting the initial, final, and intermediate electronic and vibrational states of the molecule. Upon absorption of a photon of energy $h\nu_{ex}$, a molecule in the electronic and vibrational ground state $|0,0\rangle$ is promoted to a vibrational sublevel $n = 0, 1, 2, \ldots$ of a higher electronic excited state, here assumed to be state $|1,n\rangle$: $h\nu_{ex} = E_{1,n} - E_{0,0}$. Non-radiative, fast (sub-ps) relaxation brings the molecule down to the lowest vibrational level of the excited electronic state (the emission state): $|1,0\rangle$. Radiation of a photon of energy $h\nu_{em} = E_{1,0} - E_{0,n} < h\nu_{ex}$ takes place from that emission state within nanoseconds (ns) and brings back the molecule to one of the vibrational sublevels of the electronic ground state $|0,n\rangle$, from which fast (sub-ps) non-radiative relaxation brings it back to the electronic ground state $|0,0\rangle$.

Since the vibrational states associated with the electronic states $|0\rangle$ and $|1\rangle$ are similar, the emission spectrum—while shifted to lower frequencies (higher wavelengths—is the mirror image of the absorption spectrum, see Fig. 3.3. The difference between the wavelengths at the peak of emission and peak of absorption is known as the Stokes shift. The larger that shift, the easier it is to separate (filter out) the excitation light from the emitted fluorescence. This is crucial for the observation of single fluorophores, as the

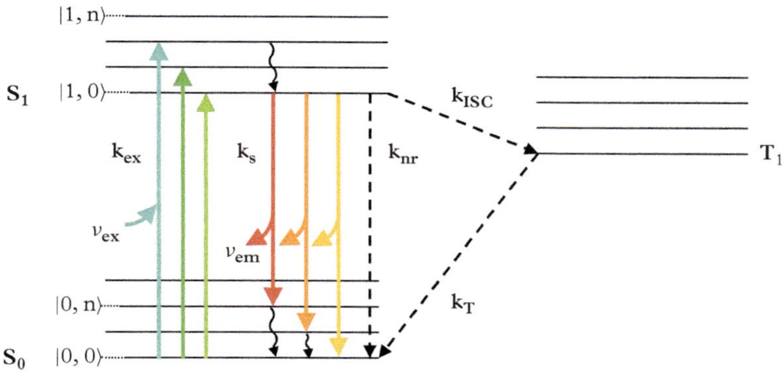

Fig. 3.2 Jablonski diagram for fluorescence absorption and emission. *Upon absorption of light at frequency ν_{ex}, an electron is excited with rate k_{ex} from the electronic and vibrational ground state $|0,0\rangle$ (spin-singlet state S_0) to a higher electronic/vibrational state, such as state $|1,n\rangle$ (where n labels the vibrational quantum state). Within a sub-ps timescale, it decays non-radiatively to the vibrational ground state $|1,0\rangle$ (spin-singlet state S_1). From emission state S_1, it returns within ns (at rate k_s) by radiation of a photon of frequency ν_{em} to one of the vibrational states associated with the electronic ground state $|0,n\rangle$. In the emission state, non-radiative de-excitation is possible with rate k_{nr}, as well as inter system-crossing to a long-lived triplet state T_1 with rate k_{ISC}. This triplet state decays back to the ground state within a time τ_T (i.e. at rate $k_T = 1/\tau_T$). Notice that due to quick non-radiative decay channels the energy levels are broadened and may overlap as shown in Fig.3.3. In such a case the excited state is a quantum mechanical superposition of the pure states $|1,n\rangle$.*

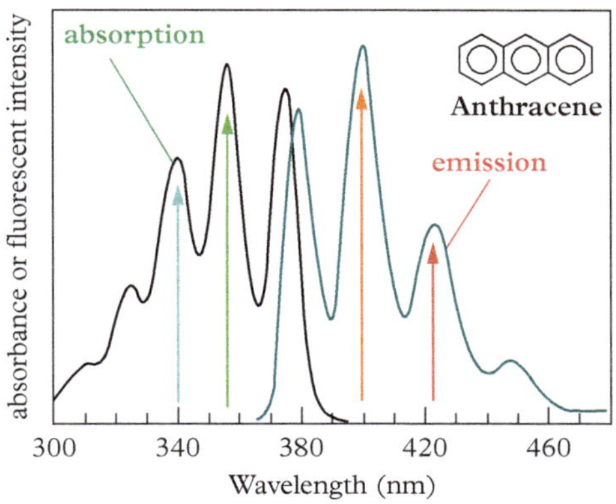

Fig. 3.3 Absorption and emission spectra of anthracene. *Notice the mirror symmetry (about 380 nm) of the two spectra. Due to the similarity of the vibrational energy states associated with the lowest electronic states, the absorption and emission spectra are a mirror image of one another, with the highest absorption line corresponding to the lowest emission line. Adapted from C.M. Byron and T.C. Werner [123].*

intensity of the back-scattered excitation light is usually much higher than the intensity of the emitted light, as shown in Eq. 3.13.

Non-radiative decay. Once in the emission state, a fluorophore often relaxes to a lower energy state $|0,n\rangle$ by radiation of a photon of energy $h\nu'$ ($h\nu' = E_f - E_n$) but it can also decay via the transfer of its energy to other atoms through non-radiative effects (e.g., collisions, dipole–dipole interactions, etc.). If the rate of these non-radiative processes is k_{nr}, the probability Φ of fluorescent emission, also known as the quantum efficiency of the fluorophore, is:

$$\Phi = \frac{k_s}{k_s + k_{nr}} \tag{3.19}$$

and the emitted power is $P_{em} = \Phi P^0_{em}$. Good fluorophores achieve quantum efficiencies $\Phi > 0.5$, with the best ones getting close to 1 in appropriate solvents. The quantum yield is, however, dependent on the molecule's environment, as new non-radiative relaxation channels can be opened, for instance, by the proximity of electron acceptor groups. One should, therefore, measure the quantum yield of a fluorophore in each new environment, e.g., at each new attachment site on target molecules that it labels. This measurement can be done at the ensemble level via comparison with a fluorophore of known quantum yield (or with the same fluorophore in a known environment) [112].

A convenient figure of merit allowing a simple comparison of fluorophores is provided by their brightness $Br = \epsilon \cdot \Phi$, which measures the amount of emitted fluorescent photons given an incident excitation intensity. Good fluorophores (e.g., fluorescent proteins such as Venus or Tomato) have typical brightness $Br > 50000\ M^{-1}cm^{-1}$.

Because of non-radiative processes, the fluorescence lifetime τ_F of the emission state is shortened:

$$\tau_F = (k_s + k_{nr})^{-1} < \tau_s \tag{3.20}$$

The fluorescence lifetime is, thus, sensitive to environmental variables influencing the total non-radiative decay rate k_{nr}, such as the proximity of another fluorophore (see section 3.2.4) or a dielectric/metallic surface [103, 124, 125, 126]. Changes in τ_F are, therefore, indicative of an alteration of the molecular environment. Due to the short lifetime of most fluorophores (ns scale), single-molecule lifetime measurements necessitate the use of a pulsed laser source with picosecond or femtosecond pulse-width and time-correlated single-photon counting (TCSPC) and analysis techniques. Once an expensive and sophisticated technical challenge, the recent apparition of relatively affordable pulsed laser diodes and integrated electronics associated with software tools [127, 128] has significantly expanded its use in single-molecule analysis.

3.2.2 Triplet states, blinking, and bleaching

In a molecule, the electrons in the various energy levels are usually paired with opposite spin states (due to Pauli's exclusion principle). The ground state $|0,0\rangle$ of the molecule is, thus, a spin-singlet state (with spin $S_z = 0$). Since the EM field couples weakly to the electron spin, the molecule's spin state has low probability of changing upon absorption or emission of EM radiation. Hence, upon excitation an electron jumps from a singlet ground state to a excited singlet state in which its spin is opposite to that of its paired electron in the ground state. If the spin-orbit coupling in the excited singlet state is strong (thus, coupling the electron angular momentum and spin) and if the energy of the triplet state (with spin $S_z = -1, 0, 1$) overlaps with the energy of the excited singlet state, the spin of the electron in the excited state may flip (and its angular momentum change by one unit) at a rate k_{ISC} so that the molecule ends up in a triplet state, a phenomenon known as intersystem crossing. Because the triplet state has a low probability of radiative decay to the singlet ground state, it is very long-lived (lifetimes τ_T in the μs–s range) and results in reduced fluorescence quantum efficiency. The radiative decay of the triplet state (if and when it happens) is the cause of phosphorescence.

By sequestering a fraction of the excitations for a long time the presence of triplet states reduces the intensity, I_s, required to saturate the emission state in their absence, Eq.3.9:

$$I_s^T = \frac{I_s}{1 + k_{ISC}\tau_T}$$

The product $k_{ISC}\tau_T$ is about 2 for Rhodamine 6G in water and about 12 for FITC in water (with a reduction in the saturation intensity I_s by a factor 3 and

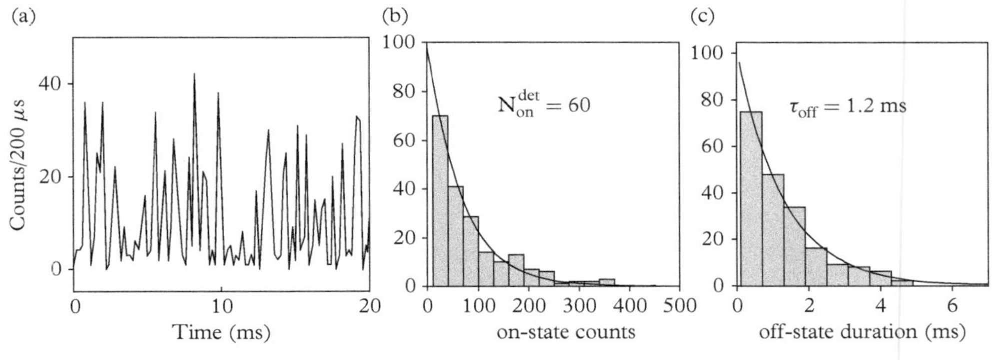

Fig. 3.4 Single-molecule blinking. *(a) Fluorescence time trace (20 ms excerpt) of a Texas Red (TR) molecule attached to a 20-nucleotide long single-stranded DNA molecule immobilized on a silanized glass coverslip [129] observed using a confocal setup (see section 3.3). Excitation: CW 514 nm (circularly polarized)* 6 kW/cm^2*, binning:* 200 μs*. On- and off-states (the latter corresponding to the molecule being in the triplet state) alternate stochastically. Histograms of on-state count (b) and off-state duration (c) for the partial time trace shown in (a). The decay parameters of the exponential fits (solid line) are indicated.* $\langle\tau_{off}\rangle = 1.2$ *ms is the triplet state lifetime. The number of emitted photons is* $N_{on} = N_{on}^{det}/\eta = 600$*, where* N_{on}^{det} *is the number of detected photons and* $\eta \sim 0.1$ *is the setup detection efficiency. Adapted from T. Ha et al. [129].*

13 respectively) [101]. The transition to triplet states thus significantly reduces the saturation intensity estimated from transitions among singlet states only, Eq.3.9.

The existence of a triplet state, which is generally non-emitting, can be observed directly by the disappearance of fluorescence emission for periods of microseconds to milliseconds, i.e., by the existence of *dark* or off-states. Figure 3.4 shows an example of this phenomenon observed on a Texas Red (TR) molecule attached to a 20-nucleotide ($\sim$7 nm) long single-stranded DNA molecule tethering it to a silanized glass coverslip [129]. Statistical analysis of the distribution of off-time τ_{off} and number of photons N_0 emitted in a burst of excitation allows an estimation of the mean triplet state lifetime $\tau_T = \langle \tau_{off} \rangle = 1.2$ ms (an overestimate since short lifetimes are difficult to measure). The probability of intersystem crossing Φ_{ISC} for this molecule was found to be: $\Phi_{ISC} = N_0^{-1} = 0.17\% = k_{ISC}\tau_F$, which for $\tau_F = 4.2$ ns yields: $k_{ISC} = 4\ 10^5 s^{-1}$. Due to intersystem crossing the saturation intensity is thus reduced by a factor $k_{ISC}\tau_T \sim 500$, which is more than is measured probably due to the over-estimation of τ_T [129]. Other (if not all) fluorescent single molecules, such as the Green Fluorescent Protein (GFP) [130] or quantum dots (QD) [131, 132, 133], exhibit similar dark state intervals, although for different reasons (different long-lived dark states for GFP, Auger ionization, or surface trapping of carriers for QD).

The ground state of oxygen (O_2) being a triplet state, it is highly reactive with other molecules in a triplet state (and much less so with molecules in a singlet state, due to spin number conservation). Hence, as a result of intersystem crossing, an excited fluorophore can be destroyed by its oxidation, which is the main cause of photobleaching, namely the irreversible loss of fluorescence due to photo-activated alteration of the fluorophore. Thus, removal of oxygen (if it does not interfere with the studied system) is essential for long-term observation of fluorophores. Various protocols exist to achieve that goal [62, 134, 135].

3.2.3 Fluorescence polarization spectroscopy

For the dipolar interaction considered above, the efficiency of photon absorption is proportional to $(\vec{p} \cdot \vec{E})^2 \propto \cos^2\theta$ and maximal when the molecular dipole is oriented along the oscillating electric field $\vec{E}$ [112, 136, 137]. Consequently, the emitted fluorescence intensity is sensitive to the orientation θ of the molecule with respect to the electric field polarization. Depending on the respective time scales of the electric field and molecular rotational dynamics and the experimental temporal resolution, different types of information on the molecular orientation can be extracted.

Consider the case where a fluorophore is labelling a molecule bound to a surface, as in Fig. 3.5. The analysis of the emitted fluorescence polarization allows us to determine the average orientation of the fluorophore absorption dipole moment $\vec{p}_a(t=0)$, its freedom of rotation, and rotational diffusion constant [138]. For an excitation field propagating along the z-axis and whose electric field $\vec{E}$ is linearly polarized at an angle ϕ_e with respect to the x-axis, the absorption probability P_a is:

$$P_a \propto (\vec{p} \cdot \vec{E})^2 = p^2 E^2 \sin^2\theta \cos^2(\phi - \phi_e) \tag{3.21}$$

where θ, ϕ are the angles the absorbing dipole makes with the z- and x-axes. These angles are assumed to be distributed uniformly within a given cone, i.e., ϕ is uniformly

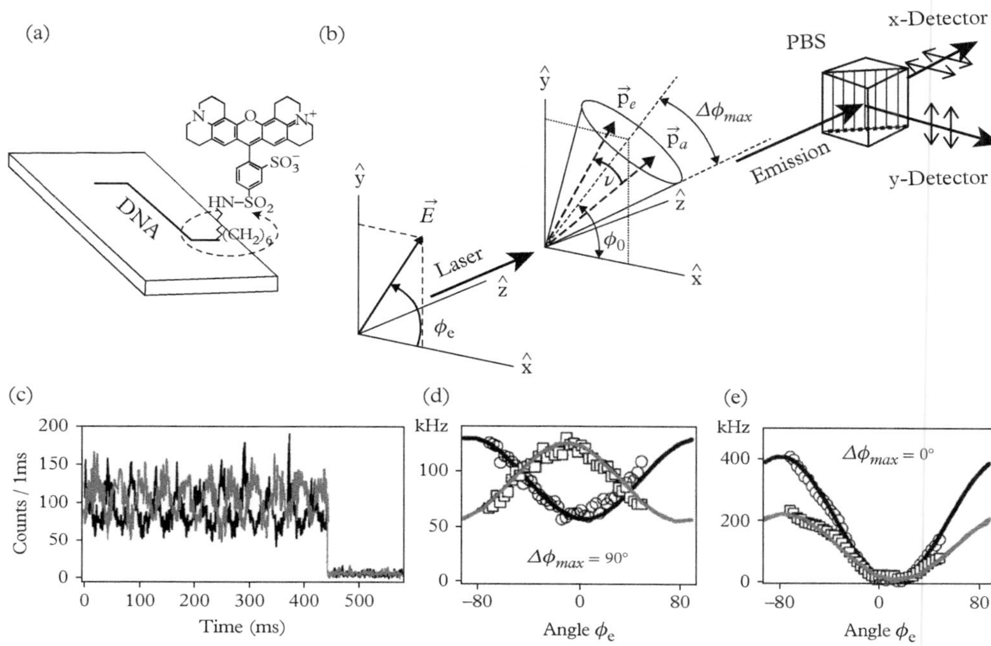

Fig. 3.5 Single-molecule polarization study of a Texas Red molecule *linked to a DNA molecule (a) absorbed on a surface. (b) Experiment schematic: an electric field* $\vec{E}$*, making an angle* ϕ_e *with the x-polarization axis, propagates along the z-axis and excites the fluorophore with absorption and emission dipole moments* $\vec{p}_a$ *and* $\vec{p}_e$*, initially aligned;* ν *represents the rotational diffusion of the emission dipole during the excited state lifetime. The dipole is supposed to be confined in a cone, which is positioned at an angle* ϕ_0 *projected on the (x,y) plane and has a half-angle* $\Delta\phi_{max}$*. A polarizing beam-splitter (PBS) splits the collected emission in two signals along the two polarization directions* I_x *and* I_y*, which are simultaneously recorded by avalanche photodiodes (APDs). (c) Recorded fluorescent intensities* I_x *(grey) and* I_y *(black) of the tethered fluorophore molecule as the polarization of the exciting field is rotated. Notice the sudden disappearance of the signals at* ~ 430 *ms due to bleaching of the single fluorophore. (d) Same data as in (c), but averaged over the eleven periods when the fluorophore was "on" (i.e., before bleaching). The fit corresponds to a* $\Delta\phi_{max}$ *close to 90° (freely rotating molecule). (e) Average data as in (d) for a different molecule with a fixed dipole orientation (*$\Delta\phi_{max} = 0$*. Adapted from T. Ha et al. [138].*

distributed in the interval: $\phi_0 - \Delta\phi_{max} < \phi < \phi_0 + \Delta\phi_{max}$. During the excitation time t (distributed with probability density $f(t) = e^{-t/\tau}/\tau$) the dipole rotates diffusively to new angles θ',ϕ' with probability $P_r(\theta',\phi',\theta,\phi,t)$ within the allowed rotational cone, so that the mean intensity $\langle I_x \rangle$ measured along the x-axis becomes:

$$\langle I_x \rangle \propto \int d\theta d\theta' d\phi d\phi' dt P_a P_r f(t) \sin^2\theta' \cos^2\phi' \tag{3.22}$$

and similarly for $\langle I_y \rangle$. By varying the polarization angle ϕ_e of the exciting field, the dependence of the average measured signals $\langle I_x \rangle$ and $\langle I_y \rangle$ on ϕ_e, (see Fig. 3.5d,e) can be used to determine the best fit parameters [138]: $\phi_0, \Delta\phi_{max}, D$ (where D is the rotational diffusion constant of the fluorophore). If $D\tau \ll 1$, then $P_r \approx \delta(\theta - \theta')\delta(\phi - \phi')$ and the average signal can be computed analytically. When $\Delta\phi_{max} = 0$, it is easy to show that the ratio $\langle I_x \rangle / \langle I_y \rangle = \tan^2 \phi_0$, i.e., constant independent of ϕ_e, see Fig. 3.5e. In the opposite limit, where the fluorophore is freely rotating ($\Delta\phi_{max} = 90°$), one can show that $\langle I_x \rangle \propto 1 + 0.5 \cos 2\phi_e$ while $\langle I_y \rangle \propto 1 - 0.5 \cos 2\phi_e$ (compare with Fig. 3.5d).

Another frequent case of interest is the study of fluorescently labelled molecules diffusing in the focal point of a laser beam, as in Fluorescence Correlation Spectroscopy (FCS, see section 3.3.1). For an absorption dipole $\vec{p}_a$ making an angle θ with the electric field (which sets the z-axis), the probability of absorption, $P_a(\theta) \propto \cos^2 \theta$. The mean components of the fluorescent emission, which are parallel ($\langle I_\parallel \rangle$) and perpendicular ($\langle I_\perp \rangle$) to the electric field $\vec{E}$, are:

$$\langle I_\parallel \rangle \propto \int P_a(\theta)(\vec{p}_{em} \cdot \hat{z})^2 d\Omega$$

$$\langle I_\perp \rangle \propto \int P_a(\theta)(\vec{p}_{em} \cdot \hat{x})^2 d\Omega$$

where we assume that the probability distribution of dipole orientations is uniform (assuming ergodicity the mean corresponds to an average measured over many molecules or on a single molecule over a timescale $t \gg \tau_F$). It is customary to define the mean polarization anisotropy:

$$r = \frac{\langle I_\parallel \rangle - \langle I_\perp \rangle}{\langle I_\parallel \rangle + 2\langle I_\perp \rangle} \tag{3.23}$$

Consider fluorophores moving so slowly that their dipole orientation does not change much between excitation and emission (i.e., within a time $\tau_F \sim 1$ ns). If the molecular emission dipole is at an angle γ to the absorption dipole, the mean polarization anisotropy is: $r_0 = (3\cos^2 \gamma - 1)/5$. The polarization anisotropy is maximal ($r_0 = 0.4$) when $\gamma = 0$ (i.e., when the emission dipole is parallel to the excitation one), it is minimal ($r_0 = -0.2$) when $\gamma = \pi/2$, i.e., when the emission dipole is orthogonal to the excitation one.

However, if, as a result of diffusion, the rotational correlation time of the molecule $\tau_R = \eta V / k_B T \ll \tau_F$ (where η is the viscosity of the solution and V is the volume of the molecule [112],[4] the orientation of the excited molecule is completely randomized before it fluoresces and, thus, $\langle I_\parallel \rangle = \langle I_\perp \rangle$ ($r = 0$). Between these extremes, the polarization anisotropy is given by Perrin's equation:

$$r = \frac{r_0}{1 + \tau_F/\tau_R} \tag{3.24}$$

[4] For a 1 nm molecule in water at 20 °C, ($\eta = 10^{-3}$ SI), $\tau_R \sim 0.1$ *ns* $< \tau_F$. By increasing the viscosity of the medium, for instance by adding glycerol, a regime where $\tau_R \gg \tau_F$ can be easily reached.

Due to the dependence of τ_R on the molecular volume (which increases the drag and, thus, the diffusion time), the observation of changes in the polarization anisotropy has been used to infer the interaction between a fluorescently labelled protein and a non-fluorescent one. Also, by comparing the value of r for a single fluorophore labelling a bound molecule to that for an ensemble of labelled but freely moving molecules, one can deduce how much of the rotational phase space is accessible to a single fluorophore (whose rotation may be sterically hindered by its binding to a bigger molecule, such as a protein).

3.2.4 Fluorescence Resonance Energy Transfer (FRET)

A special case of non-radiative decay is encountered in fluorescence resonance energy transfer (FRET, see Fig. 3.6), described theoretically by Förster in 1948 [139, 140, 112, 141, 142, 143, 61]. It consists in the transfer of energy between the excited (or donor) fluorophore and a nearby (acceptor) fluorophore via dipole–dipole interactions. For donor and acceptor molecules in quantum states $|D\rangle$, $|A\rangle$ and dipole moments $\vec{p}_D$, $\vec{p}_A$ respectively separated by a distance r, the interaction Hamiltonian is:

$$H_{FRET} = \frac{3(\vec{p}_A \cdot \hat{r})(\vec{p}_D \cdot \hat{r}) - \vec{p}_A \cdot \vec{p}_D}{4\pi \epsilon_0 \epsilon_r r^3} \tag{3.25}$$

where ϵ_r is the local dielectric constant, which is related to the local index of refraction n via: $\epsilon_r = n^2$. By analogy with Eq. 3.14, the transition rate k_{FRET} depends on the product

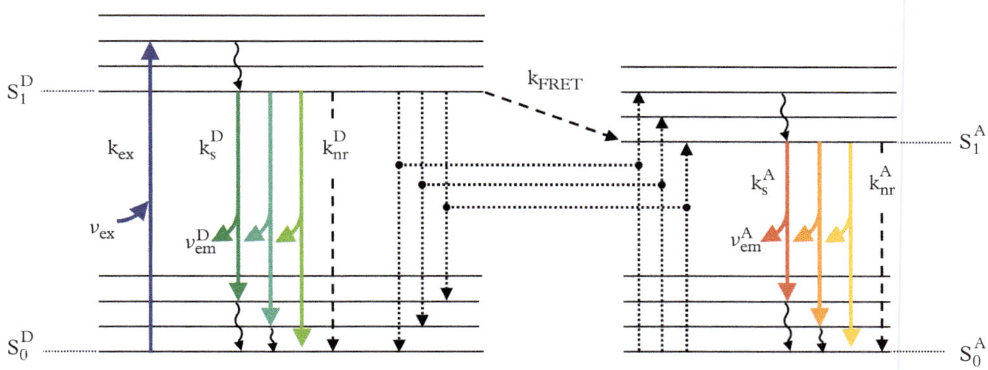

Fig. 3.6 Jablonski diagram for Fluorescence Resonance Energy Transfer (FRET) between a donor molecule (D) and acceptor molecule (A). *The triplet states of the two molecules are not represented. Upon illumination, the donor is excited from its singlet ground state S_0^D to its first excited state S_1^D. In addition to non-radiative and radiative decay of the donor emission state, donor relaxation transitions in resonance with acceptor absorption transitions (connected dotted arrows) can result in non-radiative transfer of energy to the acceptor molecule (inducing its transition from $S_0^A \rightarrow S_1^A$). On average, the phenomenon can be modelled by a transfer rate k_{FRET}.*

of the square of the transition matrix element $\langle A|H_{FRET}|D\rangle$, and the density of states $\rho_A(E)\ \rho_D(E)$, where the donor emission at energy E overlaps the acceptor absorption at the same energy:

$$k_{FRET} = \int \frac{2\pi |\langle A|H_{FRET}|D\rangle|^2}{\hbar} \rho_A(E)\ \rho_D(E) dE$$
$$\approx \frac{2\pi \kappa^2}{\hbar n^4 r^6} \int \rho_A(E)\ \rho_D(E) dE \tag{3.26}$$

Where the orientation factor $\kappa^2 = \left\langle \left[3(\vec{p}_A \cdot \hat{r})(\vec{p}_D \cdot \hat{r}) - \vec{p}_A \cdot \vec{p}_D \right]^2 \right\rangle$ depends on the mean orientation of the acceptor and donor dipoles (for freely rotating dipoles $\kappa^2 = 2/3$). Computing the overlap integral $\int \rho_A(E)\ \rho_D(E) dE$ is a task seldom performed.[5] However, due to the distance dependence of the interaction Hamiltonian, the transition rate can be defined as:

$$k_{FRET} \equiv \frac{1}{\tau_D} \left(\frac{R_0}{r} \right)^6 \tag{3.27}$$

where τ_D is the fluorescence lifetime of the donor molecule (in absence of acceptor) and R_0 is known as the Förster radius. From Eq. 3.26, one notices that R_0 (which varies between 3 and 7 nm) depends on the pair of donor–acceptor molecules, which determines the overlap integral, their orientation (via the factor κ^2), and the local index of refraction (via the factor n^{-4}). One should be aware that changes in κ^2 (via the interaction of the fluorophores with proteins or nucleic acids) or n during an experiment may, thus, alter the value of R_0 [61]. With $k_D = 1/\tau_D$, the lifetime of the donor in presence of an acceptor molecule, τ_{AD} becomes:

$$\tau_{AD} = (k_D + k_{FRET})^{-1} = \frac{\tau_D}{1 + (R_0/r)^6} \tag{3.28}$$

The FRET efficiency E_{FRET}, which measures the ratio between the energy emitted by the acceptor and that adsorbed by the donor, can be written as (see Eq. 3.19):

$$E_{FRET} = \frac{k_{FRET}}{k_D + k_{FRET}} = 1 - \frac{\tau_{AD}}{\tau_D} = \frac{1}{1 + (r/R_0)^6} \tag{3.29}$$

The FRET efficiency is, thus, very sensitive to the donor–acceptor distance, which makes it a useful molecular ruler at the nm scale [145]. However, due to the complex dependence of R_0 on a number of hard-to-measure parameters (such as the local refractive index n or the orientation factor κ^2), precise distance measurements are challenging to obtain [146], and FRET measurements are, therefore, more commonly used to assess relative distance changes [61]. FRET efficiency can be measured via the ratio of fluorescence lifetimes (see Eq. 3.29), but this requires pulsed laser excitation and

[5] This overlap integral can be cast as an integral over the product of the donor emission spectrum by the acceptor-absorption spectrum, which can be measured separately [144].

Time Correlated Single Photon Counting (TCSPC) hardware, which are not always available. Therefore, the FRET efficiency is more often measured from the ratio of (corrected) fluorescence emission F_A, F_D of acceptor and donor molecules:

$$E_{FRET} = \frac{F_A}{F_A + \gamma F_D} \tag{3.30}$$

where $\gamma = \eta_A \Phi_A / \eta_D \Phi_D$ takes care of the difference in quantum efficiency $\Phi_{A,D}$ and detection yield $\eta_{A,D}$ of the acceptor and donor fluorescence.

Single-molecule FRET (smFRET) is often used to investigate the conformation of, or interactions between, proteins or protein and nucleic acid. Each molecule can be labelled with a donor or acceptor molecule, or, in some cases, both. Changes in the conformation of one or both molecules, or of their respective position or distance, affect the amount of FRET between donor and acceptor. Since the FRET efficiency is exquisitely sensitive to the distance between the donor and acceptor, this approach is well suited to studying molecular conformational changes, e.g., during the initial stages of transcription (see section 7.6.3 and Fig. 7.13). However, since the average relative orientation of the donor and acceptor dyes, the local index of refraction, and individual molecular parameters entering in the overlap integral affect the value R_0, it is of paramount importance to assess any departure from known or typical values of these parameters for each system and experiment: in particular, the standard assumption of complete rotational freedom of both dyes and averaging of their relative orientation during the excited lifetime of the donor, which results in a value of $\kappa^2 = 2/3$, needs to be verified by polarization anisotropy measurements [140, 147]. Moreover, some dyes, such as Cy3 or Cy5, have a marked tendency to adopt stacked conformations when used as labels attached to the end of DNA molecules, resulting in a very strong modulation of the FRET efficiency as a function of the distance between donor and acceptor molecules [148]. Similarly, modulation of the donor or acceptor quantum yields (due to transient interactions of either fluorophore with a local quenching environment or isomerization) may affect the overlap integral and average value of the FRET efficiency or broaden its distribution [149].

Studies using smFRET have been done on immobilized molecules (see section 8.5) and molecules in solution (see section 7.6.3). Studies on molecules immobilized to a surface are usually done with TIRF illumination (see section 3.3), where tens to a hundred molecules can be monitored simultaneously. They usually require illumination and environmental conditions, allowing continuous monitoring of the FRET signal over a significant time (many seconds).

Alternatively, smFRET studies can be done in solution with freely diffusing molecular complexes [61]. In such a case, confocal illumination (as in FCS, see section 3.3.1) is used and the smFRET signal consists of fluorescent bursts corresponding to labelled complexes diffusing (or transported by a flow) through the illumination spot. Typical diffusion constants of single proteins or oligonucleotides in water are of the order of a few 100 $\mu m^2/s$, whereas laminar flow rates of the order of mm/s are achievable with microfluidic devices [150]. When point-like (diffraction-limited) excitation and detection are used, the total observation time is limited by the residence time in the excitation volume (of the order of 10^{-4} to 10^{-2} s for an excitation volume of $\sim 1\ \mu m^3$). Since the rate of emission of a fluorophore is also limited (by saturation) and only a fraction of the emitted photons are collected (a detection efficiency of 10% is typical), the total number of photons collected during the transit of a molecule is of the order of, at most, a few hundred photons. Hence, the quantities extracted from these experiments are usually limited by small number statistics (shot noise).

Finally, even though the duration of individual molecule transits is short, they are repeated several times before a molecule goes away permanently (over the time scale of a measurement). To prevent (or at least limit) the simultaneous occurence of two molecules in the vicinity of the excitation volume, it is necessary to work at low concentrations. Whereas the exact criterion depends on experimental details, such as the size of the excitation volume and diffusion coefficient of the observed molecules, a practical rule of thumb is to ensure that the average separation between individual bursts is at least one order of magnitude larger than the average burst duration. Typical concentrations of 100 pM or less are common, which, in a confocal geometry, lead to a burst duration of a few ms and burst separation of tens of ms. The result is data acquisition times of several minutes to hours, to obtain the thousands of bursts needed for good statistics (most bursts have low photon counts and result in poor signal/noise ratio and, thus, are generally discarded from the final analysis).

In standard smFRET experiments, each burst is comprised of two signals: the donor and acceptor signals, corresponding to photons emitted by either the donor upon donor excitation or acceptor upon transfer of the donor energy. In more sophisticated schemes, the polarization of the emitted signal can be recorded in each spectral band (two photon streams per fluorophore) [128]. Finally, in Alternating Laser Excitation (ALEX) approaches [151], the photon streams from each channel (donor and acceptor) may be further split into sub-streams, which correspond to different laser excitations (at the donor and acceptor absorption bands). In all cases, computation of the exact FRET efficiency from these raw signals requires a few corrections:

- Background subtraction for the donor and acceptor channels.
- Correction of the acceptor channel for donor-emission spectral leakage.
- Direct excitation of the acceptor by the donor-excitation laser.
- Correction for detection efficiency and quantum yield differences in both channels.

These parameters can be measured relatively easily on test samples.

3.2.5 Fluorescent labelling

There exist many naturally occurring fluorescent molecules, and even more have been synthesized by chemists, with new ones being developed constantly. In addition to these typically organic fluorophores, a variety of fluorescent nanoparticles have appeared in recent years, increasing the single-molecule experimentalist's toolkit to:

- Organic fluorescent dyes, such as fluorescein, rhodamine, or the various cyanine dyes (Cy3, Cy5, etc.).
- Fluorescent proteins, such as the Green Fluorescent Protein (GFP) and its variants.
- Fluorescent nanoparticles such as quantum dots (Qdots).

All biological molecules can be coupled (with various degrees of ease and success) to any one of these probes [77, 134, 61]. Using chemical reactions (based on cysteins, unnatural amino acids, Click^{TM} chemistry, etc.) biological molecules can be specifically labelled *in vitro* by small organic fluorophores or nanoparticles. Using genetic engineering, any cellular protein can be fused to a fluorescent protein, often without impairing the function of the target protein. Proteins can also be fused to a small peptide that will bind to a cell-permeable fluorophore, resulting in fluorescence increase. These genetic engineering methods can be easily implemented *in vivo*.

When deciding which fluorescent labelling to use, the first criterion is biological: which method is more appropriate for the system studied? Is the labelling method applicable *in vivo* or just *in vitro*? Is the function of the labelled bio-molecule affected or not (e.g., by its coupling to a rather large Qdot or charged or hydrophobic dye)? This last question is best answered by ample biochemical characterization of the labelled species at the ensemble level.

The second criterion is photophysical: which fluorophore is the brightest and has the more stable properties [153]? If oxygen scavenging protocols are not detrimental to the system studied, a good choice is to use the brightest source and get rid of oxygen. If long-term observations *in vivo* are desired, choosing a less bright but more stable fluorophore can be a good strategy. Alternatively, using the non-covalent binding of a fluorophore to a labelling peptide may provide a non-bleaching, albeit fluctuating, fluorescent source. In view of separating the emitted fluorescence signal and the much stronger excitation light, the larger the Stokes shift (the difference between the peak absorption and emission wavelengths), the larger the achievable signal/background ratio.

When studying the interaction between proteins or protein conformational changes, FRET is often the method of choice. In this case, two fluorophores (forming a donor and acceptor pair) have to be chosen. Besides the aforementioned criteria for the selection of fluorescent sources, the FRET pairs have to be chosen for their Förster radius R_0, which determines their efficiency and should be matched to the expected distance between the fluorophores labelling the protein(s).

3.3 Illumination methods

Once the fluorophores have been selected, they need to be illuminated by light at the appropriate excitation wavelength(s). To reduce background noise and minimize the photobleaching of out-of-focus fluorophores, the illumination method with the smallest possible excitation volume for the system studied should be chosen [60, 61], see Fig. 3.7.

- Unless the studied fluorophores are localized on a plane, regular epifluorescence illumination, see Fig. 3.7a, is in general not a good choice of illumination. Wide field illumination may excite many fluorophores along the optical axis, which will decrease their lifetime and generate a large background fluorescence that is incompatible with single-molecule studies.

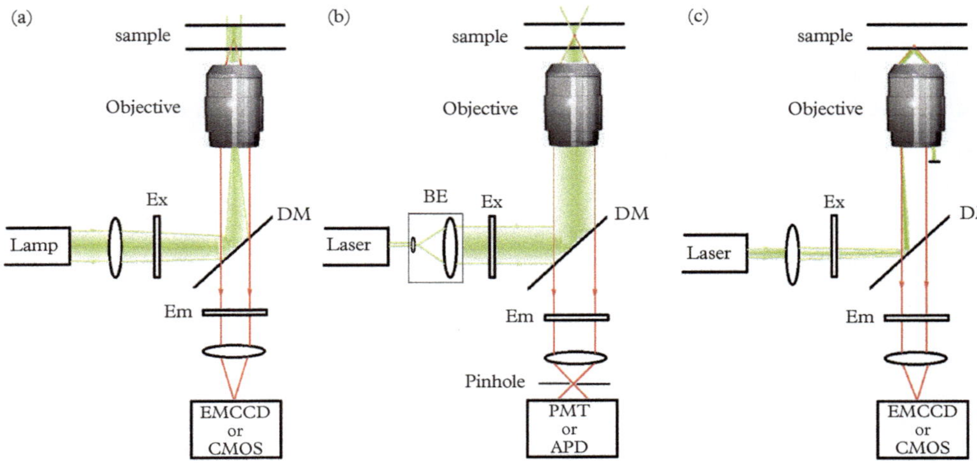

Fig. 3.7 Common illumination methods. *(a) In epifluorescence illumination, the light from a lamp (or LED) is passed through an excitation filter (Ex), reflected by a dichroic mirror (DM) and focused in the back-focal plane of a microscope objective, thus creating a uniform illumination on a sample (placed on top of an inverted microscope). The fluorescent light emitted from sources in the front-focal plane of the objective is filtered by an emission filter (Em) and imaged on an EMCCD or CMOS camera. (b) To overcome background noise (from light emitted by out-of-focus sources in thick samples) confocal illumination is often used. It utilizes the focused light from a beam-expanded (BE) laser to illuminate the sample and a pinhole conjugated to the focal plane of the objective to filter out the light emitted from out-of-focus sources that is spread on an area larger than the pinhole's. The intensity of the emitted light is measured on a Photomultiplier Tube (PMT) or avalanche photodiode (APD). By scanning both the illumination beam and conjugated pinhole, an image of the sample in the focal plane is obtained. (c) For sources close to the glass/water interface, an alternative to confocal illumination is Total Internal Reflection (TIR) microscopy. It uses illumination with a laser beam focused on the rim of the objective back-focal plane which is impinging on the glass/water interface with an angle* $\theta > \sin^{-1}(n_{water}/n_{glass}) \approx 60°$, *so that the beam is fully reflected at the interface (and blocked at its exit from the objective). An evanescent field is created in the vicinity (100–200 nm) of the glass surface, which can excite nearby fluorescent sources. Thus, the emitted light, results only from sources close to the interface e.g., fluorescently labelled DNA molecules anchored to a glass coverslip.*

- If the investigated fluorophores are localized close to a flat glass interface, evanescent wave illumination (also known as Total Internal Reflection Fluorescence (TIRF) microscopy) is the method of choice, since only fluorophores within $\sim$200 nm of the surface are excited. To generate the evanescent field, one often uses epi-illumination via a laser beam focused on the periphery of the back-focal plane, as shown in Fig. 3.7c. Alternatively, the glass coverslip (or a layer deposited on top of it) can be used as a planar waveguide [153, 154] into which light is coupled using an appropriate prism or grating.

If the concentration c of fluorescent molecules in solution is very large (higher than 1 molecule per 200 × 200 × 200 nm, i.e., $c > 200$ nM), background fluorescence may still prevent single-molecule observations. In such cases, one may use zero-mode waveguides (small holes (< 100 nm) in a conducting layer deposited on the glass surface) to generate an excitation illumination limited to the hole only. This may allow single-fluorophore observations, even if the concentration of fluorescent molecules outside the hole is > 1 μM [155], see section 7.3.3.

- For molecules in thick (but less than ~ 500 μm) samples, confocal illumination is often used. It consists of scanning a focalized laser beam on the sample and conjugating its focus with a pinhole in the emission path, which filters out the light from out-of-focus sources, see Fig. 3.7b. The minimal illumination cross section is set by the diffraction limit of the objective which, by Abbe's criterion [156], cannot be focused to a diameter smaller than $d_{Abbe} = 1.22\lambda/\text{NA}$. The longitudinal spread of the beam (twice the depth of field) is $w_z = 2\ n\ d_{Abbe}/\text{NA}$ (for water $n = 1.33$). Thus, for a wavelength $\lambda = 500$ nm, the confocal illumination volume is $V_{confocal} \sim 1\ \mu\text{m}^3 = 1$ fl. Even though the illuminating beam can excite molecules along the beam and outside the focal plane (thus reducing their lifetime), the effective detection volume is so small ($\sim V_{confocal}$) that background noise is greatly reduced. To eliminate out-of-focus excitation, an alternative is to switch a fluorophore from its ground state to an excited state using two-photon absorption. Each photon has to carry half the energy difference between the states, i.e., their wavelength is twice the wavelength of normal one-photon excitation. Since the probability of absorbing two photons almost simultaneously is negligible unless the photon density is high (i.e., at the beam focal point), two-photon excitation is often a method of choice for confocal imaging and single-molecule study (it does not require a pinhole conjugated with the focal point). Its drawback is the need for high energy laser pulses, which may be detrimental to the biological sample.

  To achieve optical imaging with arbitrary resolution, a revolutionary modification on the confocal method has been proposed by S. Hell [157, 158, 159]. In this approach, known as Stimulated Emission Depletion (STED) microscopy, a doughnut-shaped beam, colinear with the (Gaussian-shaped) excitation beam and tuned to a slightly higher wavelength, is used to deplete the excited state via stimulated emission. Due to the nonlinear dependence of stimulated emission on the intensity of the depletion beam, fluorescence emission is limited to a narrow region near the centre of the depletion beam. The extent of the emission spot can be set to dimensions (a few tens of nm) that are way below the diffraction limit, see Fig. 3.8. The major drawback of STED is the need for a high intensity depletion beam ($\gg 1$ MW/cm^2), which can be detrimental to the biological sample (although the use of fluorophores with longer excitation lifetimes and consequently smaller saturation intensities can alleviate that drawback).

- To minimize out-of-focus background noise and reduce the photobleaching effect of the illumination beam in thick samples, light-sheet microscopy (also known as Single Plane Illumination Microscopy (SPIM)) is the best current method. It consists of illuminating the sample from the side by a sheet of light generated by a cylindrical lens (or by scanning the illumination beam) and observing the excited fluorophores via an objective whose focal plane coincides with the illumination plane, see Fig. 3.9a. For technical reasons (steric hindrance between the illumination and observation objectives), the approach is difficult to implement with the

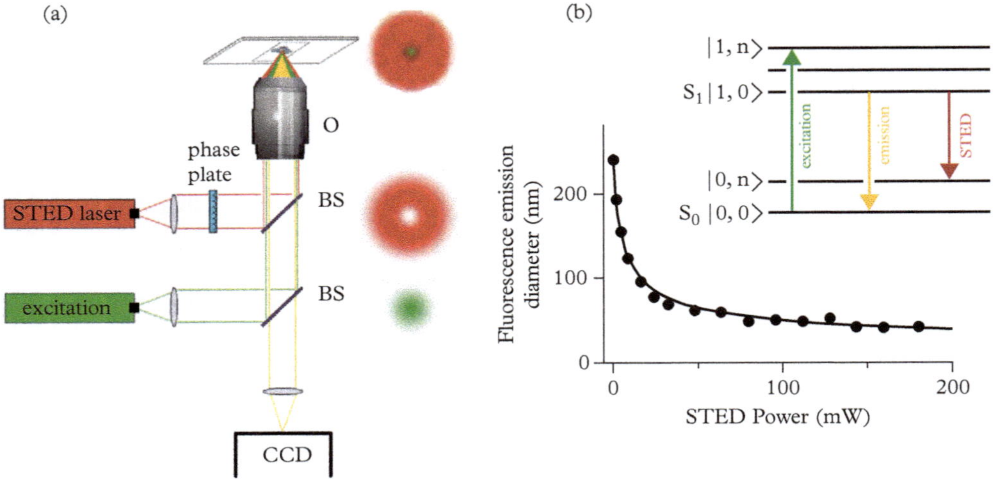

Fig. 3.8 Stimulated Emission Depletion (STED) microscopy *is a super-resolution microscopy approach that breaks the diffraction limit by relying on the strong non-linearity of stimulated emission. (a) Two co-linear beams combined with appropriate dichroic mirrors (BS) are used to illuminate the sample via the objective (O): a Gaussian-shaped excitation beam (whose intensity is maximal at the centre) and a doughnut-shaped stimulation (STED) beam (whose intensity is null at the centre and can be obtained from a Gaussian beam with an appropriate phase plate). Due to the non-linearity of the stimulated emission, the stronger the STED beam, the more the excited electrons are driven to the ground state S_0 by the stimulated beam, except in the narrower region where the intensity of the doughnut beam is zero, see (b). Hence, the region, where spontaneous (fluorescence) emission (in contrast to stimulated coherent emission) is possible, gets smaller and smaller as the intensity of the STED beam increases, allowing for theoretically unlimited optical resolution (in practice, resolution down to a few tens of nanometres has been achieved). Adapted from A. Honigmann et al. [160].*

high numerical objectives required for the efficient detection of single-molecule fluorescence, even though microscope manufacturers have recently started to offer compact systems, which implement a SPIM on a confocal microscope, see Fig. 3.9b.

With the development of super-resolution methods, the field of microscopy is undergoing a revolution. New microscopes implementing the latest developments (STED, PALM (see section 3.4.1), light sheet, etc.) are constantly appearing. Covering this rapidly evolving field is outside the scope of the present monograph. Our goal here was to point to some general guidelines when selecting an illumination method for the study of single-molecule fluorescence.

3.3.1 Fluorescence Correlation Spectroscopy (FCS)

To study the dynamics (diffusion and interaction) of free molecules (labelled with appropriate fluorophores) in solution, Fluorescence Correlation Spectroscopy (FCS) is a method often used [102, 60, 61]. It consists of monitoring the fluorescence intensity

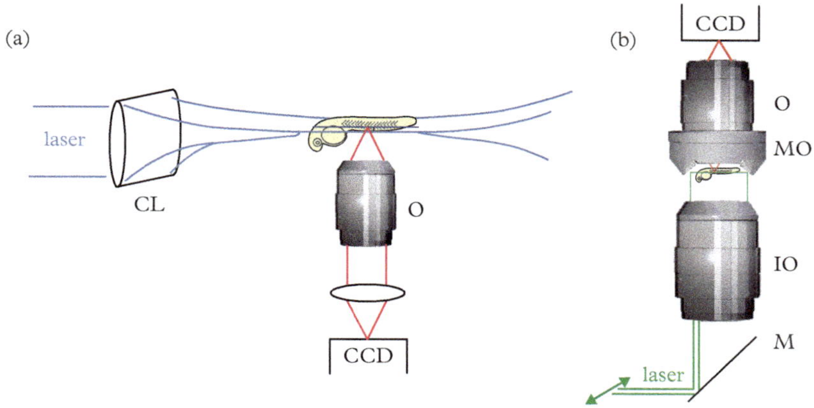

Fig. 3.9 Single Plane Illumination Microscopy (SPIM). *A sample is illuminated at 90° to the optical axis of the objective (O) by a 2D light sheet, generated by focalizing a laser beam with a cylindrical lens (a) or by scanning the beam focused with a regular lens (b). The illuminated plane is observed with an objective that collects the emitted light from all the sources in the field of view. The advantage of this method is that only observed sources are illuminated, thus minimizing photobleaching and background noise. Additionally, microscope manufacturers are offering add-ons to regular confocal microscopes to convert them into SPIM. These add-ons consist of a small contraption (MO) screwed on the objective (O), which contains two (or more) mirrors that are used to scan the sample with the confocal beam from the illumination objective (IO). The light emitted from all the sources in the illumination plane is collected by the objective (O).*

$I_{em}(t)$ as the molecules diffuse through the focal point of an excitation beam whose effective intensity I_{ex} at the focal point is assumed to vary as a Gaussian beam with transverse and longitudinal widths w_{xy}^2, w_z^2:

$$I_{ex}(x,y,z) \sim \exp\left\{-\frac{2(x^2+y^2)}{w_{xy}^2}\right\}\exp\left\{-\frac{2z^2}{w_z^2}\right\}$$

One computes the correlation function $G(t)$ defined as:

$$G(t) = \frac{\langle I_{em}(t) I_{em}(0)\rangle}{\langle I_{em}\rangle^2} - 1 \tag{3.31}$$

For freely diffusing fluorophores, it can be shown [102] that the correlation function decays with time as:

$$\begin{aligned} G(t) &= \frac{1}{N}\left(1+\frac{2\langle\vec{r}^{\,2}(t)\rangle}{3w_{xy}^2}\right)^{-1}\left(1+\frac{2\langle\vec{r}^{\,2}(t)\rangle}{3w_z^2}\right)^{-1/2} \\ &= \frac{1}{N}(1+t/\tau_d)^{-1}(1+t/\tau_d')^{-1/2} \end{aligned} \tag{3.32}$$

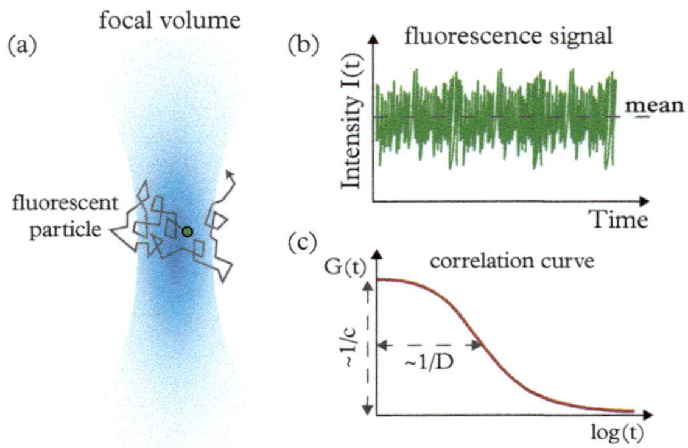

Fig. 3.10 (a) Principle of Fluorescence Correlation Spectroscopy (FCS). *The fluorescence signal $I(t)$ emitted by a particle is monitored (b) while it diffuses through the focal point of the excitation beam. (c) The correlation $G(t) = \langle \delta I(t)\delta I(0)\rangle$ of the fluorescence ($\delta I = I - \langle I\rangle$) is computed. Its amplitude is inversely proportional to the particle density (c) while its width is proportional to the diffusion time of the particle in the focal volume, i.e., the inverse of its diffusion constant: D.*

where N is the mean number of fluorophores in the excitation volume[6] $V_{ex} = \pi^{3/2} w_{xy}^2 w_z$ and $\langle \vec{r}^{\,2}(t)\rangle$ is the mean square displacement of the fluorophore within a time t. For a freely diffusing molecule, $\langle \vec{r}^{\,2}(t)\rangle = 6Dt$ and $\tau_d = w_{xy}^2/4D$ is, thus, the mean diffusion time across the beam (see Fig. 3.10), while $\tau_d' = (w_z/w_{xy})^2\tau_d \gg \tau_d$ is the mean diffusion time along the beam. Notice that for N fluorophores emitting independently, the average fluorescent intensity is $\langle I_{em}\rangle = NI_0$ (where I_0 is the average emission from a single source). The intensity fluctuations of the various fluorophores being uncorrelated, $\langle \delta I_{em}^2(0)\rangle = \langle (I_{em}(0) - \langle I_{em}\rangle)^2\rangle = NI_0^2$, from which one readily sees that: $G(0) = 1/N$. Hence, the FCS signal is larger the smaller the fluorophore concentration, being largest when there is about 1 fluorophore per excitation volume.

By inverting Eq. 3.32, the behaviour of $\langle \vec{r}^{\,2}(t)\rangle$ can be deduced. This approach has been used [161] to study the different sub-diffusive behaviours of fluorophores, which label double- and single-stranded DNA molecules, see section 4.2.

FCS has also been used to study the formation between a labelled protein (A) and a partner protein (B) [60]. If the relaxation time of the complexation reaction is longer than the diffusion times of the species in the beam, τ_A, τ_{AB} (typically < 1 ms), the correlation function is the weighted sum of the correlation function of the freely diffusing protein $1/G_A(t) = N_A(1 + t/\tau_A)$ and the complexed one $1/G_{AB}(t) = N_{AB}(1 + t/\tau_{AB})$, where N_A, N_{AB} are the number of free and complexed molecules in the excitation volume:

$$G(t) = \frac{N_A^2}{(N_A + N_{AB})^2} G_A(t) + \frac{N_{AB}^2}{(N_A + N_{AB})^2} G_{AB}(t) \tag{3.33}$$

[6] Notice that the confocal volume defined earlier is smaller than the excitation volume, as it results from a product of excitation and emission probabilities: $V_{confocal} = V_{ex}/2^{3/2}$. The same remark is valid for the excitation volume with a two-photon excitation.

Hence, by fitting the measured correlation function $G(t)$ to Eq. 3.33, one can deduce the relative concentration of the species and their diffusion coefficient (i.e., their size). For molecules doubly labelled with FRET pairs, using Alternating Laser Excitation (ALEX) at the donor and acceptor wavelengths and measuring the intensity of the emitted fluorescence in the two channels (donor and acceptor) allows one to deduce not only the concentration and diffusion of the molecules, but also the various subpopulations [151]: singly labelled (donor or acceptor) and doubly labelled with different FRET efficiencies, since they appear as isolated clusters in the 2D plot of the probability of observing a given donor and acceptor fluorescence intensity.

3.4 Detection techniques

The detection of the fluorescence emitted by single fluorophores requires an understanding of the various sources of noise and an optimization of the signal/noise ratio (SNR). When detecting the emitted photons, there are three sources of noise [60]:

- The ultimate source of noise is the so-called shot noise associated with the stochasticity of fluorescence emission. For a fluorescent source of power P_{em}, the number of photons N_{em} detected with efficiency η during a time interval Δt is (see Eqs. 3.13 and 3.19):

$$N_{em} = \eta P_{em}\Delta t/h\nu = \eta \Phi \sigma_{ex} I_{ex}\Delta t/h\nu$$

  The fluctuation in that number (the associated shot noise) is $N_s = \sqrt{N_{em}}$.
- A second source of noise is the background noise caused by extraneous fluorescent sources (e.g., Raman scattering, auto-fluorescence of the sample, etc.). This noise is proportional to the density c_b of background sources, their effective brightness b, and the effective excitation volume of the sample V_{eff}, which can be minimized by using appropriate illumination methods, such as Total Internal Reflection (TIR), two-photon, confocal, or light-sheet illumination (see section 3.3). These sources contribute a number of background photons: $N_b = bc_b V_{eff} I_{ex}\Delta t/h\nu$. To differentiate between signal and background photons, the signal/background ratio must be bigger than 1: $SBR = N_{em}/N_b \gg 1$. Alternatively, one can use time-gated techniques to reject photons emitted immediately after the excitation pulse as a way of eliminating many of the background photons not rejected by filters (e.g., Raman scattering) [32].
- The third source of noise is due to electronic noise contributed by the detector. Ideally, a photon detector (such as a single-photon avalanche diode (SPAD)) should yield a signal (e.g., an electronic current) only when illuminated. However, because of thermal fluctuations, a photon detector displays an electronic current even in the absence of illumination. This dark current i_d is indistinguishable from a background source of photons: $N_d = i_d\Delta t/e$. In addition in CCD cameras, the charges accumulated on the camera pixels (proportional to the sum $N + N_b + N_d$)

are read out at every frame with an accuracy σ_r, known as the readout noise, which, in the best cameras, is of the order of 1 electron per readout (i.e., frame).

As a consequence of all these noise sources, the signal/noise ratio of a photon detector is:

$$\mathrm{SNR} = \frac{N_{em}}{\sqrt{N_{em} + N_b + N_d + \sigma_r^2}} \sim \frac{\sqrt{N_{em}}}{\sqrt{1 + 1/SBR}} \tag{3.34}$$

where we assume that the background noise is larger than the dark current and readout noise. Notice that a large SBR results in an optimal SNR, which is larger the longer the integration time Δt. Both ratios can be increased by improving the collection efficiency η (increasing the signal). A larger excitation power or longer integration time will improve the SNR without affecting the SBR.

Typical values for the above parameters, in the case of a confocal microscopy study of freely diffusing fluorescein isothiocyanate (FITC) in water, using a silicon APD, are:

$$\lambda = 525\mathrm{nm}, \eta = 0.01, \Phi = 0.85, \sigma_{if} = 2.8 \cdot 10^{-16}\mathrm{cm}^2,$$
$$\sigma_b c_b V_{eff}/h\nu = 100\ \mathrm{Hz}/\mu\mathrm{W},\ \ i_d/e = 100\ \mathrm{Hz}$$

With an excitation power of 10 μW focused on a spot of 1 μm^2 and with an integration time of $\Delta t = 10$ ms, a mean count rate of 74 counts (per 10 ms) can be obtained, of which ten come on average from background sources, leading to an SNR ~ 7.3 and SBR ~ 6.3. Although these are good figures, they have to be moderated by the fact that single molecules have a finite life span. Single molecules in an oxygen-rich environment typically emit of the order of 10^6 photons before photobleaching. With the above parameters, this amounts to a total emission time of ~ 1.6 s. This is enough time to observe freely diffusing molecules during their transit time of a few hundred ms in the excitation beam. For an immobilized molecule, however, that sets a stringent limit on the total duration of single-molecule observations. Long-term observations of single fluorophores require lower background noise, smaller excitation volumes, reduced excitation intensity, and increased collection efficiency.

3.4.1 Criteria of singleness

In addition to SBR and SNR issues, care has to be taken to ensure that the collected signal originates from a single molecule. This is not just a purist's concern. As well designed as an experiment may be, there is always the possibility of two or more fluorescent molecules occupying the diffraction-limited excitation volume (typically, several hundred nanometres across, 1 femtolitre in volume). If intensity fluctuations (or variations of any other spectroscopic characteristics) are observed, they can be due to single-molecule dynamics or environmental changes, but they could also reflect the random mixture of emissions from nearby molecules. It is, thus, critical to eliminate this unneeded source of uncertainty. Three different strategies can limit this risk:

- Work at a low concentration, such that, at most, one molecule is present in the excitation volume or, equivalently, minimize the excitation volume. When studying diffusing fluorophores the common strategy is to use a dilute solution of fluorescent molecules that yield separate bursts of emitted light, corresponding most of the time to transits of individual molecules through the diffraction-limited focal spot of a microscope objective [162].
- Use a selective excitation or emission detection protocol, such that only one molecule is excited or detected within the sampled volume. This strategy requires the use of fluorophores having either separable excitation or emission properties [63, 64]. This approach has been used to study the fluorescence of pentacene molecules embedded in a solid matrix. Each pentacene molecule had a different, narrow peak absorption frequency that varied with its microscopic neighbourhood, and each could be selectively excited by fine-tuning the excitation laser beam, in spite of the large area covered by the beam. Alternatively, nearby fluorophores can be distinguished by their emission properties (e.g., fluorescence lifetime [163] or spectra [164, 165]).
- Use photo-activable fluorophores. Some molecules (such as the proteins Kaede and Eos) may be switched from a dark non-fluorescent metastable state into a fluorescent one by illumination at an appropriate wavelength (usually around 400 nm). By tuning the intensity of this switching beam, an average of one molecule per excitation volume may be switched on (on average). That scheme has been used to monitor the diffusion of appropriately labelled membrane proteins. It is also the foundation of stochastic super-resolution methods, such as Photo-Activable Light Microscopy (PALM [166]) and Stochastic Optical Reconstruction Microscopy (STORM) [167]. In these methods, the 2D localization (x, y) of single fluorescent sources is determined with nm precision (3D localization schemes have also been developed). The localization accuracy $(\pm\delta x, \pm\delta y)$ is roughly equal to the diffraction-limited radius of the emitting light source ($\sim \lambda/2\mathrm{NA}$) divided by the square root of the number N_{em} of collected photons (of wavelength λ): $\delta x, \delta y \approx \lambda/2\mathrm{NA}\sqrt{N_{em}}$. By stochastically switching on single fluorophores and bleaching them (thus, extracting the maximal number of emitted photons N_{em}), a pointillist image of the fluorophore distribution is obtained.

Once experimental conditions favouring single-molecule detection are satisfied, a number of criteria have to be met to ascertain that the observed signal actually comes from a single emitter [83, 65]. These criteria, summarized in the following list, are direct consequences of the photophysical properties of fluorophores (see section 3.2.5):

- The observed density of emitters is compatible with the known concentration of individual molecules and scales with the original bulk concentration.
- The observed fluorescence intensity level is consistent with that of a single emitting molecule.
- Each immobilized emitter has a well defined absorption or emission dipole.

- Fluorescence emission exhibits only two levels (on/off behaviour due to blinking or photobleaching) over time scales where no changes in the environment are expected.
- If there are two or more emission levels, the photophysical property changes are correlated with blinking [168].
- The emitted light exhibits antibunching, which is characteristic of a single quantum emitter [66]. This term refers to the negative correlation of emitted light at short times (ns), since a single source cannot emit two photons at once.

The main source of ambiguity comes from blinking. A fluctuating, non-blinking intensity recording might come, for instance, from a single molecule that undergoes rotation, or whose environment changes over time. It might also come from the superposition of the blinking or fluctuating behaviour of a few molecules. This latter case is easily detected: for several molecules, coincident emission of photons by the different molecules is likely, and will result in an autocorrelation function that does not decay at a short (ns) time-delay (as for a single source). However, for a small number of blinking molecules (say, two), the total number of coincident emissions might be too low to reject the single-molecule hypothesis with only the antibunching data. In this case, monitoring a photophysical property (spectrum, lifetime) simultaneously with the intensity fluctuations may help resolve the issue.

3.5 Single-molecule tracking

Single-molecule tracking is a natural extension of single-particle tracking (SPT), a field whose status as a quantitative science began with the early twentieth century pioneering work of Jean Perrin and his students [169], following in the footsteps of atomistic philosophers (such as Lucretius, 60 BC, in "De Rerum Natura"), and more qualitative observers of the random motion of small particles (e.g., Brown [170]). Reintroduced in the 1980s as a tool to track individual molecules to which different kinds of probes were specifically attached [171], the technique is connected to another topic of this book, namely single-molecule manipulation, as both optical traps and magnetic tweezers use similar high-resolution particle-tracking principles.

SPT can be performed with any type of optical contrast mechanism, but using fluorescent (or, more generally, luminescent) probes has several advantages, starting with the possibility of using molecular size probes. This is of particular importance for cellular targets, both to facilitate delivery of the probe and minimize its interference with the motion of the attached target. Another advantage is that tracking can be combined with simultaneous observation of other sample parameters in different (or identical) spectral channels, or even tracking of several types of probes visible in different spectral channels. However, in many cases, signal level and stability can be an issue, especially in the case of organic fluorophores or genetically engineered fluorescent proteins. Due to the fundamental limits of localization, weak signal results in low resolution, but more importantly, a limited total number of emitted photons forces the experimentalist to trade between temporal resolution (short integration time, hence weak signal) and spatial resolution (requiring strong signal). As we will see in Fig. 3.11, this is in fact an artificial dilemma, as reliable extraction of dynamical parameters from a single molecule's random trajectory, in general, requires a long integration time.

Single-molecule tracking or nm-localization, made popular by its application to super-resolution imaging, originally came as a natural by-product of the scanning imaging techniques (including near-field scanning optical microscopy) used in the early days of single-molecule fluorescence studies [67, 172, 164]. Determining the nm-localization of the intensity peak of a molecule's emission was relatively straightforward when using piezo-scanning stages moving in nm steps. The problem with these original approaches was their slow speed and invasive nature (preventing their use in live cell imaging). Before they were abandoned in favour of the easier camera-based, far-field approaches, they had revealed the wealth of information to be obtained from knowing the relative localization of different molecules separable by their spectroscopic properties (e.g., nm-resolution of co-localization [165, 173]). Such measurements helped determine the stoichiometry of molecular complexes [174] and decipher the complex arrangement of proteins in supra-molecular assemblies [175].

Extending these types of measurements to live imaging, or simply speeding them up, required one to abandon the slow piezo-scanning confocal geometry and opt for a wide field illumination, high sensitivity, camera-based detection geometry, similar to what had been used for many years in biology and biophysics to image and track polystyrene beads or bright fluorescent beads attached to molecular targets. Adopting this approach was only possible with the advent of scientific cameras of increasing sensitivity, starting with intensified CCD cameras, followed by electron-multiplying CCD cameras and, lately, scientific CMOS cameras [176]. Indeed, as mentioned before, spatial resolution and speed (temporal resolution) are conflicting goals in single-molecule fluorescence imaging because of the limited number of photons, which can typically be detected from a given molecule.

While the expression for the achievable spatial resolution given in the previous section provides a reasonable estimate of the expected performance of a system, a number of factors make the actual resolution slightly more complex to compute. Different analyses have been presented in the literature [177, 178, 179, 180, 181], all of them formalizing the idea that a single molecule's image in a microscope has a predictable theoretical intensity profile, whose centre can be computed with a precision affected by [182, 183]:

- The number N of detected photons (the precision increases as $1/\sqrt{N}$).
- The size of the Point Spread Function (PSF) of the microscope, i.e., its resolution: $\sim \lambda/2\text{NA}$, see section 3.4.1.
- The PSF sampling, i.e., pixel size, which should be smaller than the PSF itself.
- Readout and other noise factors of the camera, which should be smaller than the photon shot noise, see section 3.4.

A rule of thumb is that a single fluorophore imaged with the best possible setup (using a high-numerical aperture objective lens and a camera that samples each PSF by 4–8 pixels in each dimension) can be typically localized with a spatio-temporal resolution of about 10 nm and 10 ms, while the performance of most setups can be worse by up to a factor 10.

Being able to localize a target labelled with a single molecule rather than a large bead has obvious advantages in biology. The smaller size of the probe means that:

- The uncertainty on the localization of the *target* is smaller (if the number of detected photons is large enough).
- Its interference with the target and its environment is minimized.
- It can get into places where larger probes cannot.

The last property is crucial for the study of molecular organization in cells or organelles, but the second one is just as important for investigating dynamic processes *in vitro* or *in vivo*. By this, we mean not only dynamic conformational changes or molecular interactions, but also the 3D motion revealed by single-molecule tracking. Such studies are exemplified by the search and binding dynamics of transcription factors (or DNA-processing enzymes) on DNA and RNA *in vitro* and *in vivo* [184], motion of messenger RNAs in the cell nucleus or cytoplasm [185], and displacement of chromatin domains [186].

Single-molecule tracking has been used ever since Perrin investigated the Brownian motion of particles [169] to extract their diffusion constant D from an analysis of their trajectory $\vec{r}(t)$. Indeed, as shown by Einstein [45], the variance of a trajectory in the d dimension relevant to the studied motion (the mean square displacement or MSD) increases with time t as: $\langle[\vec{r}(t) - \vec{r}(0)]^2\rangle = 2dDt$. With the possibility of monitoring single fluorescently labelled molecules, the study of their diffusion in various environments has experienced a renewal. However, in contrast with decades-old tracking studies involving bright or contrasted beads, single-molecule fluorescence studies generally deal with poorer signal/noise or signal/background ratios, which result in limited localization accuracy. Moreover, the photophysical phenomena associated with single molecules, such as blinking and bleaching, severely constrain the total number N of images from which to construct a trajectory. While these two issues had been identified early on [187, 188], they only recently received a proper analytical treatment that allows a precise quantification of their effect on the estimation of the diffusion coefficient D of a molecule. Here, we briefly summarize these results, referring the interested reader to a growing literature on the subject [189, 190], including the increasingly popular use of Monte Carlo simulations to assess the validity of the methods of analysis.

The problem of determining the diffusion coefficient of a moving particle, based on a potentially interrupted series of noisy images, can be rigorously stated as a probabilistic parameter estimation problem [191, 192, 193, 194]. The important observables in this problem are not so much the successive molecules' locations $\vec{r}_i$, measured with an uncertainty σ_i, but rather the observed *displacements* from frame to frame, $\Delta_i = \vec{r}_{i+1} - \vec{r}_i$. These observables can be described by a probability distribution depending on the parameters of the experimental setup (the microscope resolution, camera noise, integration time, and frame rate), as well as unknown parameters characterizing the

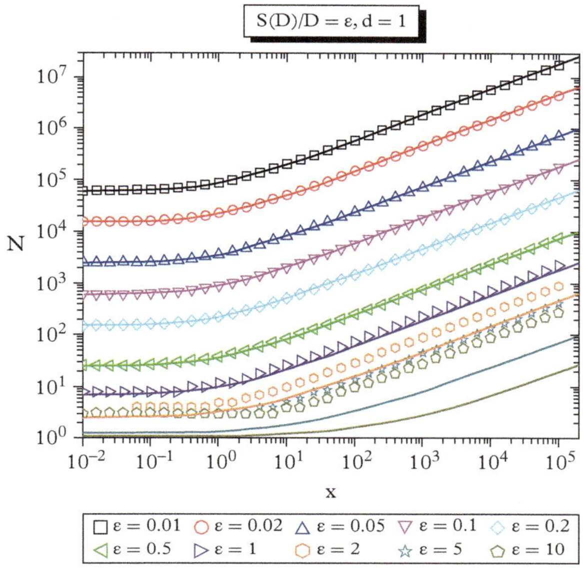

Fig. 3.11 Uncertainty in the diffusion constant estimate from single-molecule tracking experiments. *Contours of constant Cramér–Rao lower bounds for the relative uncertainty $\epsilon \equiv \sigma_D/D$ in the diffusion constant D of a fluorophore in $d = 1$ dimension. These contours are plotted as a function of the number of acquisition points N, ratio ($x = \sigma_i^2/D\Delta t$) between the precision of the particle localization σ_i^2, and the blur $D\Delta t$ resulting from the diffusion of the particle during the acquisition (frame duration) time Δt. In $d > 1$, the value of N reported on the left is related to the number N' of acquisition points: $N = d(N' - 1) + 1$. For example, in $d = 2$, to reach a relative uncertainty $\epsilon = 0.1$ when $x = 1$, the number of aquisition points N' should be larger than 500. With usual trajectories in 2D, which have at best a few tens of points, the relative uncertainty in D is of order 1 (i.e., 100%). Adapted from X. Michalet and A. J. Berglund, [189].*

diffusing molecule, such as its blinking pattern (more generally, the emission intensity time trace), and the actual trajectory of the molecule. Due to the finite integration time Δt necessary to acquire each image of the molecule, the resulting image is not only a noisy rendering of the microscope's PSF, but a motion-blurred one. Consequently, due to the standard localization uncertainty created by Shot noise and other background and noise sources (see section 3.4), an additional uncertainty is contributed by diffusion. The relative importance of these sources of uncertainty can be quantified by the ratio: $x = \sigma_i^2/D\Delta t$.

The overall *dynamic* localization uncertainty in frame i, σ_i contributes to the uncertainty on both the displacement between position i and $i-1$ and between position $i+1$ and i. This dual contribution results in a non-zero covariance between successive displacements, which is key to deriving the correct probability distribution of successive displacements. While computing the best fit diffusion parameter D requires an appropriate numerical procedure, an analytical expression of the minimum variance of D, σ_D

(the so-called Cramér–Rao lower bound or CRLB) can be computed in the simple case where the localization uncertainty σ_i is normally distributed [189, 191]. While the exact result is plotted in Fig. 3.11 for N that is not too small, a good approximation of the relative error $\epsilon \equiv \sigma_D/D$ is given by:

$$\epsilon \geq \sqrt{\frac{2\left(1+2\sqrt{1+2x}\right)}{d(N-1)}} \tag{3.35}$$

The uncertainty on D increases with the relative localization uncertainty x, but is also strongly dependent on the number of trajectory points N used to estimate the diffusion coefficient. For instance, for a typical membrane protein with diffusion coefficient $D = 10^{-2}\mu\text{m}^2/\text{s}$, a localization uncertainty $\sigma_i = 30$ nm and integration time $\Delta t = 30$ ms, a value $x = 3$ is obtained, which results in a relative error in the estimate of D: $\sigma_D/D \sim 2.5/\sqrt{N}$. For instance, estimating D with thirty frames (one-second trajectories) cannot be done with better than a 50% relative uncertainty. This result, therefore, calls for caution when extracting diffusion coefficients D from short and noisy trajectories (as most single-molecule trajectories are): numerical procedures will always yield a value but rarely a reliable estimate of the associated error, which can be so large that it renders the value of D meaningless. Obviously, the situation is even less favourable for motion regimes necessitating additional fitting parameters, as is the case with the popular models of confined diffusion or anomalous diffusion which are used to analyse single-molecule motion inside cells, cell compartments (such as the nucleus), or cell membranes [195].

3.6 Perspectives

In this chapter, we have presented an overview of the main spectroscopic methods that are used to investigate single molecules via the fluorescence emission of appropriate labels. While the theoretical background for these methods was laid by Einstein as far back as 1917, it is only within the past twenty years, with the development of single-molecule fluorescent techniques, highly efficient dyes, and discovery of fluorescent proteins, that the use of these methods for the study of single molecules has taken off. With the development of super-resolution methods, these techniques are revolutionizing the field of optical microscopy, allowing real-time observations of cellular processes with unprecedented spatial resolution (a few tens of nm). In parallel, the recent development of optogenetic methods [196, 197, 198], which use light to control and readout [199, 200, 201, 202] physiological processes are revolutionizing the field of neurophysiology. The combination of these approaches (single-molecule spectroscopy, super-resolution microscopy, and optogenetics) to control and study processes *in vivo* in live organisms will open fascinating new fields of research.

3.7 Problems

Measuring the photon flux on a CCD camera. In a CCD camera, every pixel (of size 5×5 μm) outputs at a rate of 60 Hz a signal proportional to the illumination and varying between 0 and 255 (8 bits). The mean value and variance of the signal for a uniform illumination by an LED are displayed in Fig. 3.12. Assume the camera to be ideal so that the signal S is proportional to the number N of photons converted into electrons in every pixel: $S = \alpha N$.

1. What is the source of the measured noise (variance in signal)? From the data shown in Fig. 3.12, determine α.
2. For a CCD with quantum efficiency of 40%, what is the number of incident photons at an illumination intensity of 100? How many photons per pixel does the camera noise correspond to? Is that bigger than the discretization noise?
3. That CCD is used to visualize a single fluorophore emitting at 535 nm through a 100x, NA = 1.4 microscope objective. How many pixels are illuminated by that fluorophore? What is the level of pixel signal required to reach a resolution of 10 nm?

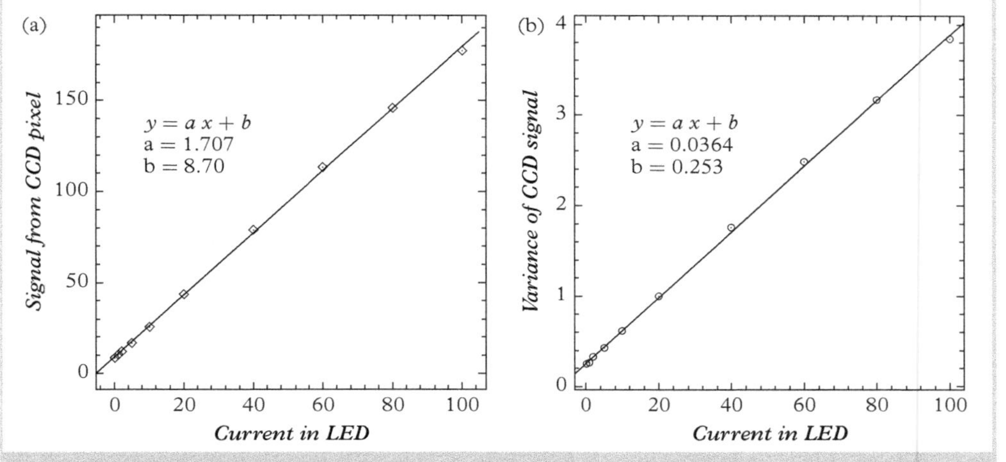

Fig. 3.12 Photon flux on CCD camera. *(a) Mean signal per pixel measured as a function of the current intensity in the illuminating LED. (b) Variance of that signal.*

Doppler effect and spectral broadening. The Doppler effect describes the observed frequency f of the EM field when a radiation source oscillating at frequency f_0 is moving at velocity v: $f = f_0/(1 + v/c)$. The frequency is lower when the source is moving away from the observer ($v > 0$) and higher when it is moving towards the observer ($v < 0$). Since, in solution, fluorophores are free to move around, the radiation they emit is subject to Doppler broadening. Particles of mass m at a temperature T have an average kinetic energy: $\langle m\vec{v}^{\,2}/2\rangle = 3k_BT/2$ (with $k_BT = 4\ 10^{-21}$ J at $T = 300°$K). Express the standard deviation $\Delta f = \sqrt{\langle f^2\rangle - \langle f\rangle^2}$ in the observed emission frequency f as a function of temperature. Compute the broadening Δf resulting from this Doppler effect in the emission lines at 570 nm of CY3 (MW: 767). How does it compare with the broadening deduced from the typical lifetime of fluorescent emission?

Measuring the absorption cross section of a single fluorophore. In a recent paper (see Kastrup and Hell, Angew.Chem.43,6646 (2004)), the stimulated emission (= absorption) cross section for a single fluorophore was measured through the time decay of fluorescence in the presence of a Stimulated Emission Depletion (STED) beam of intensity I_{STED} (in photons/cm^2), which forces the lowest vibrational excited state $|1,0\rangle$ to emit coherent stimulated radiation, rather than isotropic fluorescence, while returning to a lower vibrationally excited energy state $|0,n\rangle$, see Fig. 3.2. Assuming that the vibrational relaxation rate k_{vib} is much higher than the stimulated rate $k_{st} = \sigma_{10,0n}I_{STED}$, which is much larger than the spontaneous rate of emission k_{fl}, show that the probability $P(t)$ of fluorescence emission, a time t after the STED beam is turned on, decays as $P(t) = P(0)\exp(-\sigma_{10,0n}I_{STED}t)$. Explain how $\sigma_{10,0n}$ was measured, and verify that the simplifying assumption $k_{st} > k_{fl}$ is valid (using the published experimental parameters). Estimate the absorption of a beam at the STED wavelength passing through 1 cm of a 1 μM solution of fluorophores that are absorbing with the deduced cross section.

Super-resolution microscopy. A new microscope uses the stimulated emission from an excited state to obtain super-resolution images of fluorescent particles. In that scheme, see Fig. 3.13, an excitation beam at $\lambda_{ex} = 480$ nm of amplitude I_{ex} is used to drive the ground state $|1\rangle$ to an excited level $|4\rangle$ with an absorption cross section $\sigma_{14} = 2\ 10^{-16}$ cm^2. From this state, the system relaxes rapidly to state $|3\rangle$ from which it can relax with rate $k_{31} = 10^9$ s^{-1} back to level $|1\rangle$ with emission of a fluorescent photon at $\lambda_{em} = 550$ nm. A second beam of intensity I_{STED} and wavelength $\lambda_{st} = 600$ nm is used to stimulate the decay from state $|3\rangle$ to state $|2\rangle$ with a stimulated emission cross section $\sigma_{32} = 10^{-16}$ cm^2. (From state $|2\rangle$, the system returns rapidly to state $|1\rangle$).

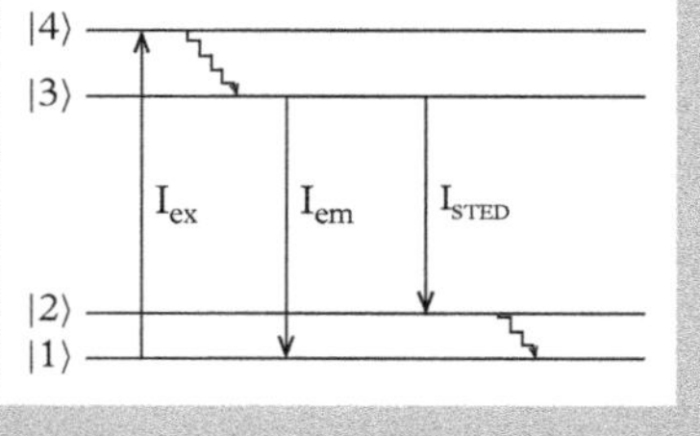

Fig. 3.13 STED example. *Scheme for the depletion of fluorescent emission from state $|3\rangle$ to state $|1\rangle$ by using a beam that stimulates the coherent emission from state $|3\rangle$ to state $|2\rangle$.*

1. Write the steady state equation for the population in state $|3\rangle$, N_3. Show that the probability of being in state $|3\rangle$ ($P_3 = N_3/N_1$) decreases with the increasing intensity I_{STED} of the stimulating beam.
2. Show that the fluorescence power emitted per molecule: $I_{fluo} = h\nu_{em}k_{31}P_3$ varies as:

$$I_{fluo} = \frac{I_{fluo}^{max}}{1 + I_{STED}/I_{sat}}$$

where the maximal fluorescence power: $I_{fluo}^{max} = (\nu_{em}/\nu_{ex})\sigma_{14}I_{ex}$. Express I_{sat} in terms of the various rates and cross section of the problem and find its value. Is it detrimental to the observation of biological tissues?
3. If the stimulating beam is modulated: $I_{STED} = I_{STED}^0 \sin^2 \pi x/a$, with $I_{STED}^0 \gg I_{sat}$ show that the fluorescence signal mainly arises from sources in a region of size:

$$\delta x \sim \frac{a}{\pi}\sqrt{I_{sat}/I_{STED}^0}$$

(which can be $\ll \lambda/2\text{NA}$) around $x_n = na$. Hence, the resolution of a microscope can be increased by increasing the power of the stimulating beam.

Estimating the diffusion constant from 2D particle tracking, neglecting noise, and camera integration. Consider a fluorophore diffusing on a 2D membrane and its position $\vec{x}_n(t)$ determined at times $t = n\Delta t$. Let the distance travelled during a time step Δt be: $\Delta r_n = |\vec{x}_n - \vec{x}_{n-1}|$. Let $P(D_n)$ be the probability distribution of the diffusion constant D_n estimated from n steps via: $4D_n t = \sum_{i=1}^{n} \Delta r_i^2$. Assume that the probability distribution of step sizes $f(\Delta r_n)$ is purely diffusive:

$$f(\Delta r_n) = \frac{1}{4\pi D\Delta t}\exp\left(-\frac{\Delta r_n^2}{4D\Delta t}\right)$$

Show that the distribution $P(D_1)$ is exponential. Assume that the successive steps are uncorrelated and show that the distribution $P(D_n)$ is then distributed according to the Erlang distribution of order n:

$$P(D_n) = \frac{n^n D_n^{n-1}}{D^n(n-1)!}e^{-nD_n/D}$$

Show that $\langle D_n\rangle = D$ and find the variance of D_n [187].

Saturation intensity in presence of Triplet states. Refer to the Jablonski diagram Fig. 3.2. Show that the probability of occupation of state S_1 is:

$$P_1 = \frac{1}{1+\theta}\frac{I/I_s^T}{1+I/I_s^T}$$

with $\theta = k_{ISC}/k_T$ and $I_s^T = I_s/(1+\theta)$, where I_s is the saturation intensity in absence of intersystem crossings (Eq.3.9).

4

The Mechanical Properties of Nucleic Acids

As any polymer, DNA in a good solvent (e.g., water) adopts a random coil conformation which maximizes its entropy [5]. Pulling on the molecule reduces this entropy and costs energy. The associated entropic forces result from a reduction of the number of possible configurations of the system consisting of the molecule and its solvent, so that at full extension there is only one configuration left: a straight polymer linking both ends. To reach this configuration, work has to be done against entropy, a force has to be applied. Entropic forces are rather weak, typically less than 10 pN. Beyond this force and up to about 70 pN, DNA stretches like any spring: it is in an enthalpy-dominated regime.

The first measurements of the entropic elasticity of a single DNA molecule were reported in 1992 by S. Smith, L. Finzi, and C. Bustamante [15]. They used a combination of magnetic fields and hydrodynamic drag to pull on small super-paramagnetic beads tethered to a surface by a single DNA molecule. The precision of their measurement was good enough to invalidate the theoretical prediction of the Freely-Jointed-Chain (FJC) model, which treats DNA as a chain of freely jointed rods whose entropy is reduced upon pulling. A physically more sensible approach that considers DNA to be a continuous flexible (worm-like) chain, see Fig. 4.1, was later solved by J. Marko and E. Siggia. This Worm-Like Chain (WLC) model computes the chain entropy change more realistically than the FJC model, in particular at high forces, where the entropy is underestimated in the latter model since it neglects the bending fluctuations of the discretized rods. The predictions of the WLC model were shown to agree remarkably well with the data measured by Smith *et al.*, see Fig. 4.2. A particularly stringent test of this model is to use magnetic tweezers to pull on a DNA molecule and measure simultaneously the extension of the molecule and its stiffness (deduced from its longitudinal fluctuations). The ratio of force to stiffness satisfies the prediction of the WLC model with no adjustable parameters. In fact, single-molecule measurements of DNA elasticity have turned out to be the most precise method of deducing its persistence length and studying its dependence on ionic conditions [15]. The excellent agreement between the elastic theory of an ideal polymer and the measurements on DNA has provided a benchmark, calibration, and framework for all single-molecule work on nucleic acids. It sets a solid

Single-Molecule Studies of Nucleic Acids and Their Proteins. David Bensimon, Vincent Croquette, Jean-François Allemand, Xavier Michalet, Terence Strick 2019. © Oxford University Press 2019. Published in 2019 by Oxford University Press.
DOI: 10.1093/oso/9780198530923.001.0001

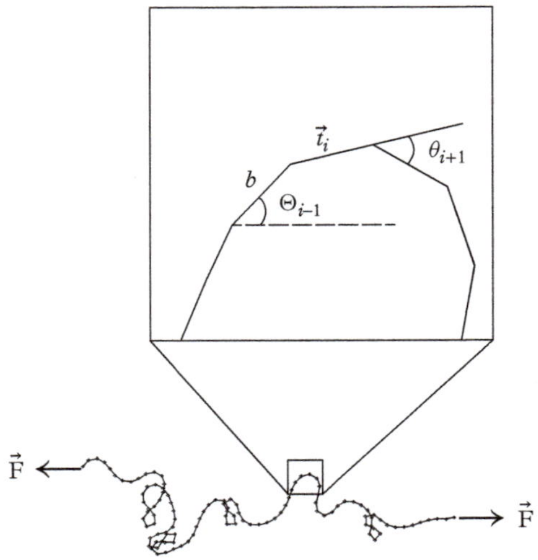

Fig. 4.1 The Kratky–Porod model. *A continuous polymer chain can be simulated by a chain of segments of size b and orientation vector $\vec{t}_i$. The direction of the stretching force F defines the z-axis.*

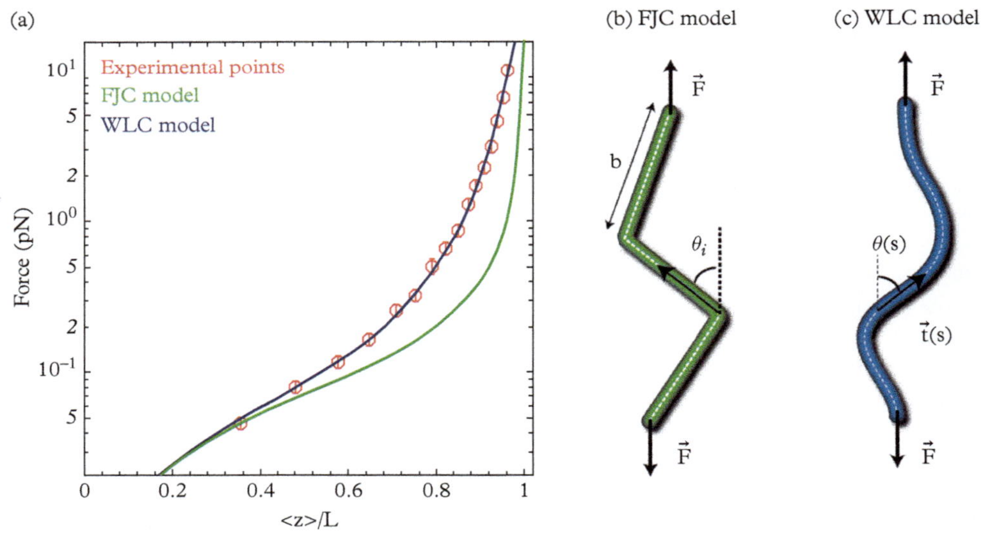

Fig. 4.2 Force vs extension for dsDNA and polymer models for DNA. *(a) Experimental force–extension curve for one dsDNA of* 16 μm *(red squares). The fit using WLC model (blue curve) gives* $\xi = 51.6$ *nm and* $L = 15.6$ μm*. The plot of the FJC (green curve) model using the same value of* ξ *shows an obvious discrepancy with the experimental data. (b) The FJC model: the DNA chain is modelled as a sequence of segments of size* $b = 2\xi$*;* θ_i *is the angle between the i-segment and the direction of force. (c) The WLC model: the DNA chain is treated as a continuous curve with a given bending rigidity* $B = k_B T \xi$ *(see section 4.1.3).*

foundation for understanding the structural transitions observed in DNA under high force (or high torque), and for studying its interactions with proteins. In the following, we will review first the theoretical models describing polymer elasticity and the dynamics of polymers. Because DNA is one of those rare polymers that can bear a torque, i.e., be twisted, we will review in section 4.3 the behaviour of DNA under twist. We will then describe the behaviour of braided DNA molecules (a configuration encountered during DNA replication) and briefly describe how the behaviour of stretched, twisted, and braided DNA molecules can be numerically simulated in section 4.7. We will end this chapter with a discussion of the elastic behaviour of single-stranded nucleic acids (DNA and RNA) where inter-strand interactions (self-avoiding or complementary base-pairing) strongly affect their elastic behaviour.

4.1 Models of stretched polymers

4.1.1 The Kratky–Porod model

For simplicity, let us first consider a polymer chain with no torsional stress. Such a chain is often described by the Kratky–Porod model [203]: a succession of N segments of equal length b and orientation vector $\vec{t_i}$, see Fig. 4.1. The energy $\mathcal{E}_{KP}$ of a chain configuration (the ensemble of segment orientations $\{\vec{t_i}\}$) is the sum of the bending energies of successive segments:

$$\mathcal{E}_{KP} = -\frac{B}{b}\sum_{i=2}^{N}\vec{t_i}\cdot\vec{t}_{i-1} = -\frac{B}{b}\sum_{i=2}^{N}\cos\theta_i \tag{4.1}$$

where θ_i is the angle between successive orientation vectors and B is the bending modulus. This model is formally similar to that of a classical 1D magnetic (spin) system, a model which has been solved exactly [204]. The angular correlation decays exponentially with distance along the chain:

$$\langle\vec{t_i}\cdot\vec{t_j}\rangle = e^{-b|i-j|/\xi} \tag{4.2}$$

where $\xi = B/k_BT$, the decay length of the angular correlation, reflects the stiffness of the chain and is known as the persistence length. The chain end-to-end mean square distance R_g satisfies:

$$R_g^2 \equiv \langle\vec{R}^2\rangle = (b\sum_{i=1}^{N}\vec{t_i})^2 \simeq 2Nb\xi = 2l_0\xi \tag{4.3}$$

where $l_0 = Nb$ is the chain length. A DNA molecule in solution is, thus, predicted to adopt a fluctuating random coil configuration of typical size R_g, known as the radius of gyration, see Fig. 1.1d. For many years, the measurement of R_g by various means (sedimentation, light scattering, etc.[203, 205]) was the only way to estimate the persistence length of DNA (or any polymer).

Stretching a single DNA molecule now provides a much more precise way of measuring ξ. To model the behaviour of a polymer chain under tension, a term representing the work $W = -\vec{F}\cdot\vec{R} = -Fb\sum t_{i,z} = -Fb\sum \cos\Theta_i$ done by a force F acting on the chain along the z-axis (Θ_i is the angle between $\vec{t_i}$ and the z-axis) is added to Eq. 4.1:

$$\mathcal{E}_{KP} = -\frac{B}{b}\sum_{i=2}^{N}\vec{t_i}\cdot\vec{t}_{i-1} - Fb\sum_{i=1}^{N} t_{i,z} \tag{4.4}$$

$$= -\frac{B}{b}\sum_{i=2}^{N}\cos\theta_i - Fb\sum_{i=1}^{N}\cos\Theta_i \tag{4.5}$$

Unfortunately this model can be solved only for small forces [204], where the mean extension of the chain $\langle z\rangle(\ll l_0)$ is:

$$\langle z\rangle = \frac{2F\xi}{3\,k_BT}l_0 = \frac{F}{3\,k_BT}R_g^2 \tag{4.6}$$

To compute the elastic response of a chain at higher forces, one has to resort to numerical calculations (e.g., transfer matrix methods) or to various approximations of the Kratky–Porod model.

4.1.2 The Freely Jointed Chain model

An interesting limit is the Freely Jointed Chain (FJC) model, which consists of setting $B = 0$ in Eq. 4.5. This assumption models a polymer chain of total length l_0 by a series of N freely rotating rigid segments of orientation $\vec{t_i}$ and length $b = 2\xi$ (the so-called Kuhn length): $l_0 = Nb$. In this FJC model, the energy of a given chain configuration $\{\vec{t_i}\}$ is, thus, $\mathcal{E}_{FJC} = W = -Fb\sum^N\cos\Theta_i$. The partition function $\mathcal{Z}$ is:

$$\mathcal{Z} = \sum_{\{\vec{t_i}\}} e^{-\mathcal{E}_{FJC}/k_BT} = \sum_{\{\vec{t_i}\}}\prod_{i=1}^{N} e^{Fb\cos\Theta_i/k_BT} \tag{4.7}$$

$$= \left[\int d\Omega e^{Fb\cos\Theta/k_BT}\right]^N = \left[\frac{2\pi k_BT}{Fb}\sinh\frac{Fb}{k_BT}\right]^N \tag{4.8}$$

From the free energy $\mathcal{F} = -k_BT\log\mathcal{Z}$, one can compute the mean extension of the chain $\langle z\rangle$:

$$\langle z\rangle = -\frac{\partial\mathcal{F}}{\partial F} = l_0\left(\coth\frac{Fb}{k_BT} - \frac{k_BT}{Fb}\right) \tag{4.9}$$

Notice that at small forces, one recovers our previous result, Eq. (4.6), using the approximation $\coth(x) \sim 1/x + x/3$. However, as shown in Fig. 4.2a, the FJC model is too crude and is **not** a good approximation of the elastic behaviour of a DNA molecule at large extensions ($l > R_g$).

4.1.3 The Worm-Like Chain model

A much more precise description is provided by the Worm-Like Chain (WLC) model, the continuous ($b \to 0$) limit of Eq. 4.5:

$$\mathcal{E}_{WLC} = \lim_{b\to 0}\left[\frac{B}{2}b\sum_{i=2}^{N}\left(\frac{\vec{t}_i - \vec{t}_{i-1}}{b}\right)^2 - F\sum_{i=1}^{N} b\cos\Theta_i + \frac{(N-1)B}{b}\right]$$

Neglecting the irrelevant last (constant) term and replacing the sum by an integral yields:

$$\frac{\mathcal{E}_{WLC}}{k_BT} = \frac{\xi}{2}\int_0^{l_0}\left(\frac{d\vec{t}}{ds}\right)^2 ds - \frac{F}{k_BT}\int_0^{l_0}\cos\Theta(s)ds \tag{4.10}$$

where s is the curvilinear coordinate along the chain and $\xi = B/k_BT$. The calculation of the chain's mean extension can be done using path integral techniques. It is similar to the calculation of the ground state of a quantum mechanical dipole in an electric field as shown by Marko and Siggia [206, 207, 208, 209, 210]. For the reader not familiar with these techniques, we briefly sketch this calculation.

Let $P(s,\vec{t})$ be the probability of the chain segment at position s pointing in the direction $\vec{t}$. For the next segment at position $s+b$

$$P(s+b,\vec{t}) = A\int d^2t' \exp\left[-\frac{\xi}{2b}(\vec{t}-\vec{t}')^2 + \frac{Fb}{k_BT}\vec{t}'\cdot\hat{z}\right]P(s,\vec{t}') \tag{4.11}$$

The integration is over all directions of the tangential vector (i.e., the surface of the unit sphere). The term in the exponential is the Boltzmann factor associated with the bending energy between two consecutive segments as in Eq. 4.10 (or Eq. 4.5) and A a normalization constant. In the limit $b \to 0$, setting $\vec{t}' - \vec{t} = \sqrt{b/\xi}\vec{u}$, Eq. 4.11 becomes:

$$P(s+b,\vec{t}) = \frac{Ab}{\xi}\int\int_{-\infty}^{\infty} d^2u e^{-u^2/2}(1 + \frac{Fb}{k_BT}\cos\Theta)P(s,\vec{t}+\sqrt{b/\xi}\vec{u}) \tag{4.12}$$

Expanding the probability distributions on both sides of Eq. 4.12 to first order in b implies that the normalization constant $A = \xi/2\pi b$. Performing the integration yields the diffusion (Schrödinger-like) equation:

$$\frac{dP}{dx} = \frac{1}{2}\Delta_S P + \alpha\cos\Theta P \tag{4.13}$$

where $x \equiv s/\xi$, $\alpha = F\xi/k_BT$, and Δ_S is the Laplacian on the sphere ($\Delta_S = -\hat{L}^2$, the angular momentum operator in quantum mechanical notation). The solution of Eq. 4.13 can be written as:

$$P(x,\Theta,\Phi) = \sum_\beta a_\beta \Psi_\beta(\Theta,\Phi) e^{-\epsilon_\beta x} \tag{4.14}$$

where the eigenstates $\Psi_\beta(\Theta,\Phi)$ satisfy the Schrödinger equation:

$$\left[\frac{\hat{L}^2}{2} - \alpha\cos\Theta\right]\Psi_\beta = \epsilon_\beta\Psi_\beta$$

with eigen-energy ϵ_β. For long chains, $x \to \infty$, the smallest eigenvalue ϵ_0 (the ground state) dominates the sum in Eq. 4.14. The free energy of the polymer is then: $\mathcal{F} = (k_BTl_0/\xi)\epsilon_0$, and by Eq. 4.9, the mean extension of the chain becomes:

$$\langle z\rangle = -\frac{\partial\mathcal{F}}{\partial F} = -l_0\frac{\partial\epsilon_0}{\partial\alpha} \tag{4.15}$$

Thus, to compute the mean extension of the chain one simply has to find the ground state energy, ϵ_0. At low forces ($\alpha \ll 1$), perturbation theory readily yields $\epsilon_0 = -\alpha^2/3$, and, thus, $\langle z\rangle/l_0 = 2\alpha/3$, as shown earlier in Eq. 4.6. At large forces, i.e., when $\alpha \gg 1$, the chain is almost straight (i.e., $\Theta \ll 1$) and the $\cos\Theta$ term in Eq. 4.13 can be expanded to $\cos\Theta \sim 1 - \Theta^2/2$. Eq. 4.13 is then identical to the Schrödinger equation of a 2D harmonic oscillator whose ground state energy: $\alpha + \epsilon_0 = \sqrt{\alpha}$. Therefore, at high forces the free energy of the chain satisfies:

$$\mathcal{F} = -Fl_0 + l_0\sqrt{Fk_BT/\xi} \tag{4.16}$$

The mean extension is, thus, $\langle z\rangle/l_0 = 1 - 1/(2\sqrt{F\xi/k_BT})$. Between those two extremes, there is no analytic formula (equivalent to Eq. 4.9) for the force–extension behaviour of a WLC, but a simple and efficient numerical solution was provided by Bouchiat *et al.* [211], who gave an approximation better than 0.1%:

$$F = (k_BT/\xi)\, g(\zeta) \tag{4.17}$$

where $\zeta = \langle z\rangle/l_0$ and:

$$g(\zeta) = \zeta - \frac{1}{4} + \frac{1}{4(1-\zeta)^2} + \sum_{i=2}^{7} a_i\zeta^i \tag{4.18}$$

with $a_2 = -0.5164228$, $a_3 = -2.737418$, $a_4 = 16.07497$, $a_5 = -38.87607$, $a_6 = 39.49944$, $a_7 = -14.17718$ [211]. Notice that at small relative extensions $\zeta \ll 1$, we recover Eq. 4.6, as for the FJC model. However, when compared over the whole extension range, the WLC model is a much better description of the behaviour of DNA than the FJC model. As shown in Fig. 4.2 the WLC model fits the measured data extremely well and allows for a very precise estimate of the DNA's persistence length: $\xi = 52 \pm 2$ nm in near-physiological conditions (10 mM phosphate buffer (pH = 8.0), 10 mM NaCl).

4.1.4 Self-avoidance effects

In the theoretical models described previously, the fact that a real polymer cannot intersect itself was not taken into consideration. For example, in the computation of the partition function $\mathcal{Z}$, Eq. 4.8, these unphysical configurations are included. In this section, we shall try to justify this approximation. A treatment of the self-avoidance effects of a stretched polymer exists only at low extensions ($l \ll l_0 = Nb$), where a heuristic (though wrong) argument due to Flory [212] works remarkably well: due to self-avoidance, one expects a real chain to occupy a larger volume than the so-called Gaussian (intersecting or phantom) chains considered previously, which occupied a volume R_g^3. This, of course, costs some entropic energy which is compensated by a reduction in the probability of self-interaction. In this approach, the free energy of an FJC polymer is:

$$\frac{\mathcal{F}}{k_B T} = \frac{3l^2}{2R_g^2} + \frac{vN^2}{l^3} \tag{4.19}$$

The first term on the right-hand side (RHS) describes the energy cost associated with the swelling of the polymer ($\sim \int F dl$, with F given by Eq. 4.6). The second term accounts for self-avoidance. It is proportional to the probability that two segments of the FJC (assumed cylindrical with a radius r and length b) sweeping a volume $v = \pi b^2 r$ will share the same location in the volume l^3 occupied by the chain. This repulsive term decreases as the chain swells. By minimizing $\mathcal{F}$ with respect to l, one obtains the equilibrium (Flory's) radius of a self-avoiding polymer:

$$R_F = (vb^2)^{1/5} N^{3/5} = l_0^{3/5} (v/b)^{1/5} = l_0^{3/5} (\pi br)^{1/5} \tag{4.20}$$

Excluded volume interactions will become important when the second term of the RHS of Eq. 4.19 becomes non-negligible, i.e., when $vN^2 > R_g^3 = b^3 N^{3/2}$. Self-avoidance, thus, becomes non-negligible when $N > (b/\pi r)^2$. For a DNA molecule with $r = 1$ nm and $b = 100$ nm, this corresponds to $N \sim 1000$, i.e., a molecular length: $l_{0,F} = Nb \sim 100$ μm. Almost all manipulations of DNA molecules so far have been in a regime $l_0 < l_{0,F}$, where self-avoidance is totally negligible (except during twisting experiments). However, for single-stranded DNA and proteins (in denaturing conditions) with persistence length $\xi \sim 1$ nm $\sim r$, self-avoidance plays a crucial role and alters the predictions based on FJC or WLC models.

4.2 Relaxation dynamics of DNA

The well-defined properties of DNA (its exact length and composition) and the possibilities offered by the tools of molecular biology for labelling the molecule at precise locations have opened the way for a detailed study of its dynamics. Thus, the diffusion of DNA can be monitored by fluorescently labelling one DNA extremity and using FCS techniques (see section 3.10) to measure its mean square displacement $\langle r^2(t)\rangle$ as a function of time [161].

For a fluorophore not bound to single- or double-stranded DNA, $\langle r^2(t)\rangle$ increases linearly with time, as expected from regular diffusion: $\langle r^2(t)\rangle = 6D_0 t$, where D_0 is the fluorophore's diffusion constant. For a typical fluorophore (of radius $r_0 = 1$ nm) in water (viscosity: $\eta = 10^{-3}$ N s/m^2) the hydrodynamic drag experienced during diffusion is $\gamma_0 = 6\pi\eta r_0$, and therefore by Einstein's relation:

$$D_0 = \frac{k_B T}{\gamma_0} \sim 200\ \mu\text{m}^2/s$$

When the fluorophore is bound to single-stranded DNA (in dilute and denaturing conditions), diffusive behaviour associated with diffusion of the whole ssDNA molecule occurs on long time scales (i.e., when $\langle r^2(t)\rangle > R_g^2$). At shorter times, subdiffusive behaviour is observed: $\langle r^2(t)\rangle \propto t^{2/3}$, see Fig. 4.3 (where the symbol $\propto$ stands for 'proportional to'). This behaviour is expected from the Zimm model [213] of polymer dynamics, which assumes complete hydrodynamic coupling between the various polymer segments: the drag γ_r on a polymer region of size r is assumed equivalent to that of a solid particle of similar radius, $\gamma_r = 6\pi\eta r$. Thus, diffusion of the fluorophore on a length scale $r < R_g$ is associated with a diffusion constant $D_r = k_B T/\gamma_r$ and happens on a time scale t:

$$t = \frac{r^2}{6D_r} = \frac{\pi\eta}{k_B T} r^3$$

and therefore $\langle r^2(t)\rangle \propto t^{2/3}$, as observed.

When the fluorophore is bound to double-stranded DNA (in a dilute solution), a different subdiffusive behaviour is observed on short time scales (i.e., for $\langle r^2(t)\rangle \ll R_g^2$: $\langle r^2(t)\rangle \propto \sqrt{t}$, see Fig. 4.3). This behaviour is predicted by the Rouse model [214] of polymer dynamics, which assumes no hydrodynamic coupling between the polymer segments. In other words, it presumes that the motion of one segment does not drag along the others. In this model, the drag on a polymer of total length l is equivalent to the sum of the drag γ_b on each Kuhn segment of size b and diameter d: $\gamma_b = 3\pi\eta b/\log(b/d)$. The total drag γ_r on a polymer blob of size r, comprising $N_r = r^2/b^2$ Kuhn segments, is, thus,

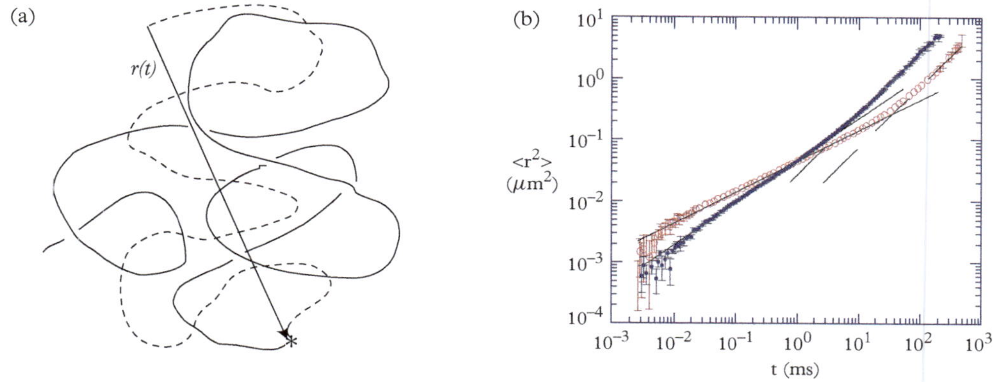

Fig. 4.3 Diffusion of a fluorophore bound to a DNA extremity. *(a) A DNA molecule (continuous line) is labelled at one end with a fluorophore (shown by a *). The average distance $r^2(t)$ travelled by the bound fluorophore (dashed line) is measured by FCS techniques. (b) The average distance $r^2(t)$ covered by a fluorophore bound to ssDNA (dots) and dsDNA (open circles). At long times (t > 10 ms) both display the typical diffusive behaviour ($r^2 \propto t$) associated with the random motion of the whole molecule (dashed lines). At short times (t < 1 ms) both display subdiffusive behaviour, characteristic of random motion within the polymer. For a 6700-base ssDNA (blue): $r^2 \propto t^{2/3}$. For the equivalent 6700-bp dsDNA (red): $r^2 \propto t^{1/2}$. Adapted from R. Shusterman et al. [161].*

$$\gamma_r = N_r \gamma_b = \frac{r^2 \gamma_b}{b^2} = \frac{3\pi \eta r^2}{b \log(b/d)}$$

Diffusion of a bound fluorophore over a length scale r (implying diffusion of a polymer blob of that size) occurs on a time scale t:

$$t = \frac{r^2}{6D_r} = \frac{\gamma_b}{6k_B T b^2} r(t)^4$$

and therefore $\langle r^2(t) \rangle \propto \sqrt{t}$, as observed.

Rouse behaviour is valid as long as the polymer segments do not interact via their hydrodynamic drag. This assumption breaks down when the total drag on the polymer blob of size $r(t)$: $\gamma_r = N_r \gamma_b$ is larger than the drag computed, assuming full hydrodynamic coupling (i.e., the Zimm model), $6\pi\eta r$. Rouse behaviour is, thus, valid, as long as $r(t) < r_c \sim b \log(b/d)$. For ssDNA, for which $b \sim d$, this regime is non-existent, whereas for dsDNA, with $b/d \sim 100$, the Rouse regime can be observed over distances as large as $r_c \sim 0.5\ \mu$m.

On long time scales, motion of the fluorophore implies motion of the whole chain of size $R_g \propto l_0^{1/2}$ (or $\propto l_0^{3/5}$ for long chains $l_0 \gg 40\ \mu$m, where self-avoiding interactions increase the coil size). It is, thus, controlled by diffusion of the DNA molecule with diffusion constant: $D_{dsDNA} = k_B T / 6\pi \eta R_g \sim 0.66\ \mu m^2/s$ (for a 10 μm long dsDNA [33]).

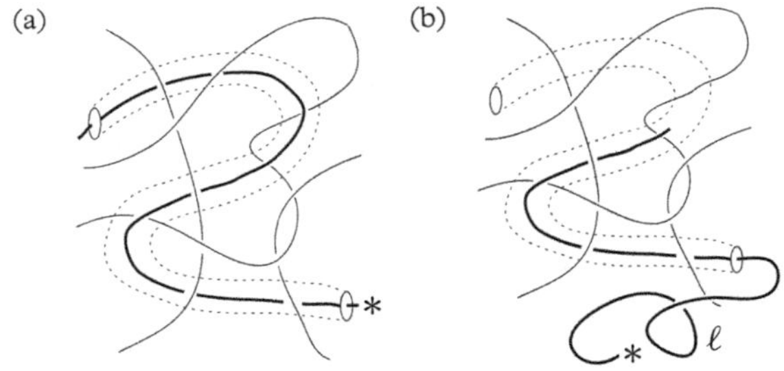

Fig. 4.4 Schematics of DNA reptation in an entangled solution of molecules, *as might be the case in the nucleus of eukaryotes. (a) A particular DNA molecule (thick line) of length l_0 is constrained by its interactions with neighbouring molecules to diffusing along a tube with a diffusion constant $D_{rep} \propto l_0$. (b) After a time $t = l^2/D_{rep}$ a length l of the molecule has disentangled. Total disentanglement, thus occurs on a time $t_{rep} \propto l_0^3$ which can be extremely long (years!). To study this reptation in a DNA solution, some of the molecules can be labelled at one end with a fluorescent dye (shown by a *).*

The results just described correspond to diffusion of DNA segments in dilute solutions, where the distance between two chains is larger than the coil size R_g. In a typical eukaryotic cell, however, the concentration of DNA is so large (about 1 μg/μl, which corresponds to roughly one Kuhn segment of DNA per excluded volume) that the molecules are highly entangled. The motion of a DNA segment involves reptation of the whole chain of length l_0 along the virtual 1D "tube" which surrounds it [5], see Fig. 4.4, with a drag $\gamma_{rep} \sim 3\pi\eta l_0/\log(b/d)$ and a diffusion constant, $D_{rep} = k_BT/\gamma_{rep} \propto l_0^{-1}$. Two molecules will typically disentangle in the time t_{rep} it takes them to diffuse in 1D along their whole length:

$$t_{rep} = \frac{l_0^2}{2D_{rep}} = \frac{3\pi\eta}{2k_BT\log(b/d)} l_0^3 \approx \frac{1.2\eta}{k_BT} l_0^3 \tag{4.21}$$

For a 10 μm long DNA (at a concentration of 1 μg/μl): $t_{rep} = 300$ s [215]. For a typical chromosome of length $l_0 \sim 10$ cm: $t_{rep} = 3\ 10^{14}$ s ~ 10 million years! If nature had to rely on reptation to disentangle the chromosomes prior to cell division, life would have been limited to simple organisms with very short chromosomes. Chromosomes disentangle due to the activity of a fascinating class of enzymes, the topoisomerases (see Chapter 9) which are capable of undoing knots and other topological barriers to chromosome segregation by a mechanism that is still not fully understood.

4.3 The elastic behaviour of twisted DNA

Double-stranded DNA is one of the few natural polymers (with actin and collagen) which can store torsion. Living cells take advantage of this fact, since in nearly all living organisms, DNA is in a slightly untwisted state. This is probably needed in order to initiate local unravelling of the double helix. Bacteria living at high temperatures (thermophiles) have an opposite problem: to prevent the denaturation of their DNA, the molecule is overwound. Opened DNA regions (denaturation bubbles) may serve as a starting point for the proteins whose job it is to replicate DNA. As we will see, by twisting DNA, various structures can be formed, including such denaturation bubbles. Before describing these structures, we will review the topological formalism used to study DNA twisting.

4.3.1 DNA topology

Two topological quantities are needed in order to describe a twisted rope or DNA, see Fig. 4.5. First, the *twist* Tw of the double helix measures the number of times the two constituent strands of the molecule wrap about each other. For a linear DNA in the absence of any external constraints, the natural twist Tw_0 is equal to the number of base pairs (bp) n divided by the helical pitch h of the structure: $Tw_0 = n/h$ ($h = 10.4$ bp for B-DNA).

The second topological quantity, the *writhe* Wr of the molecule, counts the number of times the axis of the molecule crosses itself. On average, a linear DNA molecule, in the absence of external constraints, will have neither writhe nor macroscopic curvature, and thus $Wr_0 = 0$.

The sum of these two components is known as the *linking number* Lk, and represents the sum total of crossings (strand–strand and axis–axis) in the molecule. Thus, the natural linking number Lk_0 of a linear DNA in the absence of external constraints is equal to the number of helical turns of the molecule $Lk_0 = Tw_0$.

A mathematical theorem derived by White [218] shows that for a circular DNA (where the extremities of the molecule are connected), the linking number is a topological invariant (which can be modified only by cutting the molecule):

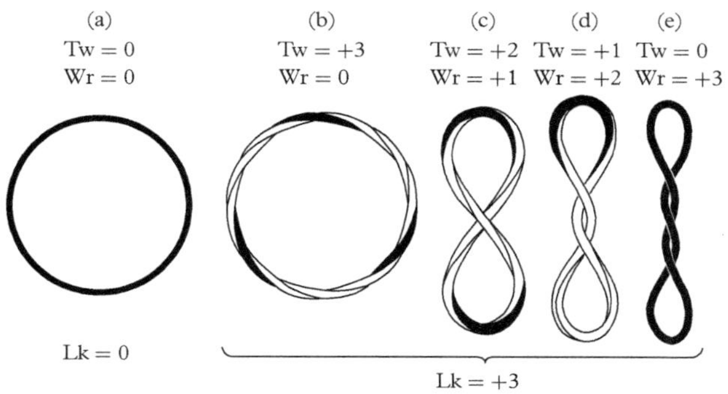

Fig. 4.5 Representation of a supercoiled plasmid where ΔLk is partitioned in different ways. *The plasmid is represented by a tube with a square cross section where one of the faces has been coloured black. (a) $\Delta Lk = 0$, the plasmid is torsionally relaxed. (b–e) $\Delta Lk = +3$, the plasmid is overwound and may partition this torsional constraint in several ways. (b) Supercoiling is stored entirely by a change in the twist Tw of the DNA. (c) and (d) The torsional constraint is partitioned between the formation of supercoils (plectonemes) (Wr) and twisting (Tw) of the molecule's axis. (e) Supercoiling is entirely stored in plectonemes and the torsion of the axis is negligible. Adapted from [216].*

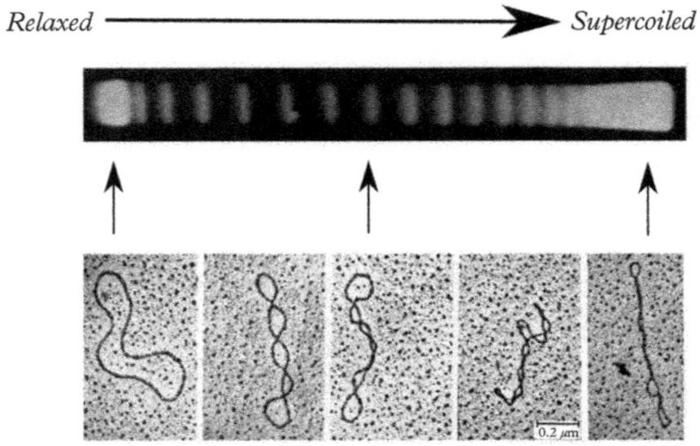

Fig. 4.6 Supercoiled plasmids. *Examples of a 5 kbp plasmid of varying degrees of supercoiling: from the relaxed form on the left to the highly supercoiled ($\sigma \sim -0.06$) form on the right. Theses various "topoisomers" can be separated on an electrophoretic gel where the more compact the DNA, the faster it migrates: the relaxed form is the slowest migrating species. Adapted from [217].*

$$Lk = Tw + Wr = constant \tag{4.22}$$

This is also valid for linear DNA molecules, whose extremities are prevented from rotating freely by being attached to macroscopic surfaces at multiple points at both extremities.

A DNA molecule is supercoiled when $Lk \neq Lk_0$, that is to say, when the molecule has an excess or a deficit of linking number relative to its torsionally relaxed state: $\Delta Lk = Lk - Lk_0 = Tw - Tw_0 + Wr = \Delta Tw + Wr$. The degree of supercoiling σ is defined as:

$$\sigma = \frac{\Delta Lk}{Lk_0} = \frac{\Delta Tw + Wr}{Lk_0} \tag{4.23}$$

Plasmids and phages extracted from bacteria are typically underwound, i.e. negatively supercoiled, with $\sigma \sim -0.06$, see Fig. 4.6. As an example, a 50 kb bacteriophage-λ DNA molecule with this degree of negative supercoiling is underwound by roughly 300 turns.

4.3.2 The buckling instability in DNA

Much of the elastic behaviour of coiled DNA can be understood by calling upon the common experience of twisting a rubber tube pulled on with a force F, (see Fig. 4.7). When one begins to twist the tube, its extension initially remains unchanged and the constraint accumulates as pure torsion. The tube's torque Γ increases linearly with the twist angle Ω, and its twist energy increases quadratically: $\Gamma = \frac{\mathcal{C}}{l_0}\Omega$ and $E_{torsion} = \frac{1}{2}\frac{\mathcal{C}}{l_0}\Omega^2$, where $\mathcal{C}$ is the tube's twist stiffness (usually written in the DNA context in units of k_BT: $\mathcal{C} \equiv k_BTC$, where C is the torsional persistence length). As one continues to twist the tube, after a certain number of turns n_b (corresponding to a torque Γ_b) the system

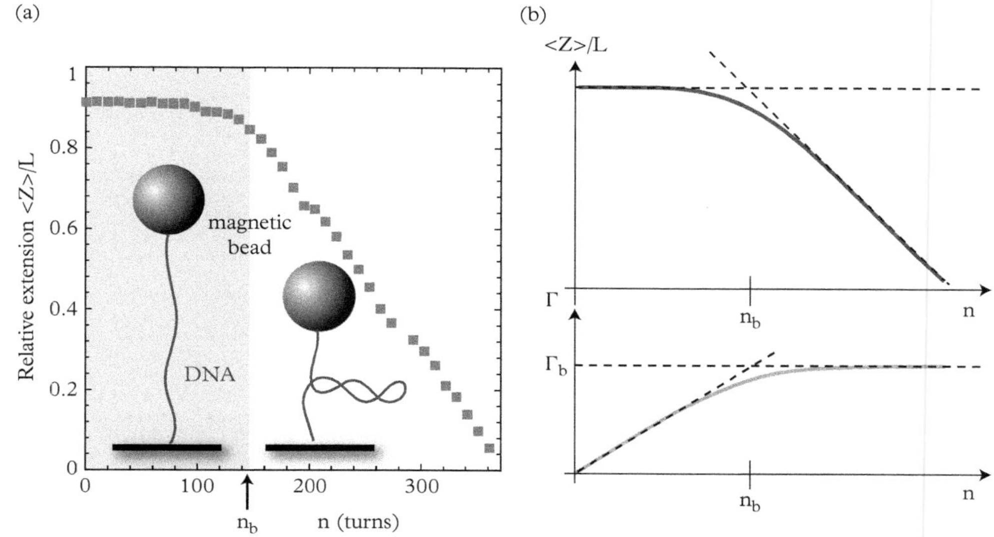

Fig. 4.7 Extension vs σ for one dsDNA. *(a) The experimental squares show the normalized extension vs number of turns of a single DNA molecule (50 kb) at $F = 1.2$ pN. Two regimes can be identified: at low n, change in DNA extension is small, while the molecule stores torsional energy; at $n = n_b \approx 140$, the molecule buckles and starts forming supercoils (plectonemes, as seen on a phone cord). After buckling, the extension decreases almost linearly with the addition of further twisting. (b) As can be understood for an elastic rubber tube, initial twisting does not change the system's extension, but the torque stored in the tube increases linearly with the number of turns n applied. At a given number of turns $n = n_b$, forming a loop costs less energy than increasing the torsional energy. Each additional turn leads to the formation of another loop, so that the extension decreases linearly with n, but the torque $\Gamma = \Gamma_b$ remains constant. Adapted from G. Charvin et al. [219].*

buckles and a loop of radius R is formed. The twist energy is thus transferred into bending energy. Upon further twisting the tube coils upon itself to form plectonemes but the torque no longer increases, see Fig. 4.7b. As a result, and in-spite of the stretching force F, the system's extension decreases by $2\pi R$ for every turn. Balancing the torsional energy against the work done and the increase in bending energy results in the relation:

$$2\pi\Gamma_b = 2\pi RF + 2\pi R\frac{1}{2}\frac{B}{R^2} \tag{4.24}$$

The radius which minimizes the bending energy is given by $R = \sqrt{B/2F}$, so that:

$$\Gamma_b = \sqrt{2BF} = \sqrt{2\xi F k_B T}; \quad n_b = L\sqrt{2BF}/(2\pi C) \tag{4.25}$$

Thus, the greater the force, the larger the critical buckling torque Γ_b and the number of turns n_b and the smaller the loop radius. Upon further twisting, the tube coils around itself but the torque Γ_b no longer increases.

As shown in section 4.3.3 below, this simple model qualitatively describes the behaviour of DNA under torsion. The buckling transition is not as sharp as expected for a macroscopic transition because of thermal fluctuations, but increasing the force tends to make the transition sharper.

4.3.3 The Rod-Like Chain (RLC) model

Although the buckling instability appears abruptly on a rubber tube, in the case of DNA thermal fluctuations, which are particularly important at low forces ($F < 0.5$ pN) and close to the mechanical instability, tend to smoothen the transition. A detailed understanding of twisted DNA thus requires a full statistical mechanics analysis of its buckling in the presence of thermal fluctuations. This analysis has been performed by Moroz and Nelson [220] and Bouchiat and Mézard [221] in the framework of the Rod-Like Chain (RLC) model. This approach consists of describing a DNA molecule not as a chain free to rotate along its axis (as in the WLC model), but as an isotropic homogeneous rod with a finite torsional modulus $\mathcal{C}$. In this approach, the helical structure of the molecule is neglected: it is expected to behave similarly whether wound or unwound. As we shall see, this approximation is partially valid at low forces ($F < 0.5$ pN). At higher forces, the torque built into the DNA molecule before it buckles can be large enough to induce local structural transitions. The energy of the rod-like chain, $\mathcal{E}_{RLC}$, is obtained by adding to the bending energy of the WLC model, Eq. (4.10), an energy of twist $\mathcal{E}_T$:

$$\mathcal{E}_{RLC} = \mathcal{E}_{WLC} + \mathcal{E}_T = \mathcal{E}_{WLC} + \frac{\mathcal{C}}{2}\int_0^{l_0} ds\, \Omega^2(s) \tag{4.26}$$

where $\Omega(s)$ is the local excess of the twist. Since the molecule cannot pass through itself as it is twisted, self-avoidance is essential for a proper description of twisted DNA. Although an analytical solution of this model including self-avoidance effects is not available, two approaches have been taken to deal with this issue. One group [220, 222] solved it at high forces, where plectonemes are suppressed and self-avoidance can be neglected. The result of this computation is that twisted DNA can be treated before buckling as an elastic tube characterized by the bending rigidity of DNA and an effective torsional modulus $\mathcal{C}_{eff}$, which, owing to molecular fluctuations, is smaller than the bare DNA torsional modulus $\mathcal{C}$ (appearing in Eq. 4.26) and varies with the tension on the molecule as:

$$\mathcal{C}_{eff} = k_B T C\left[1 - (C/4\xi)\sqrt{k_B T/\xi F}\right] \tag{4.27}$$

A second group [221, 211], noticing that the twist integrand in Eq. 4.26 was singular, regularized it with the introduction of a short length cut-off (of the size of the DNA pitch) which allows for a solution of the model at lower forces (the cut-off presumably mimics some of the effects of self-avoidance at intermediate forces [223]).

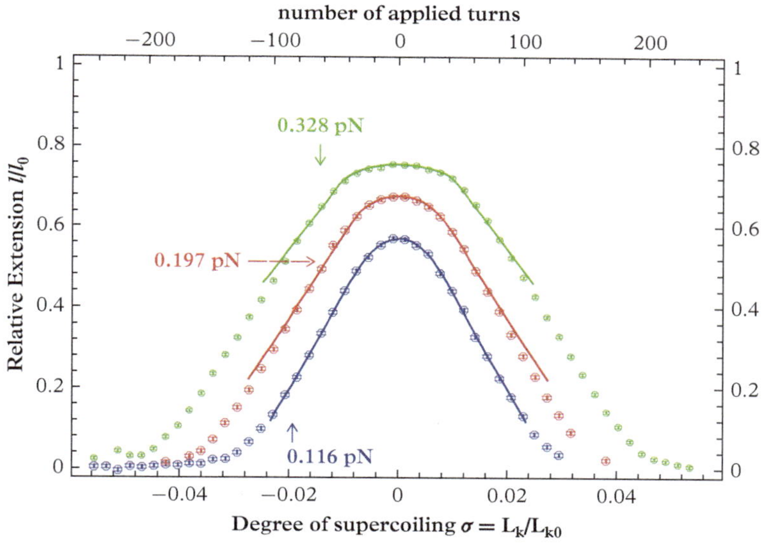

Fig. 4.8 Variation of the relative extension l/l_0 as a function of the degree of supercoiling *$\sigma = n/Lk_0$ at various (low) forces. The continuous curves are best fit to the RLC model with torsional persistence length: $C = 86$ nm, from T.R. Strick et al. [224].*

Although the solutions of the RLC model do not fully address the chain self-avoidance, which stabilizes the plectonemic state, their predictions fit remarkably well observations at high forces and low twist [220] (before buckling) and at low forces (< 0.4 pN) and moderate twist [221] (around buckling), a domain where the molecule end-to-end distance decreases as it is coiled, see Fig. 4.8. By fitting the RLC predictions to the data, one can extract a value for the torsional modulus $\mathcal{C}$ of DNA (or torsional persistence length C): $C = \mathcal{C}/k_B T = 86 \pm 5$ nm [221]. This result depends weakly on the magnitude of the cut-off, whose best-fitted value of ~ 6 nm is roughly equal to twice the DNA pitch, $p = 3.4$ nm.

4.3.4 Visualization of plectonemes on DNA

Using fluorescently labelled DNA van Loenhout *et al.* [225] could visualize the dynamics of plectonemes *in vitro*. Homogeneously labelled DNA appears in epifluorescence as a bright line with a homogeneous intensity along the molecule. However, as plectonemes appear, they contract the DNA (the end-to-end extension decreases for a given DNA contour length) and some bright spots appear where the local concentration of fluorophores is higher, i.e., at the locations of plectonemes, see Fig. 4.9. The observations show that there can be multiple plectonemes for a given supercoiling density. Their number depends on the salt concentration and the tension on the molecule. The higher the tension and the salt concentration the smaller the number of plectonemes. The dynamics of plectonemes on these stretched and labelled DNA molecules can be monitored with a video camera. Unless they are pinned by defects

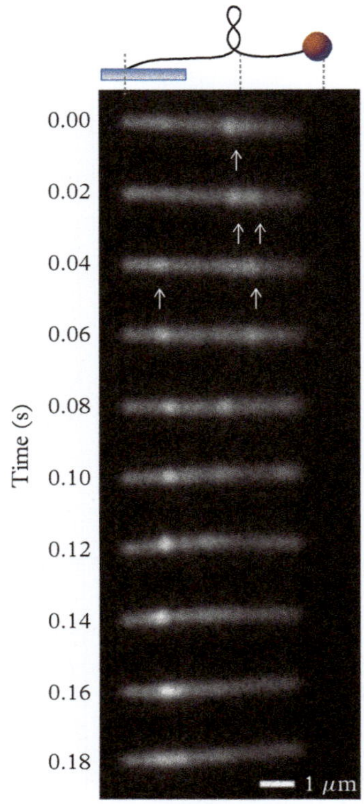

Fig. 4.9 Visualization of plectonemes *by epifluorescence microscopy. A twisted DNA molecule tethering a small magnetic bead is pulled sideways by magnets. The fluorescently labelled DNA displays fluorescent bright spots indicative of the presence of plectonemes. These fluctuate along the DNA, appearing and disappearing at various locations (indicated by arrows). Adapted from van Loenhout et al. [225].*

or other nucleation elements present along the molecule [226], plectonemes are not static objects: they may diffuse or hop from one location on DNA to another. Diffusion is observed as the random displacement of a bright spot with fixed intensity. Hopping is observed as the disappearance of a bright spot at a given location accompanied by an increase in fluorescence at a second location that can be a few kbp away.

4.4 Torque estimate from single-molecule thermodynamics

While measuring the tension applied on a DNA molecule and its extension is easily done with a number of manipulation techniques, measuring the torque applied on a supercoiled molecule is much more demanding. The most direct and elegant method uses optical tweezers to apply both force and torque on birefringent quartz cylindrical particles whose easy (extraordinary) polarization axis is perpendicular to the cylindrical axis [50, 49]. One of the cylindrical faces is treated so as to bind DNA at one end, the other DNA end being anchored to the surface of a glass coverslip, see section 2.4.4 and Fig. 2.5. Linearly polarized light is used to trap the cylindrical particle, whose axis aligns with the optical axis. Like the magnetic beads, which align with the magnetic field in a magnetic trap, the particle's polarization axis tends to align with the electric field. By rotating the field's polarization, the quartz particle can be rotated. This is essentially the optical equivalent of a magnetic trap. The big difference is that the torsional stiffness k_θ of this "optical torque wrench" [50] (about 10 pN nm/rad for every mW of laser power) is 10^5 times smaller than the torsional stiffness of a magnetic trap. The particle thus exhibits non-negligible angular fluctuations, from which the applied torque can in principle be deduced. In practice, the force on a trapped bead is deduced from the change in the beam linear momentum, and the torque on the particle is deduced from the change in the beam angular momentum. Because the right- and left-handed circular polarizations of the incoming linearly polarized beam are equal, the torque is related to the imbalance between these components in the outgoing beam resulting from its interaction with the particle:

$$\Gamma = (P_R - P_L)/\omega_0$$

where P_R (resp. P_L) is the power of the right (resp. left)-handed component of the scattered beam and ω_0 is the optical frequency. This setup has been used to measure the torque on DNA as it is twisted [227], see Fig. 4.10, confirming the intuitive picture derived from our experience with twisting ropes and allowing one to extract a direct estimate of the DNA's bare torsional modulus of $C \sim 100$ nm. The only drawback of this approach is that it is difficult to measure torques on DNA molecules under low tension (< 1 pN), as both the trap and torsional stiffnesses become very weak.

An indirect method has also been used to deduce the torque on a DNA molecule from the extension–rotation curves at various forces [228]. The interest of this method is that it exemplifies the use in a single-molecule context of a thermodynamic approach to deducing global parameters (free energies, forces, torques, rigidity moduli, etc.). It stems from the general observation that the free energy $\mathcal{F}$ of a twisted and stretched molecule depends on the force F and the applied rotational angle, $\theta = 2\pi n$. The mean extension of the molecule at a given force is therefore (see Eq. 4.9):

$$l = -\partial\mathcal{F}/\partial F|_\theta$$

and the mean torque is similarly:

$$\Gamma = \partial\mathcal{F}/\partial\theta|_F$$

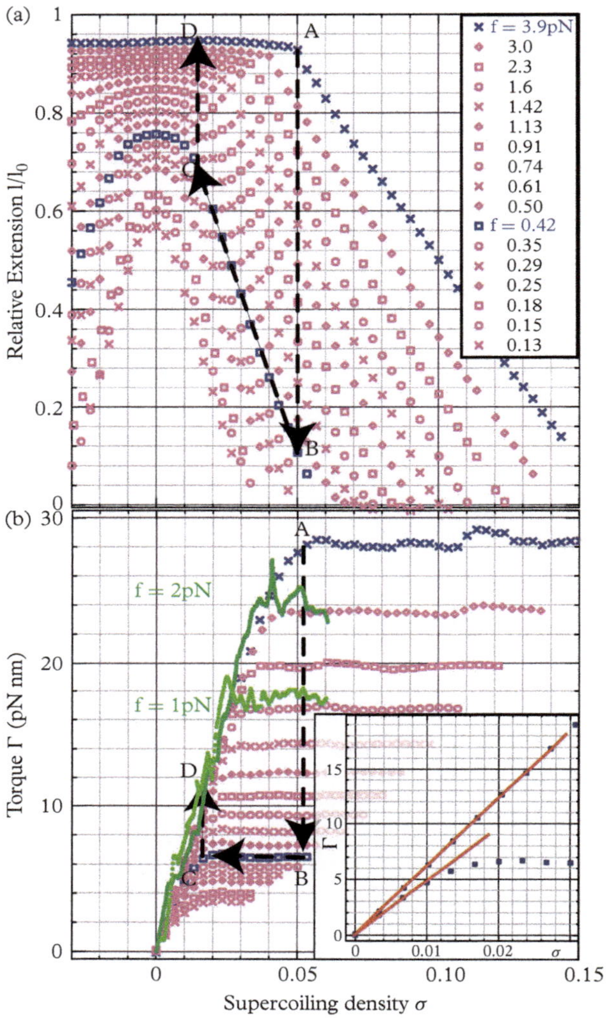

Fig. 4.10 Estimating the torque in DNA from force–extension measurements at various degrees of supercoiling. *(a) Variation of the mean relative extension l/l_0 of dsDNA as a function of its degree of supercoiling $\sigma = n/Lk_0$ in 100 mM NaCl. (b) Variation of the torque in DNA as a function of σ. These curves have been obtained by integrating, as explained in section 4.4, Eq. 4.28 from points A to B and from points C to D in (a), assuming that the torque in the plectonemic regime at a given force (i.e., at points B and C) is constant. The continuous wiggly lines in (b) are results from a direct measurement with the optical torque wrench. The inset is a zoom-in at low σ displaying two curves (at $F = 0.42$ pN and $F = 3$ pN) with different slopes, i.e., different effective torsional moduli. Adapted from F. Mosconi et al. [228].*

From these thermodynamic relations, one readily derives an expression for the mean torque at a given force $\Gamma(F,\theta)$, obtained from a measurement of the decrease in extension with increased coiling $\partial l/\partial\theta|_F$:

$$\Gamma(F,\theta) = \Gamma(F_0,\theta) - \int_{F_0}^{F} dF' \left(\frac{\partial l}{\partial\theta}\right)_{F'} \tag{4.28}$$

Since the angular rotation is known and the force and change in extension with rotation are easily measurable, the determination of the torque difference is reduced to a problem of sampling these variables precisely enough to estimate the above integral with sufficient precision. The integration constant $\Gamma(F_0,\theta)$ is ultimately set by the requirement that there is no torque on an untwisted molecule: $\Gamma(F,0) = 0$.

To compute the torque at various forces and degrees of supercoiling using Eq. 4.28, one starts from point A at the buckling transition ($\sigma_s = \sigma_A$) which is characterized by the highest stretching force ($F_A = 3.9$ pN) and the highest torque in the data shown in Fig. 4.10a. The value of the torque at A (which serves as the reference point) is initially unknown, but will ultimately be set by the requirement that $\Gamma(F,0) = 0$. The torque at point B is computed by integrating Eq. 4.28 from F_A to F_B, along path AB of constant $\sigma = \sigma_A$ but varying force, see Fig. 4.10a. Taking into account the fact that the torque in the plectonemic regime is constant, i.e., $\Gamma_C = \Gamma_B$, the torque at point D (at force F_D and $\sigma_D = \sigma_C$) is obtained by integrating Eq. 4.28 along the path CD and subtracting the result of the torque difference along path AB. Since this procedure is valid only when points B and C are in the plectonemic regime, it is effectively restricted to values of $\sigma > 0.02$, where the buckling transition can be correctly identified. To evaluate the torque at smaller values of σ, one notices that the values of $\Gamma(F,\sigma)$ at fixed but large forces increase linearly with σ. The values of $\Gamma(F,\sigma)$ for one of these curves can then be extrapolated down to $\sigma = 0$ requiring $\Gamma(F,0) = 0$, and these extrapolated values can be used to infer, from the numerical integration of Eq. 4.28, the values of the torque at other forces and for values of $\sigma < 0.02$. The results are shown in Fig. 4.10b. The results obtained by this thermodynamic approach are similar, see Fig. 4.10b, to the results obtained by a direct measurement of the torque using optical tweezers [227]. Unsurprisingly, the variation of the torque with force differs from the simple model outlined in section 4.3.2, Eq. 4.25: Γ appears to scale as $F^{0.7}$ and not as $F^{0.5}$.

The slope of the torque–supercoiling curves at low values of σ (see inset in Fig. 4.10b) yields $\mathcal{C}_{eff} = C_{eff}/k_BT$, i.e., the effective torsional stiffness of DNA. As shown in Fig. 4.11, C_{eff} decreases with the force acting on the molecule. As seen previously, this was anticipated by Moroz and Nelson [220] in their high force analysis of the RLC model, see Eq. 4.26. Their estimate of this effect, see Eq. 4.27, is shown in Fig. 4.11. This allows us to deduce a value of the DNA bare torsional modulus $C = 94 \pm 7$ nm, in agreement with the previous single-molecule estimates of $C \approx 100$ nm [229, 219, 227]. The variation of C_{eff} with force may also explain the great variability of values of the torsional modulus reported in bulk studies: from $C_{eff} \approx 50$ nm measured by fluorescence polarization anisotropy [230, 231] on supercoiled plasmids (effectively at $F \sim 0.4$ pN) to $C_s \sim C = 75$–100 nm [232, 233] in circularization assays on small DNA fragments.

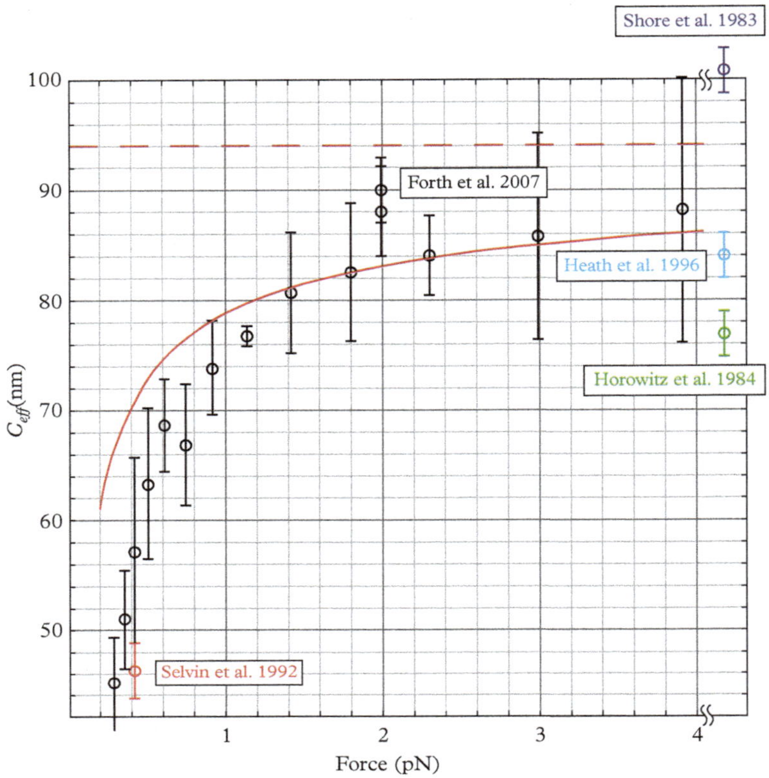

Fig. 4.11 Variation with force of the effective torsional stiffness of DNA *in* 100 *mM NaCl. The continuous curve is the best fit of high force data to the theoretical prediction (valid only at high force), with* $C = 94 \pm 7$ *nm (dashed line). The figure also displays the values of* C_{eff} *reported in other single-molecule and bulk experiments. Adapted from F. Mosconi et al. [228].*

From the data in Fig. 4.10, one can also deduce the effective tension and torque on a bare plasmid (circular DNA) unwound by 6% (as often found in nature [234]). The entropic tension is determined by the value of force at which the DNA's extension is zero at $\sigma = 0.06$, i.e., $F \sim 0.42$ pN [219] (the line passing through the points BC (at $F = 0.42$ pN) crosses the abscissa at $\sigma = 0.06$). The buckling torque at this force is obtained from Fig. 4.10b: $\Gamma \sim 6$ pN nm. This value is close to the value Γ_d where DNA denatures at negative supercoilings [235], see section 5.1.1. This may explain the variable sensitivity of gene expression to sequence and degree of supercoiling [236], see section 7.6.

4.5 Twist-stretch coupling in DNA

When previously discussing the behaviour of DNA under twist, we neglected its chiral structure and assumed it was akin to an isotropic elastic rod. This, of course,

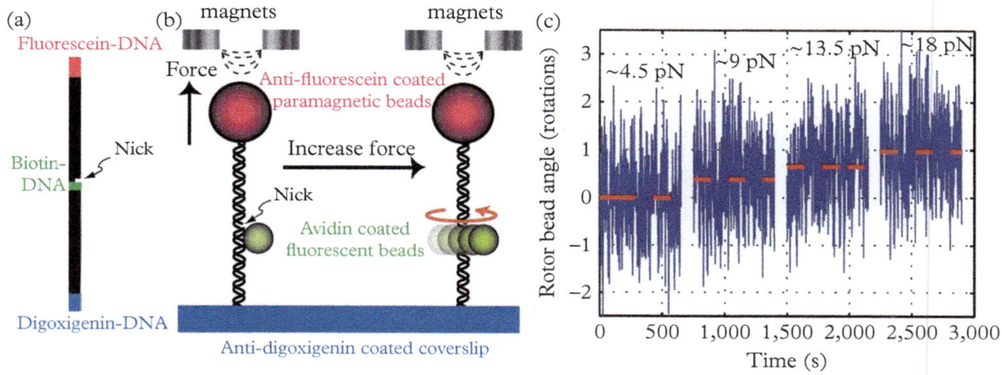

Fig. 4.12 Measurement of the twist-stretch coupling in DNA. *(a) The molecular construct for the marker bead tracking experiments contains three distinct attachment sites and a site-specific nick, which acts as a swivel. (b) A DNA molecule is stretched between the glass coverslip and a magnetic bead, while an avidin-coated marker bead is attached to the central biotinylated patch. Tension in the DNA is controlled by raising or lowering the magnets, and changes in twist are monitored via the rotation of the marker bead. (c) When the DNA molecule is held at a fixed force, the marker bead angle (blue trace) fluctuates around a mean (red dashed lines). As the tension is increased, the molecule lengthens and counterintuitively overwinds, i.e., the bead angle increases. From the ratio between the change in the twist angle and the change in the molecule's extension, Gore et al. deduced a twist-stretch coupling coefficient $g = -90$ pN nm, see text in section 4.5. Adapted from J. Gore et al. [238].*

is only an approximation whose validity needs to be tested experimentally. Intuition suggests that DNA should unwind under tension as it is pulled towards a denatured structure, see section 5.1.1, or, equivalently, that DNA should extend when unwound. The experimental evidence contradicts this naive picture: DNA actually extends (very slightly) when overwound and contracts when unwound [237, 238].

To simplify the analysis, we shall consider a DNA under a large enough tension that bending fluctuations can be neglected. In this case, the free energy of the DNA is given by:

$$\mathcal{F} = \frac{\mathcal{C}}{2l_0}\theta^2 + \frac{\mathcal{S}}{2l_0}\epsilon^2 + \frac{g}{l_0}\epsilon\theta - F\epsilon \tag{4.29}$$

The first term of the RHS is the torsional elasticity term; the second is the Hookean elasticity of a spring overextended by an amount ϵ with the stretch modulus of DNA $\mathcal{S} \sim 1100$ pN; the third term represents the coupling between twist and stretch; the last term is the work done by the external force. At equilibrium, we expect: $\partial\mathcal{F}/\partial\theta|_{eq} = 0$, from which we derive the change in the DNA twist when the molecule is stretched: $\theta_{eq} = -g\epsilon/\mathcal{C}$. If $g > 0$, the DNA unwinds ($\theta < 0$) as it is stretched ($\epsilon > 0$), and overwinds if $g < 0$. To visualize that effect, the group of C. Bustamante [238] introduced an ingenious modification to the magnetic tweezer setup. They anchored the magnetic bead via a DNA molecule possessing a single nick, below which a small non-magnetic marker bead was attached, see Fig. 4.12. As the twist in the bottom portion of the

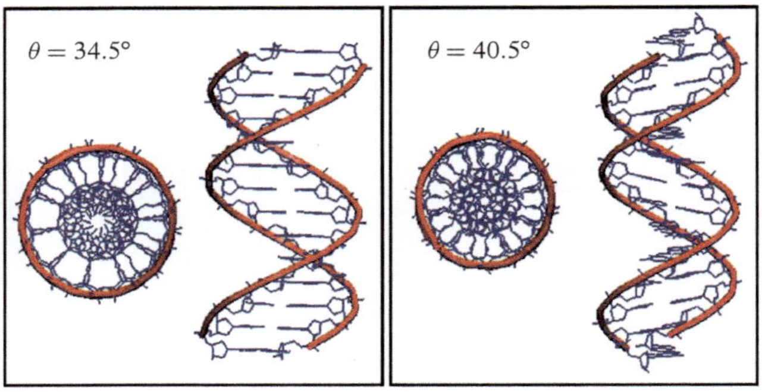

Fig. 4.13 Axial and lateral views of the structures of a slightly overwound B-DNA *obtained by energy minimization of a 14 base pair "average" sequence. Left: relaxed B-DNA state (angle between neighbouring nucleotides: $\theta = 34.5°$). Right: B-DNA under positive supercoiling ($\theta = 40.5°$) displays a smaller diameter, an increased length, and a negative base pair inclination. Adapted from T. Lionnet et al. [237].*

molecule changed, the mean angular position θ_{eq} of the marker bead changed accordingly. Surprisingly, as the molecule was stretched, the DNA overwound slightly (by a change in twist of 0.1%, i.e., $\delta\theta \sim 10^{-3} \cdot 2\pi l_0/p$ for a 1% change in extension, i.e., $\delta x \sim 10^{-2} l_0$), implying a twist-stretch coupling coefficient $g = -\mathcal{C}\delta\theta/\delta x \sim -90$ pN nm.

The complementary experiment of looking for a change in extension ϵ_{eq} upon coiling a stretched DNA was performed with a conventional magnetic tweezers setup [237, 238]. In this case, at equilibrium: $\partial\mathcal{F}/\partial\epsilon|_{eq} = 0$, from which one derives: $\epsilon_{eq}/l_0 = (F - g\theta/l_0)/S$. In this experiment, upon overwinding, the DNA extended by about 0.4 nm/turn from which one obtains a value: $g = -S\partial\epsilon_{eq}/\partial\theta \sim -70$ pN nm. Numerical calculations of the minimal energy configuration of an overwound dsDNA showed that the bases were slightly inclined with respect to the helical axis, allowing for a smaller diameter and a longer extension than relaxed B-DNA, see Fig. 4.13.

4.6 The braiding of two DNA molecules

During replication of a DNA molecule, the two newly synthesized strands may end up intertwined. Although they are decatenated by a class of enzymes known as topoisomerases (see Chapter 9), an investigation of the elastic properties of interlinked DNA molecules is important for the study of these enzymes and an understanding of the structures adopted by replication intermediates in the cell.

To study the braiding of two DNA molecules, the same magnetic trap system used to study the response of a single DNA molecule to supercoiling can be used. One needs to look for magnetic beads bound to the surface by two molecules (instead of one) and work with nicked DNA molecules to avoid the complications arising from the coupling between braiding and supercoiling. In such cases, only beads bound by two (or more) molecules respond to rotation of the magnets by a shortening of their DNA tether. One

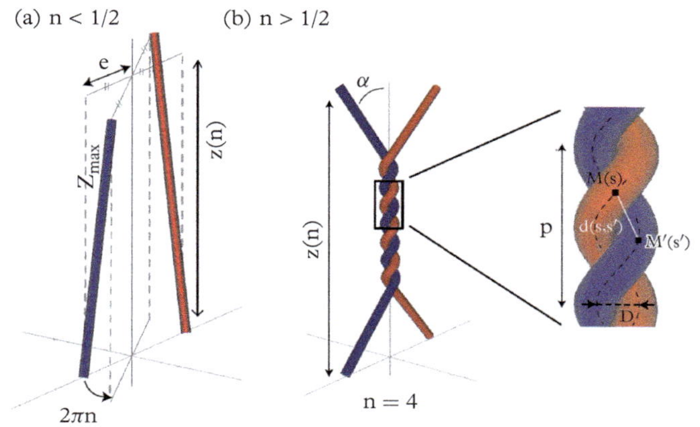

Fig. 4.14 Geometrical model for DNA braiding. *(a) Case $|n| < 0.5$: simple geometric considerations lead us to derive the extension $z(n)$ as a function of the angle $2\pi n$, the length $z_0 = z(n=0)$, and the distance $2e$ between molecules: $z(n) = \sqrt{z_0^2 - 4e^2 \sin^2 \pi\, |n|}$. (b) Case $|n| > 0.5$: further braiding leads to the formation of an helix with an effective diameter D, so that the extension $z(n)$ follows: $z(n) = z_0 \cos\alpha = \sqrt{z_0^2 - (2e + \pi D(|n| - 1/2))^2}$. This model remains valid as long as the molecules are not in close contact; this condition is reached when $d(s,s') \equiv \left\| \vec{MM'}(s,s') \right\| = D$, i.e., when the braid angle α reaches $\alpha_c = 45°$. Adapted from G. Charvin et al. [239].*

can then check that a bead is bound by only two molecules, by verifying that twice the force is required to reach a given extension as compared to a single DNA.

By rotating the magnets above the sample, it is possible to braid the molecules by n turns and to monitor the consecutive variations in their extension $z(n)$. Three different regimes can be discerned, which are simply accounted for by the twisting of a swing whose ropes of extension $z_0 = z(0)$ and diameter D_b are separated by a distance e, see Fig. 4.14 and Fig. 4.15:

- $|n| \leq 1/2$. The first half turn leads to a large decrease in DNA extension, which corresponds to the formation of the first crossing between the two dsDNA molecules. From the variation of extension $z(n)$ between $n=0$ and $n=0.5$, we can calculate the spacing e between the molecules:

$$z(n) = \sqrt{z_0^2 - e^2 \sin^2 \pi\, |n|} \tag{4.30}$$

- $1/2 < |n| < n_c$. Further rotation of the magnets leads to a smaller decrease in the system's extension per added turn. This regime corresponds to the formation of the braid. From the geometry of a twisted swing (with pitch angle α), one derives:

$$z(n) = z_0 \cos\alpha = \sqrt{z_0^2 - (e + \pi D(|n| - 1/2))^2} \tag{4.31}$$

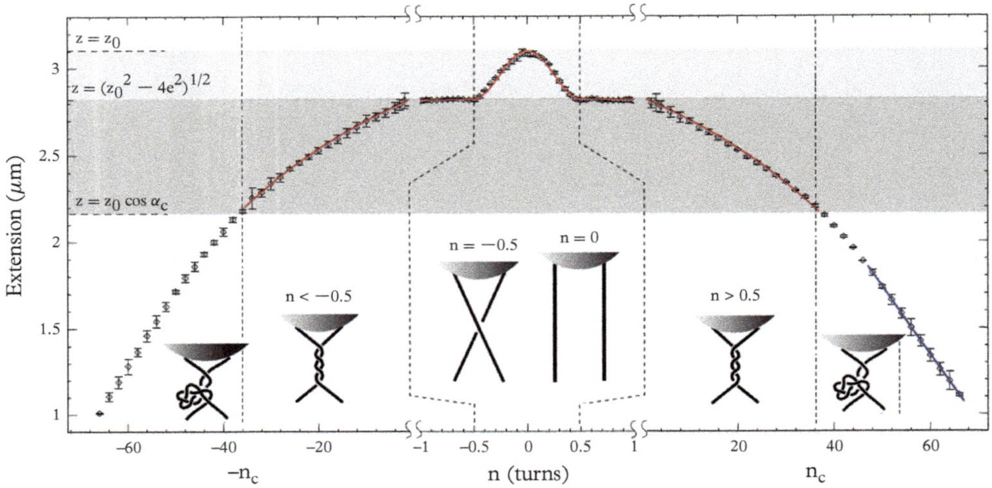

Fig. 4.15 Extension versus braiding of two DNA molecules. *The ◇ are experimental points, obtained at $F = 2pN$ in 100 mM phosphate buffer. We have distinguished three different ranges of braiding, based on the geometrical model described in the text:* $|n| < 0.5$ *(light grey shading), corresponding to the crossing of the two DNA molecules and* $0.5 < |n| < n_c$ *for further braiding (heavy grey shading). The red plain curve is a fit to the experimental points using the model described in the text with:* $z_0 = 3.10\,\mu m$, $2e = 1.28\,\mu m$, $R = 3.9$ *nm. For* $|n| > n_c$, *braids form 3D plectonemes. The extension then decreases linearly with the number of turns, by 39 nm/turn as shown by the blue line. Adapted from G. Charvin et al. [239].*

- $|n| > n_c$ At a critical number of turns n_c, the molecules are in close contact, so that the torque rapidly builds up and further rotation of the magnets leads to a buckling of the braid [240]. This transition to a different regime of braiding is manifested by a change in the slope of the extension versus n plot, see Fig. 4.15. This transition is characterized by an increase in the fluctuations in extension also observed in MC simulations of braided molecules.

4.7 Simulations of stretched, twisted, or braided molecules

Because the elastic response of DNA is described so well by a polymer model (WLC or RLC), experimental data on the stretching, twisting, and braiding of DNA molecule(s) at low forces ($F < 10$ pN) can be reproduced with high accuracy by Monte Carlo (MC) simulations [241] of the stretching, twisting, or braiding of one or two discrete chains of bending and torsional persistence length: $\xi \sim 50$ nm and $C \sim 100$ nm respectively [242, 208]. In these simulations, a DNA molecule is modelled as a discrete chain constituted by N rigid segments (typically $N \sim 100$) of fixed length $a \ll \xi$ (typically $a = 10$ nm), so that the chain length is $L_0 = Na$. The simulation then consists of:

- **Computing the elastic energy of the chain.** If θ_i is the angle between adjacent segments $i, i+1$, the bending energy E_b associated with a given chain configuration is [243]:

$$E_b = \alpha k_B T \sum_{i=1}^{N-1} \theta_i^2$$

where k_B is the Boltzmann constant, T the temperature, and $\alpha = 2.409$ the bending rigidity constant. With $a = 10$ nm, α is set so that the persistence length of the chain is 50 nm, i.e., five rigid segments [243], a discretization which has been shown to provide an accurate simulation of a DNA molecule [242]. For a stretched chain, the force F tends to align each segment along its direction (taken as the z-axis), an effect which is taken into account by an additional energy term $E_F = -Fz$, where z is the extension of the molecule. Finally, for an unnicked DNA molecule, one has to further add to these energy terms a contribution from the twisting of the molecule:

$$E_t = (2\pi^2 k_B TC/L_0)(\Delta Lk - Wr)^2$$

ΔLk is the imposed (i.e., fixed) number of turns on the molecule, the writhe Wr is a geometrical property of the chain which is computed for every chain configuration [242].

- **Comparing the energy of different chain configurations.** Chain conformations are generated using a Monte Carlo (MC) Metropolis algorithm [244] which randomly samples the configurational space by a combination of two types of moves, see Fig. 4.17: first, choose a segment i and change its angle ϕ_i with respect to the z-axis to an arbitrary value (uniformly distributed in $\cos(\phi_i) \in [-1, 1]$), while keeping the orientation of all other segments fixed. Second, randomly choose two vertices k and l in one chain and rotate the segments between these two vertices by an angle ϵ about the axis Δ_{kl} passing through the vertices. The value of ϵ is randomly chosen in an interval $[-\epsilon_{max}; \epsilon_{max}]$. ϵ_{max} is set such that the probability of acceptance is 25%. Note that this move does not modify the extension of the chain. The energy E' of a new chain configuration is computed and compared to the energy E of the current one. The new conformation is accepted, with probability $P = 1$ if $E' < E$ and probability $P = \exp(-(E' - E)/k_B T)$ if $E' > E$. This algorithm ensures that the distribution of chain conformations asymptotically reaches the equilibrium distribution as the number of trials $N_{MC} \to \infty$ (typically $10^7 < N_{MC} < 10^{10}$).

- **Taking care of topological constraints.** The moves used in an MC simulation may generate knots and alter the number of braids between two chains. To check that the topology of a new trial configuration is the same as the previous, one can compute the Alexander polynomial associated with a closed loop [245] (the chain is closed by appropriate straight segments). Configurations that do not return the same values of the Alexander polynomials are rejected.

- **Taking care of additional constraints.** Excluded volume and electrostatic interactions are taken care of by considering each segment as a cylinder of diameter D_{DNA} set by the ionic conditions [246, 247]. Configurations with intersecting segments are discarded.

When simulating the braiding of two molecules, one has to relax the constraint of a fixed spacing e between the chains at one of their extremities, in order to permit changes in the chains' conformation. Small variations of the spacing and orientation of the chains at their extremities are allowed by adding three potential energy terms: E_o, E_d, and E_a to the total energy of the chain. If $\vec{\tau}_1$ $(= e\hat{x})$ and $\vec{\tau}_2$ are the vectors joining the anchoring points respectively at the bottom and the top of the braid, and β the angle between $\vec{\tau}_1$ and $\vec{\tau}_2$, see Fig. 4.17. These terms are:

$$
\begin{aligned}
E_o &= \frac{1}{2} k_o (\vec{\tau}_2 . \hat{z}/e)^2 \\
E_d &= \frac{1}{2} k_d (|\vec{\tau}_2|/e - 1)^2 \\
E_a &= \frac{1}{2} k_a \beta^2
\end{aligned}
$$

with k_o, k_d, and $k_a \gg k_B T$. E_o ensures that $\vec{\tau}_2$ is orthogonal to the direction of the force, $\hat{z}$. E_d prevents the spacing between the molecules from deviating too much from its average value, e. E_a limits the angular fluctuations of β. Measuring the mean deviation $\langle \beta \rangle$ allows for an estimation of the torque in the braid. Note that this method also allows for the simulation of non-integer catenation numbers (in this case, β is the angle between $\vec{\tau}_2$ and the desired orientation).

The algorithm just described nicely reproduces the elastic data on stretching, twisting, and braiding of DNA molecules. Typical snapshots of the MC simulation of the braiding of two DNA molecules at various catenation numbers are shown in Fig. 4.18. The MC simulations agree very well with the data for the stretching of one or two DNA molecules, see Fig. 4.16, the supercoiling of a single DNA molecule at low forces ($F < 0.5$ pN), and the braiding of two DNA molecules, see Fig. 4.18. The success of these numerical simulations in accounting for the elastic response of DNA molecule(s) allows us to use them in situations that are difficult to reproduce or study using single-molecule techniques. Thus one can use them to estimate the torque generated when braiding two molecules [239] or to visualize the plectonemic (and braided) structures formed by twisted molecule(s) [242], and compare them to their electron microscopy images [248]. For braided replication intermediates, the simulations suggest that the molecules are mostly interlinked in the vicinity of the replication fork (a configuration that maximizes the overall entropy of the chains). This localization of the topological entanglement might play a role in the ability of the type II topoisomerases to systematically decatenate DNA molecules (i.e., reduce the global topological constraint) while acting locally on the chains, see Chapter 9.

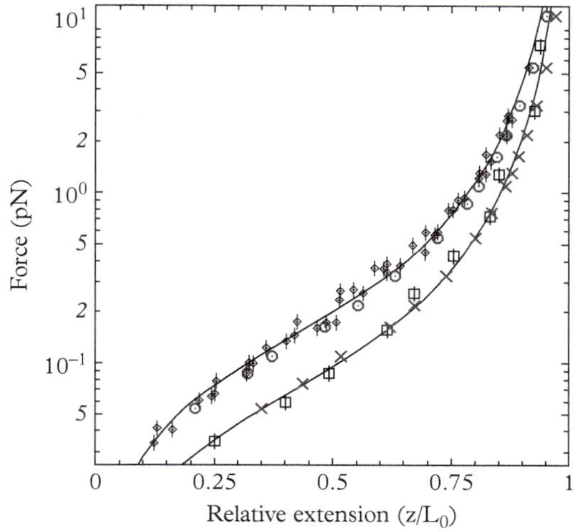

Fig. 4.16 Force–extension curve for one (right plot) and two (left plot) DNA molecules *anchored to the same bead. If F is the stretching force, the tension in the DNA is F for one tethering molecule and $F/2$ for two tethering molecules. Black □ (◇): experimental points obtained on a bead anchored by one (two) DNA molecule(s). The black lines are best fits using the WLC model with $\xi \sim 46$ nm. Gray symbols: results from a Monte Carlo simulation of the stretching of one (×) or two (○) unbraided molecules of persistence length 46 nm, adapted from Charvin et al. [239].*

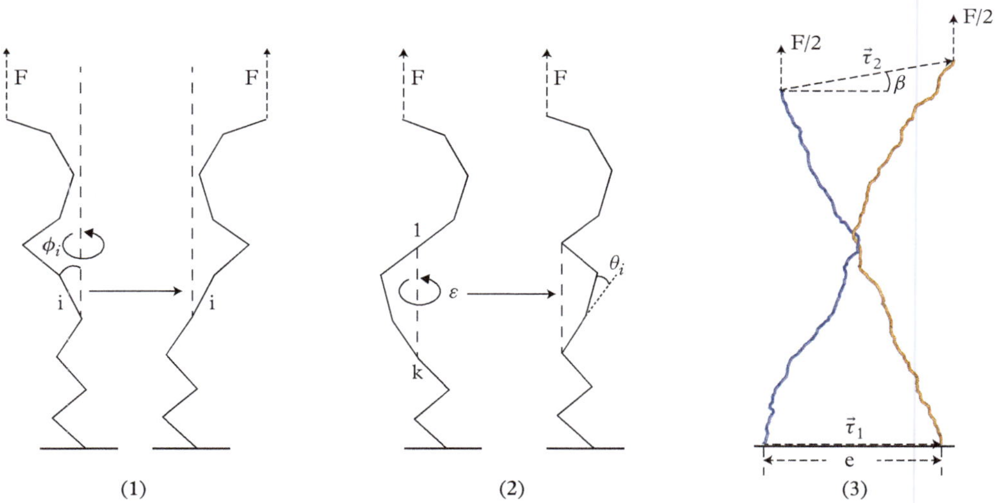

Fig. 4.17 The Monte Carlo moves used to probe the configuration space of one or two stretched polymer chains. *1) Either one segment i of the chain is chosen at random and rotated by an arbitrary angle ϕ_i about the z-axis (defined by the direction of the stretching force) or 2) two nodes k and l are chosen at random and the intervening segments are rotated by an angle ϵ about the axis passing through these nodes. The angle θ_i between adjacent segments is used to compute the elastic energy (see p. 92). (3) When two chains are intertwined, the same moves are used and the distance e between the chains' ends is a soft constraint (see equations below) on the $\vec{\tau}_2$ between the free ends and the angle β between $\vec{\tau}_2$ and $\vec{\tau}_1$ (the vectors between the anchored extremities).*

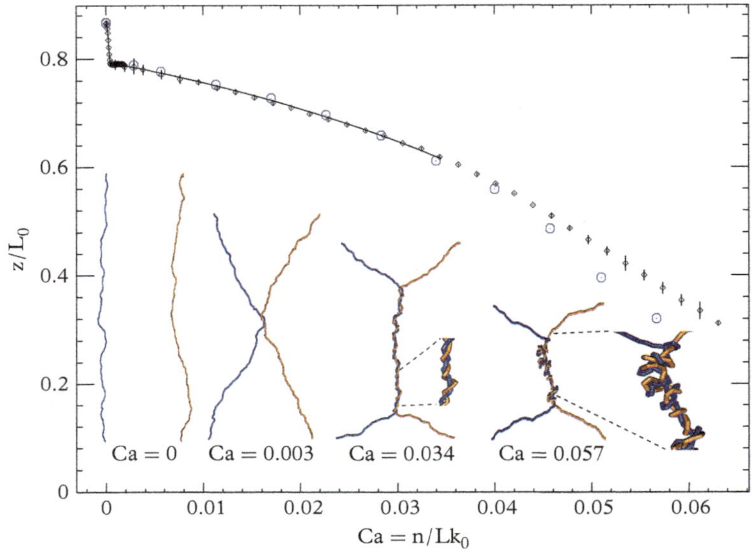

Fig. 4.18 Relative extension z/L_0 versus catenation density *$Ca = n/Lk_0$ (where n is the number of turns) in numerical simulation of the braiding of two polymer chains and comparison with the experimental data in 100 mM PB. MC simulations of two braided chains of extension $L_0 = 1.2\ \mu m$, intermolecular distance $2e/L_0 = 0.36$ at $F = 2$ pN using an effective DNA diameter $D_{DNA} = 6$ nm (blue circles, corresponding to the experimental ionic conditions). The black ◇ are experimental data. The black line is a fit of the numerical data to the geometric model (described in section 4.6) with a braid's diameter $D_b = 8.8$ nm $> D_{DNA} = 6$ nm due to entropic repulsion. The error bars in the estimates of the simulated extension are smaller than the size of the symbols. Adapted from [239].*

4.8 Single-stranded nucleic acids

Proteins acquire their biological function when they fold into a 3D structure, which is determined by the complex interactions between their various amino acid residues. Predicting this structure from the known sequence of amino acids is a major challenge of structural biology. While it is not our intention to address this very complex issue, single-stranded nucleic acids, in particular RNA, can also adopt a 3D structure that will determine their function: tRNAs adopt a clover-leaf structure which is recognized by the ribosome, whose rRNA is responsible for catalysing peptide bond formation in the growing protein chain. A large number of RNA molecules (so called ribozymes) possess a catalytic function which depends on their proper folding. Since interactions between the four different bases in nucleic acids are simpler and better understood than the interactions among the twenty different amino acids in proteins, RNA folding is better understood than protein folding. The possibility to pull on single nucleic acids offers a new method to probe their structures and their folding/unfolding pathways.

In this section, we shall discuss the effect of a stretching force on single-stranded nucleic acids with various levels of interactions. We first start with single-stranded DNA that may be thought of as a "random" sequence (or at any rate a sequence not designed to fold). It may nonetheless form transient double-helical structures known as hairpins, which are often found at palindromic sequences. The self-complementarity of these sequences allows them to pair and form a local stretch of double-helical DNA (known as the "stem" of the hairpin). After describing the elastic behaviour of single-stranded DNA (ssDNA), we shall discuss the case of small RNA molecules which present sequences that have evolved in order to fold into well-defined structures.

4.8.1 Pulling on single-stranded DNA

The elasticity of single-stranded DNA has been measured by several groups [33, 249, 250, 251]. It exhibits a complex behaviour strongly sensitive to salt conditions and *sequence* that is strikingly different from the elasticity of double-stranded DNA. The difference in extension at a given force between single- and double-stranded DNA has been used to monitor the activity of enzymes which convert DNA between single- and double-stranded states, such as DNA polymerases, helicases, and exonucleases.

The complex elastic behaviour of ssDNA results from the interplay between the elasticity of the chain (its entropy) and its self-interactions (self-avoidance and electrostatic repulsion, base-pairing interactions). For dsDNA, whose persistence length ξ is very large ($\sim$50 nm), we have seen previously that excluded volume interactions (of a typical scale determined by the Debye length, l_D ($\sim$1nm $\ll \xi$)) do not become relevant until the molecules exceed 100 μm in length, a rarely accessed experimental regime. Thus, for all practical purposes, the elastic properties of dsDNA resemble that of an ideal (non-interacting) polymer.

For ssDNA and RNA, on the other hand, as for many natural and artificial polyelectrolytes, the Debye length is never much smaller than the chain's persistence length (typically $\sim$1 nm). Therefore, electrostatic self-avoiding interactions play an essential role in the elastic response of these polymers. Moreover, since the bases in ssDNA (and RNA) have the singular ability to pair and form secondary structures, these attractive interactions cannot be neglected, and significantly complicate the energy landscape and elastic behaviour of these molecules. These pairing interactions can, however, be suppressed in denaturing conditions (e.g., in the presence of urea, guanidine, DMSO, etc.), or by a proper chemical treatment of the molecules, for instance, with glyoxal, which modifies the bases and prevents their pairing. The elastic properties of glyoxal-modified ssDNA in different salt conditions have been studied by various groups [251, 252] and their behaviour at low forces can be understood by simple scaling arguments [252]. For these non-pairing molecules, self-avoidance is the only essential interaction and can be treated in the framework of the Flory approximation.

We have seen that for a self-avoiding polymer, the radius of gyration of the chain is given by Eq. 4.20: $R_F \sim l_0^{3/5}(v/b)^{1/5}$. As this defines the characteristic size of the polymer, when pulling with a force F on the chain one expects the extension l of the polymer to scale with R_F:

$$l = R_F \Psi(FR_F/k_BT) \tag{4.32}$$

where $\Psi(x)$ is some unknown function of the dimensionless variable $x = FR_F/k_BT$. At very small forces, when $x \ll 1$, one expects $\Psi(x) \sim x$, so that linear elasticity is recovered: $l \sim FR_F^2/k_BT$. At intermediate forces, i.e., when $1 \ll x < R_F/b$, so that one is still in the entropic regime (i.e., $F < F_c = k_BT/b$), the polymer is extended to a significant fraction of its full length l_0 and we expect l to scale with l_0. This implies that $\Psi(x) \sim x^{2/3}$, since in this case Eq. 4.32 yields:

$$l = l_0(v/b^3)^{1/3}(Fb/k_BT)^{2/3} \tag{4.33}$$

As stated before, Eq. 4.33 is expected to be valid for forces $F < F_c$, i.e., extensions $l < l_c = l_0(v/b^3)^{1/3}$. This relation seems to be remarkably satisfied by glyoxal-treated ssDNA at low forces, see Fig. 4.19a,b. In fact, since the effective Kuhn length b of the molecule increases at low salt concentration c (it appears to vary as the Debye length [252]: $b \sim \lambda_D \sim c^{-0.5}$), rescaling the force–extension data by F_c and l_c collapses the curves measured in different salt conditions into a unique curve. At forces $F < F_c$ the rescaled extension varies as $(F/F_c)^{2/3}$, while at larger forces the extension surprisingly

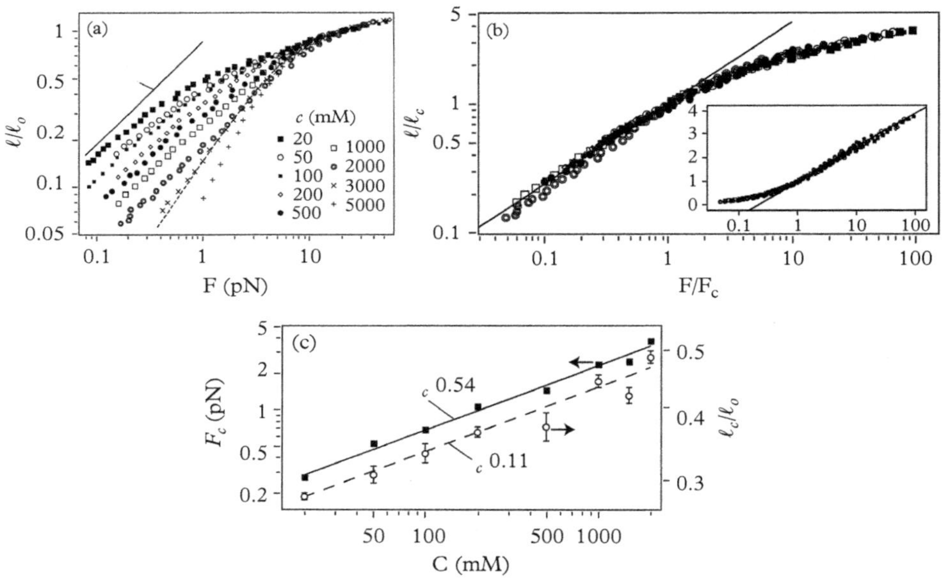

Fig. 4.19 Force versus relative extension l/l_0 of ssDNA with no base-pair interactions *(treated with glyoxal) in various monovalent salt conditions (l_0 is the value of molecular extension at 20 pN, close to its crystallographic full length). (a) The low force data seems to scale as $F^{2/3}$. (b) Same data as in (a) collapsed into a single curve by a simple rescaling of the force and the extension by best-fitted values F_c and l_c. Notice that the high force ($F/F_c > 1$) behaviour varies as the logarithm of the force (the continuous curve in the inset is $l/l_c = 0.63\log(F/F_c)$). (c) Values of the characteristic force F_c (filled dots) and relative extension l_c/l_0 (open dots) as a function of monovalent salt concentration* C*, adapted from Saleh et al. [252].*

varies as $log(F/F_c)$. This last result, which seems to be true also for proteins in denatured conditions, so far has no general explanation.

The simple scaling arguments that worked so well for glyoxal-modified (or denatured) ssDNA do not work for regular single-chain nucleic acids whose bases can pair. Nonetheless, even though a complete analytical treatment of such molecules under tension is missing, the basic physics is understood well enough that MC simulations were carried out and found to agree nicely with experimental data.

The first element in an elastic model of ssDNA (or RNA) is a description of the molecule as a chain of N freely jointed **elastic** segments of rest-size $b = 1.6$ nm (Kuhn length) and Young modulus $Y = 120\ k_BT/\mathrm{nm}^2$ [33]. The free energy of this so-called modified Freely Jointed Chain (mFJC) model is [33, 253]:

$$\mathcal{E}_{mFJC} = \frac{Y}{2}\sum_{i=1}^{N}(|\vec{r}_i - \vec{r}_{i-1}| - b)^2 - Fz_N \tag{4.34}$$

where $\vec{r}_i = (x_i, y_i, z_i)$ is the position of node i. To account for self-interaction effects, it is necessary to add electrostatic self-avoidance and base-pairing interactions to that free energy [254, 253, 255, 251], yielding:

$$\mathcal{E}_{ssDNA} = \mathcal{E}_{mFJC} + \frac{\nu^2}{\epsilon_W}\int ds_i ds_j \frac{\exp(-|\vec{r}_i - \vec{r}_j|/\lambda_D)}{|\vec{r}_i - \vec{r}_j|} + \sum_{\text{nested i,j pairs}} V_{i,j} \tag{4.35}$$

The second term on the RHS of Eq. 4.35 represents the electrostatic repulsion between DNA segments in the Debye–Hückel (low salt) approximation, where ν is the effective DNA charge density, $\epsilon_W = 80$ the dielectric constant of water, and integration is performed over the molecule's curvilinear coordinates. The value of ν is computed by solving the Poisson–Boltzmann equation for a straight chain in the same ionic concentration [253]. The last term in Eq. 4.35 represents the base-pairing energy contribution from all nested pairs (i.e., all possible hairpin structures): $V_{i,j}$ represents the pairing energy between nodes i and j. This term will in general depend on the nucleic acid sequence and be dominated by its most stable secondary structures. For ssDNA molecules which have not evolved to a defined 3D structure, it can be replaced by its mean field approximation [253]: $\sum_{\text{nested pairs}} V_{i,j} = \sum_{i=1}^{N_p} V_p$, where N_p is the number of (nested) paired nodes and V_p represents the average pairing energy over a Kuhn segment (V_p increases with GC content). The system defined in Eq. 4.35 was simulated by a Monte Carlo algorithm (see section 4.7 and [253]) to determine the mean molecular extension $l = \langle z_N \rangle$ vs F. The results are shown in Fig. 4.20, where the only fit parameter was $V_p = 4.6\ k_BT$. They agree nicely with experimental results in the various salt conditions over the whole range of forces investigated.

As can be seen in Fig. 4.20, at forces below $F \sim 10$ pN the ssDNA extension decreases as the ionic strength increases. Above 10 pN, this salt concentration dependence of the elasticity curves disappears. The most likely explanation for this observation is that the low-force behaviour of ssDNA is dominated by the formation of hairpins, whose stability

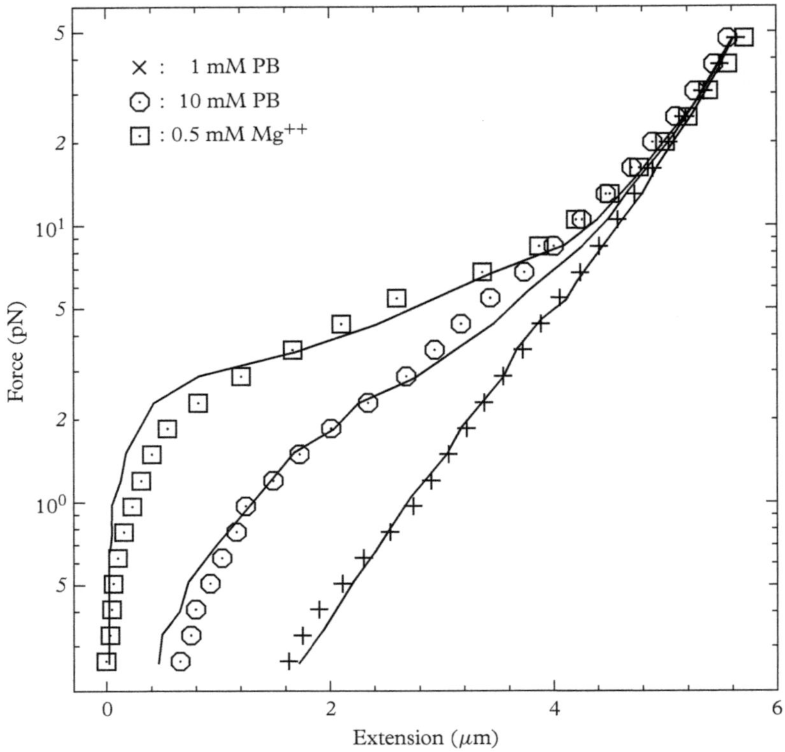

Fig. 4.20 Force–extension behaviour of ssDNA in various salt conditions: *1 mM phosphate buffer (PB) (+), 10 mM PB without (o), and with (⊡) 0.5 mM Mg^{++}. Notice the strong dependence on salt concentration and the good agreement with MC simulations (continuous line) described in Eq. 4.35. Adapted from M.-N. Dessinges et al. [251].*

is reduced (at low ionic concentrations) by electrostatic repulsion between negatively charged phosphates on the DNA backbone. Increasing the force unzips these hairpins, and effectively suppresses them at $F > 10$ pN, making the elasticity of ssDNA at high forces insensitive to sequence and only weakly dependent on ionic strength. The stability of hairpins depends upon the DNA sequence: at a given force, AT-rich molecules are comparatively longer than GC-rich ones, which form more stable hairpins.

4.9 Pulling on RNA

While base-pairing interactions are usually short-lived in ssDNA, they are essential for maintaining the stability of the 3D structure of RNA molecules whose sequence has been selected by evolution to ensure proper folding and stability. Many algorithms have been developed to predict the secondary structure of RNA molecules (the loops and hairpins created by interactions between its bases). It is not our purpose here to describe these algorithms, but rather to present the contributions of single-molecule techniques to the investigation of RNA folding and stability.

4.9.1 Equilibrium experiments

The group of C. Bustamante has pioneered the investigation of the folding/unfolding transition of small RNA molecules resulting from an applied pulling force [256]. There are two interesting regimes which these experiments can probe. When pulling slowly on an RNA molecule, its thermodynamic equilibrium properties can be studied: the existence of different conformations, their free energies, and the transition rates between those conformations can be determined. When pulling rapidly on the molecule, hysteretic effects are observed when increasing or decreasing the force, and the properties of non-equilibrium thermodynamical systems can thus be investigated. In particular, Jarzinsky's equation (see section 4.9.2) was first verified in pulling experiments performed on RNA molecules.

In one of these experiments see ref [256], the two ends of RNA molecules were attached via a 500 bp RNA/DNA linker to two handles consisting of 2 μm polystyrene beads. When optical tweezers were used in their normal mode of operation, that is, when the length of the molecule was fixed, a typical WLC extension curve was measured until the force reached $\sim$ 14 pN. At that point the molecule yields by $\sim$ 20 nm. This abrupt increase corresponds to the tension-induced unzipping of the RNA hairpin.

By an appropriate feedback loop, the optical tweezers can be made to work at constant tension (rather than fixed extension). In this mode, the fluctuations of the molecule's extension can be studied at fixed force. Below 14 pN and above 14.5 pN, the amplitude of fluctuations is small, but around 14.2 pN a distinct telegraphic noise is observed: the extension of the molecule is alternating between an extended (open) state and a short (closed hairpin) one. The probability of occupation of these states obeys a simple Boltzmann distribution:

$$P_{open}(F) = \frac{1}{1 + e^{\Delta E/k_B T}} \tag{4.36}$$

The difference in energy between the two states varies with the force as: $\Delta E = \Delta G_0 - Fl_{RNA}$, where ΔG_0 is the free energy difference at zero force and $l_{RNA} \sim 22$ nm is the size of the opened hairpin. When $P_{open} = 1/2$ (i.e., $\Delta E = 0$), the difference in free energy at $F = 0$ between folded and unfolded states: $\Delta G_0 = F_{1/2} l_{RNA} \sim 312$ pN nm = 78 $k_B T$. This value is in good agreement with the predicted value, $\Delta G_0 = 60\ k_B T$, obtained after correcting for the reduction in entropy of the stretched RNA in the open configuration.

An analysis of the folding–unfolding kinetics is very instructive, since it provides unique information on the distance of the transition state to the stable folded or unfolded configurations. Arrhenius kinetics predicts that the transition from folded to unfolded states proceeds at a rate:

$$1/\tau_f = k_{fu} = k_{fu}^0 \exp\left(F\Delta x_{fu}/k_B T\right)$$

which is the inverse of the state's lifetime, and similarly for the reverse rate:

$$1/\tau_u = k_{uf} = k_{uf}^0 \exp\left(-F\Delta x_{uf}/k_B T\right)$$

Therefore, by plotting the lifetimes of the states versus force, their distances to the transition state can be determined. For the hairpin shown in Fig. 4.21: $\Delta x_{fu} \sim \Delta x_{uf} \sim 11$ nm, i.e., the transition state is halfway between the closed and open hairpin configurations. This is very different from tension-induced unfolding of proteins, where the transition state is very close to the native state (a few Å away). This difference might reflect the greater importance of non-local 3D interactions in the stabilization of proteins.

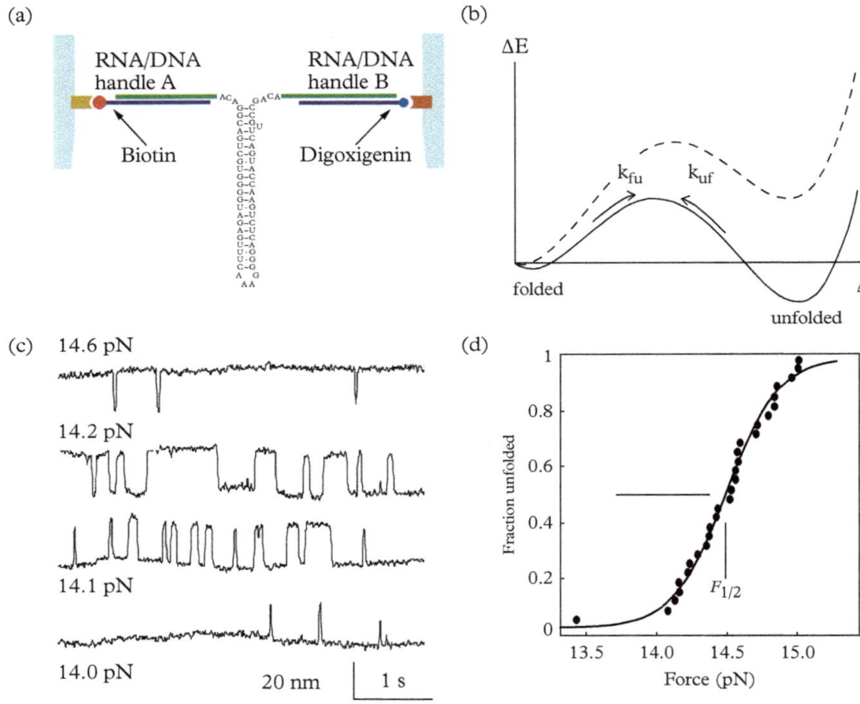

Fig. 4.21 RNA stretching experiments. *(a) The RNA molecule is held by two DNA handles bound to adequate surfaces. (b) For simple RNA hairpins, a two-state model (folded or unfolded) is often valid. Upon pulling on the hairpin, the energy landscape is tilted and the unfolded state becomes energetically more stable. The transition rates k_{uf} and k_{fu} are affected by the change in the energy landscape upon the application of a force on the RNA. (c) The telegraphic signal observed around $F_{1/2} \sim 14.2$ pN as the RNA switches randomly between the folded and unfolded states of similar free energies. (d) Fraction of unfolded hairpins as a function of force and fit to a Boltzmann two-state distribution. Adapted from Liphardt et al. [256].*

4.9.2 Non-equilibrium experiments

Experiments involving rapid pulling on RNA have been used to study theoretical predictions about systems far from thermodynamic equilibrium. Of particular interest was testing Jarzinsky's equation [258], relating the difference in free energy between two equilibrium states to the irreversible work $W = \int F dx$ done when going from one state (e.g., a folded hairpin) to the next (e.g., an unfolded hairpin):

$$\Delta G_0 = -k_B T \ln \langle e^{-W/k_B T} \rangle \; < \; \langle W \rangle \tag{4.37}$$

where the average is done over many unfolding experiments, see Fig. 4.22. When RNA unfolding is done slowly (at quasi-equilibrium), and therefore reversibly, the work is always the same and Eq. 4.37 reduces to: $\Delta G_0 = W$. However, when pulling rapidly on an RNA hairpin, unfolding happens spontaneously within a range of forces, see Fig. 4.22; the work is not always the same, and the hysteretic effects are pronounced. This is the regime where Eq. 4.37 can be tested. The main problem in testing it, however, is in evaluating the average on its RHS. The largest terms in this average come from

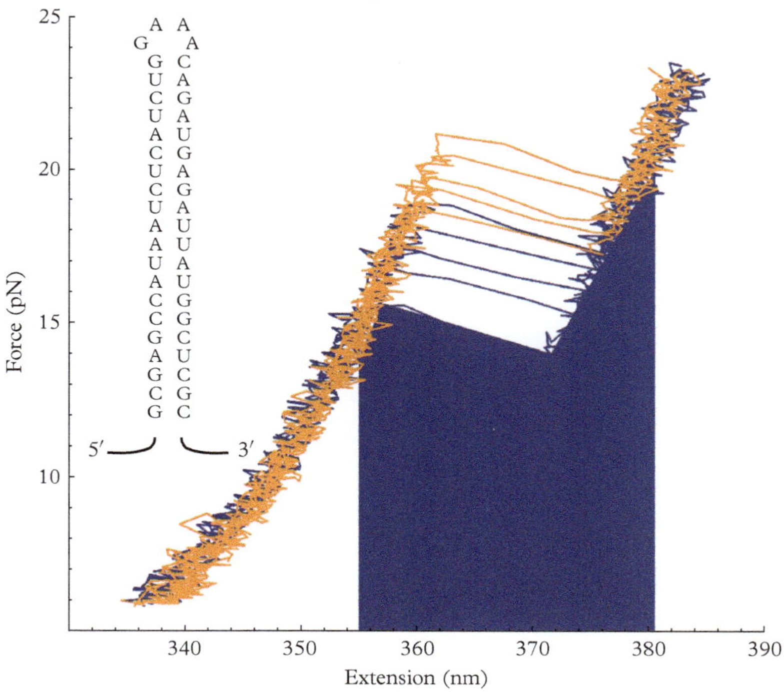

Fig. 4.22 Force–extension curves obtained during rapid pulling on a tethered RNA loop. *The work done by the force when opening the hairpin (the sequence displayed on the left) is equal to the area under one of the curves (shown here for the lowest curve), adapted from D. Collin et al. [257].*

the very rare events for which $W < \Delta G_0$. In fact, one can show that the number $N_{\mathcal{J}}$ of repetitions of the experiments necessary to obtain a good estimate of ΔG_0 (i.e., within $\pm$ 1 $k_B T$) scales as $N_{\mathcal{J}} \sim \exp(\langle W \rangle - \Delta G_0)/k_B T$ [259, 257]. Except for small systems where the dissipated work $\langle W \rangle - \Delta G_0$ is a few $k_B T$, $N_{\mathcal{J}}$ is prohibitively large. Thus, while Eq. 4.37 provides a rare example of an exact result in non-equilibrium thermodynamics, for practical purposes it is of little use. Nonetheless, Eq. 4.37 and its variants have been tested on single RNA unfolding experiments. Their outcomes agree with the value of ΔG_0 estimated from equilibrium measurements, such as the ones described previously [260]. These investigations provide a unique example of the use of a biological system to test fundamental physical principles.

4.10 Problems

DNA elasticity. At what tension would a DNA molecule at room temperature ($T = 300°$ K) extend to 20% of its full length (the DNA Kuhn length: $a = 100$ nm). Assume that the molecule obeys the Freely Jointed Chain model (a reasonable model at low extensions).

WLC and FJC models. Show that for a bead tethered by a single DNA molecule in a magnetic trap, the ratio of the transverse fluctuations $\langle \delta x^2 \rangle$ to the longitudinal ones $\langle \delta z^2 \rangle$ is a function only of the relative extension: z/L. Compute and plot that ratio for the FJC and WLC models.

Equilibrium constant and free energies from RNA unfolding experiments. Fig. 4.21c displays the telegraphic signals resulting from a transition between the folded and unfolded states of a simple RNA hairpin at various tensions. Describe how the equilibrium constant between these states (at a fixed tension) can be deduced from these measurements. Fig. 4.23 displays the variation of this equilibrium constant

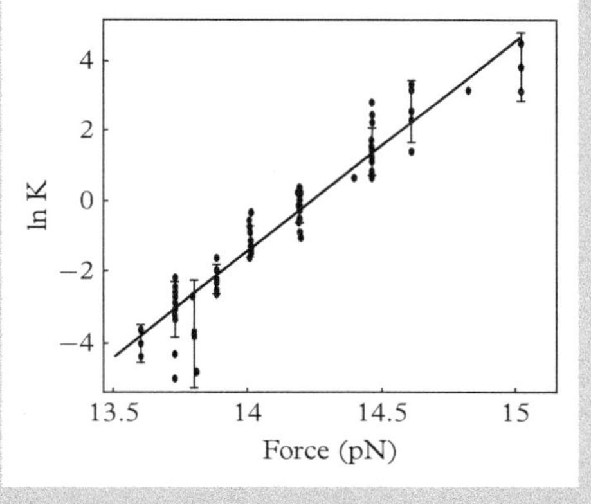

Fig. 4.23 Variation with the force of the equilibrium constant *between unfolded and folded conformations of the RNA hairpin shown in Fig. 4.21a, adapted from Liphardt et al. [256].*

with force. From this graph, estimate the free energy difference (at zero force) between the folded and unfolded RNA configurations and compared with the value mentioned in section 4.9.1. From Fig. 4.23 what is the increase in extension when one bp is unfolded?

FRET pairs and DNA beacons. The fluorophores CY3 and CY5 are often used as acceptor–donor pairs in FRET experiments. Find on the internet which is the donor and which is the acceptor, and find out the donor fluorescence lifetime and the Förster radius of the pair. In a DNA beacon, these fluorophores are labelling the extremities of a 25 bp ssDNA molecule that can self-hybridize to form a hairpin. Find the change in FRET efficiency between the open and closed conformation of the beacon (assume that in the open conformation the distance between the molecule's ends is that of a 25 bp ssDNA).

5

Structural Transitions in DNA

We have until now treated dsDNA as a structureless polymer, a continuous chain or rod whose entropy is reduced upon stretching or twisting. This approximation is only valid at low forces (less than a few pN) and torques (less than a few $k_B T$). Beyond these regimes, mechanical stresses built into the molecule can be large enough to induce structural changes, which we will now examine.

5.1 Molecular transitions induced by twisting

The chiral double-helical structure of DNA has profound implications for many cellular processes. In particular, by altering the DNA twist (and, thus, the torque), the local access of various proteins to particular sequences (e.g., DNA promoters [236]) can be controlled. In the following, we shall see that a torque of order 9.6 pN nm ($\sim 2.4 k_B T$) is enough to locally denature the molecule, favouring access of various proteins to AT-rich regions of the genome.

5.1.1 Denaturation upon unwinding

In section 4.3.2, we saw that twisting a DNA molecule while pulling on it would form supercoils past a certain number of turns, defining the buckling threshold. If, however, the molecule is unwound at a stretching force $F > 0.5$ pN, the expected decrease in extension associated with supercoil formation is never observed. Instead, the extension of the DNA varies little as it is unwound, the increase in the molecule twist is adsorbed in the opening of its two strands, and the two strands unpair by about 10.5 bases for every extra turn of unwinding. As previously shown, the buckling torque Γ_b increases with the stretching force. If the critical torque for denaturation Γ_d is smaller than the buckling torque Γ_b, the molecule denatures before it can buckle. That is what happens when $F > 0.5$ pN.

Two different experiments demonstrate that the bases in such stressed and unwound DNA are indeed unpaired:

(1) The stretched and twisted DNA molecule can be incubated with a reagent (glyoxal) that specifically reacts with unpaired bases (it was mentioned previously in

Single-Molecule Studies of Nucleic Acids and Their Proteins. David Bensimon, Vincent Croquette, Jean-François Allemand, Xavier Michalet, Terence Strick 2019. © Oxford University Press 2019. Published in 2019 by Oxford University Press.
DOI: 10.1093/oso/9780198530923.001.0001

the context of suppressing base-pairing interactions in ssDNA). The measured amount of modified bases is equal to the expected size of the unpaired region (the denaturation bubble [261]) demonstrating the existence of such a region in negatively twisted and stretched DNA.

(2) DNA probes complementary to AT-rich regions of the DNA (characterized by a weaker pairing interaction) were made. They were observed to hybridize to stretched negatively supercoiled DNA molecules [18] showing that unpaired DNA regions are created in an AT-rich region of the molecule.

These experiments support the existence of torque-induced denaturation regions in unwound and stretched DNA. In fact, their presence did not come as a surprise. Denaturation bubbles had previously been observed in negatively supercoiled plasmids with $\sigma < -0.06$. The observed unpairing of the two DNA strands at lower σ values is a result of the increased torque due to the tension in the molecule.

5.1.2 Transition to P-DNA upon overwinding

A behaviour similar to the one just described was observed upon overwinding DNA with a force $F > 3$ pN. Taking a hint from the torque-induced denaturation of unwound DNA, one can hypothesize that a positive torque similarly induces a local modification of DNA to a new structure, called P-DNA.

Two types of experiment were designed to test that hypothesis and characterize the P-DNA structure:

(1) The extension of the molecule was measured over a wide range of forces ($F < 70$ pN) and degrees of supercoiling ($\sigma < 3$). If a new structure of DNA with intrinsic twist σ_P appears locally upon overwinding, it should be possible to describe the extension $l(F,\sigma)$ of the molecule as a combination of a fraction $p = \sigma/\sigma_P$ of P-DNA phase with extension $l_P(F)$ coexisting with a fraction $(1-p)$ of stretched B-DNA phase of extension $l_B(F) = l(F,0)$:

$$l(F,\sigma) = p l_P(F) + (1-p) l_B(F) \tag{5.1}$$

In other words, if a P-DNA structure exists in positively supercoiled DNA, all extension curves $l(F,\sigma)$ (with $\sigma > 0$) can be collapsed, with a proper choice of σ_P, upon a single curve: $l_P(F)$. As shown in Fig. 5.1, this is indeed the case with $\sigma_P = 3$ (i.e., $Tw_P = 4$). This very large value of the intrinsic twist of the molecule implies that P-DNA has a pitch (expressed in number of bp/turn) four times smaller than B-DNA, i.e., about 2.6 bases per turn. Moreover, the data suggest that the maximal extension of P-DNA is about 75% longer than that of B-DNA. Numerical energy minimization of a DNA molecule under large positive twist show, that a new conformation with similar pitch and extension is indeed obtained. The molecule exhibits a helical structure, with its phosphate backbone winding at the centre and its bases exposed in solution, a structure reminiscent of that proposed by Pauling and Corey in 1953 [262].

(2) In order to demonstrate that the DNA bases in P-DNA were indeed exposed, unpaired, in solution, the stretched overwound molecule was reacted with glyoxal

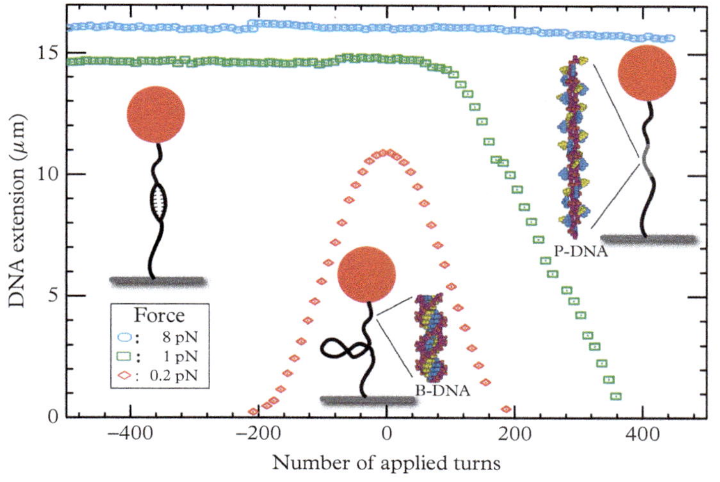

Fig. 5.1 Structural transitions in stretched and coiled DNA. *Extension vs supercoiling curves, $l(n)$, obtained on λ-DNA ($l_0 = 16$ μm) at three different stretching forces. (◇): $F = 0.2$ pN. The molecule contracts, whether over- or underwound, forming plectonemes on its classical B-DNA form (left structure). (□): $F = 1$ pN. The molecule's extension decreases as it forms plectonemes while overwound, but its extension remains essentially constant as it is underwound and denatured. (○): $F = 8$ pN. The DNA's extension (close to its maximal B-DNA contour length l_0) does not vary noticeably in the range of supercoiling shown here. At negative supercoilings the molecule denatures, while at positive supercoilings the molecule forms supercoils of P-DNA (the inside-out structure shown on the right, which is (when fully extended) about 75% longer than B-DNA).*

(as was done with unwound denatured DNA). The number of modified bases measured corresponded to the expected number of unpaired bases in the predicted P-DNA structure [261], supporting the existence of that structure in stretched positively supercoiled DNA.

These results show that the helical structure of DNA is very sensitive to the torque. When underwound the molecule can denature, while upon overwinding it adopts an inside-out helical structure with a shorter pitch than B-DNA.

5.1.3 Measuring the critical torques for the under- and overwinding transitions

While there is ample evidence for the existence of local denaturation bubbles in twisted and unstretched DNA, it is unclear whether the overstretched DNA structure, P-DNA, is of any relevance to cellular processes. To partly address this issue, the group of C. Bustamante has measured the critical torques required to induce these structural transitions [229]. This group has developed a setup which measures the torque on a twisted DNA molecule stretched between a micropipette and a dual-beam optical trap, by tracking the rotational motion of a small bead attached to the molecule. This reporter

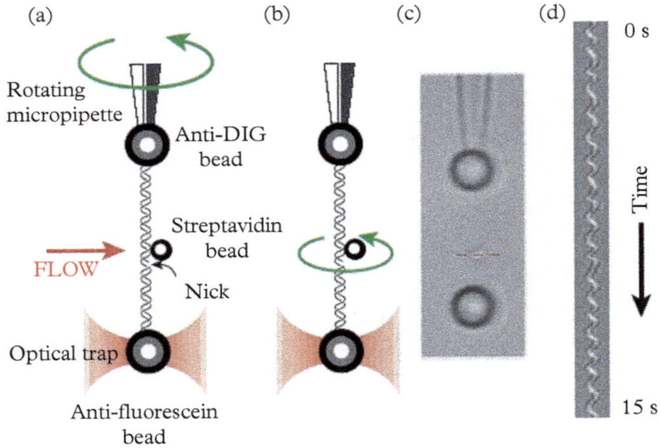

Fig. 5.2 Setup for the measurement of the critical torques at the structural transitions of coiled DNA [229]. *(a) A DNA molecule is stretched using a combination of an optical trap and a micropipette. The principle of the experiment is the following: in the presence of a flow, the small bead is unable to rotate around the DNA (the swivel is blocked), so that rotating the micropipette increases the torque in the upper part of the DNA. (b) When the flow is released, the torque stored in the DNA causes the bead to rotate along the molecule's axis with an angular velocity ω, which depends on the torque Γ and the rotational drag. (c) Picture of the different beads. (d) Space-time diagram (kymograph) of a 920 nm rotating bead. Adapted from [229].*

bead is set in motion by the torque in the molecule, and the measurement of its rotation rate yields (via the viscous drag on the bead) an estimate of the torque on the DNA.

As shown in Fig. 5.2a, a single-strand break is engineered in the DNA close to the small bead. It provides a swivel about which torsion built up in the upper part of the molecule can be relaxed. The build-up of that torsional stress is achieved by rotating the micropipette while applying a flow perpendicular to the DNA's axis, to prevent rotation of the small bead. When the flow is stopped, the torque stored in the upper part of the DNA induces the rotation of the bead, see Fig. 5.2b. The torque applied on the molecule Γ can be computed from the angular velocity $\omega = -2\pi\, dn/dt$ of the bead (where n is the number of turns of the bead, i.e., the change in the DNA linking number):

$$\Gamma = 14\pi\eta r^3\omega \equiv -\alpha\, dn/dt$$

where r is the radius of the reporter bead and η the viscosity of the medium. In Fig. 5.2d, the position of the bead's centre is shown as a function of time for a bead of known size. Clearly, the smaller the bead, the smaller the dissipation and the faster the rotation.

The force stretching the molecule ($F = 45$ pN) is high enough to prevent formation of writhe. As a consequence, all the torsional energy is stored as twist. However, at this

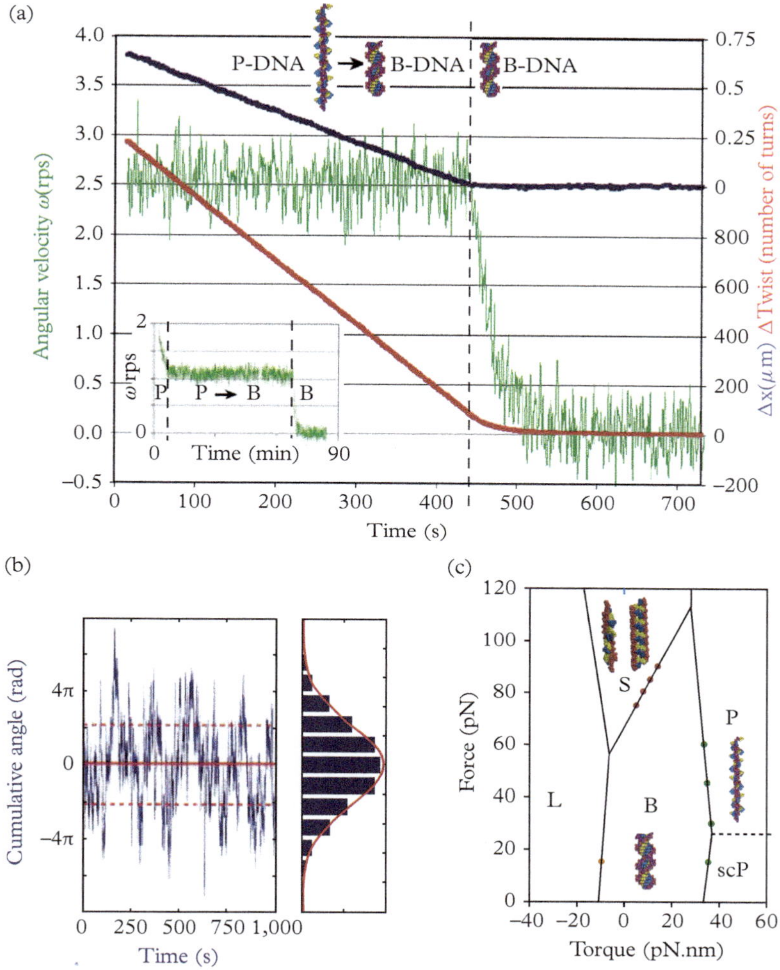

Fig. 5.3 Phase diagram of a stretched and twisted DNA. *(a) Analysis of bead motion: after rotating the micropipette by $\Delta L_k = +1{,}200$ turns, the DNA molecule consists of coexisting phases of P-DNA and B-DNA. As the flow is stopped, the bead starts to rotate, decreasing L_k. Initially, as the torque in the coexisting phase is constant, the angular velocity stays the same. Once all P-DNA has reverted to B-DNA, the angular velocity decreases to zero as the extra twists in the molecule and the torque are completely relaxed. Knowing the rotational drag on the bead, the torsional modulus, $C = 102 \pm 7$ nm, can be deduced. (b) Recording of the angular fluctuations of the bead near equilibrium (near $\Delta T_w = 0$). From the twist fluctuations $\langle(\Delta T_w)^2\rangle$ and the equipartition theorem ($1/2C/L\langle(2\pi\Delta T_w)^2\rangle = 1/2k_BT$), one deduces $C = 110 \pm 10$ nm. (c) From these measurements (and others), the phase diagram of DNA structures in the tension–torque plane can be deduced: B stands for B-DNA, P for P-DNA (and scP for supercoiled P-DNA) observed for positive torques, L for denatured DNA observed for negative torques, and S for the S-DNA phase observed at high tensions (see section 5.1.3). Adapted from [229].*

large force, a large positive degree of supercoiling (i.e., a large torque) induces the DNA molecule to undergo a structural transition to P-DNA. As long as these two phases coexist, the torque in the molecule remains constant and the bead rotates at constant angular velocity, see Fig. 5.3a. The critical torque in P-DNA can be deduced from this rate: $\Gamma_p = 34$ pN nm $\sim 8.5\ k_BT$. A similar experiment on stretched underwound DNA yields the critical torque in a denatured molecule: $\Gamma_d = 9.6$ pN nm $\sim 2.4\ k_BT$.

Once twist has relaxed by full conversion of P-DNA into B-DNA, the torque in the molecule is that expected in a regular twisted dsDNA (or tube): $\Gamma = 2\pi Cn/l$. The twist in the molecule is then relaxed exponentially:

$$n(t) = n_c \exp(-2\pi Ct/l\alpha)$$

where n_c is the residual number of turns (i.e., the value of the linking number) at the transition to P-DNA and l the length of the molecule. This experiment, thus, allows one to estimate the torsional modulus of DNA by fitting the data at low n to an exponential decay. It yields: $C = 102.5 \pm 7$ nm.

Another way to deduce C using this setup is to monitor the angular fluctuations $\delta\theta$ of the bead around its equilibrium position. The equipartition theorem for the bead angular fluctuations $\langle\delta\theta^2\rangle$ yields:

$$\frac{1}{2}\frac{C}{L}\langle\delta\theta^2\rangle = \frac{1}{2}k_BT$$

Fig. 5.3b shows a typical recording of the bead angular fluctuations obtained at $F = 15$ pN. From this measurement, one deduces $C = 110 \pm 10$ nm. The advantage of this method of estimating C over that using the relaxation of torsion is that it does not rely on the estimation of the bead's drag (which depends sensitively on the bead's radius).

5.2 Molecular transitions induced by stretching

The experiments on torsionally unstressed DNA described previously were done in the entropic regime (at $F < 10$ pN), where the molecule behaves as an ideal (WLC) polymer of persistence length $\xi \sim 50$ nm. Its elastic behaviour is due purely to a reduction of its entropy upon stretching. From 10 pN and up to about 60 pN, DNA stretches elastically, as does any material, following Hooke's law: $F \sim \Delta l/l_0$ (where $\Delta l = l - l_0$ is the increase in the molecule extension l and l_0 is its contour length in the absence of force). However, at about $F_t \sim 60$ pN, a surprising transition was discovered, where DNA stretches to about 1.7 times its crystallographic length [32, 33]. This transition (like the torque-induced ones to P-DNA and denatured DNA) is highly cooperative: a small change in force results in a large change in extension, see Fig. 5.5.

To study the possible structural modification in the molecule resulting from pulling on it, a numerical energy minimization of DNA under stress was performed by R. Lavery

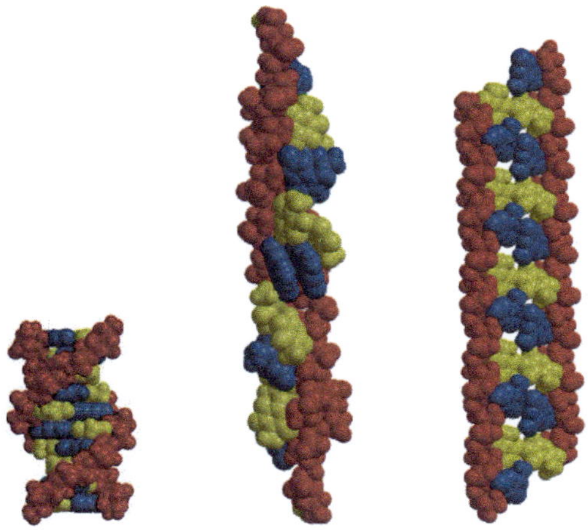

Fig. 5.4 Numerical simulations of DNA structure under high tension. *The phosphate backbone is shown in red, the complementary bases (AT and GC) in blue and green. Left: usual B-DNA structure. Middle: if the molecule is pulled by its 5′ ends, it keeps a double-helical structure with tilted bases. Right: if the DNA is pulled by its 3′ extremities, the final structure resembles a ladder. Figures adapted from [263].*

and collaborators [32, 34]. Its results reveal the existence of a conformational transition to a new form called S-DNA, about 70% longer than B-DNA, whose exact structure depends on which extremities of the DNA are being pulled (3′–3′ or 5′–5′). If both 3′ extremities are pulled, the double helix unwinds upon stretching. The final structure resembles a ladder, see Fig. 5.4. If both 5′ ends are pulled, the helical structure is preserved, but it is characterized by a strong base pair inclination, a narrow minor groove, and a diameter roughly 30% less than that of B-DNA, see Fig. 5.4. In both cases, the rupture of the molecule (by unpairing of the bases) is predicted to occur, as observed in other experiments when its extension is more than twice that of B-DNA [264, 20, 21].

These numerical results are consistent with experiments done many years ago by Wilkins *et al.* [264], before the double-helical structure of DNA was proposed. These experiments suggested that stretched DNA fibres undergo a transition to a structure with inclined bases about twice as long as the relaxed molecule.

For many years, a debate existed as to the reality of S-DNA. It was argued that when DNA is overstretched it denatures. In support of this interpretation, the existence of a strong hysteresis when the force is reduced was put forward, as well as the fact that ssDNA can be generated by overstretching a nicked DNA molecule. Many experiments [266, 267, 268] have demonstrated the existence of force-induced denaturation of DNA. This fact has been used over and over again to generate ssDNA from dsDNA. In particular, it has been used in single-molecule studies of DNA polymerases as a way of regenerating a single-stranded DNA template after its replication [249]. Hence,

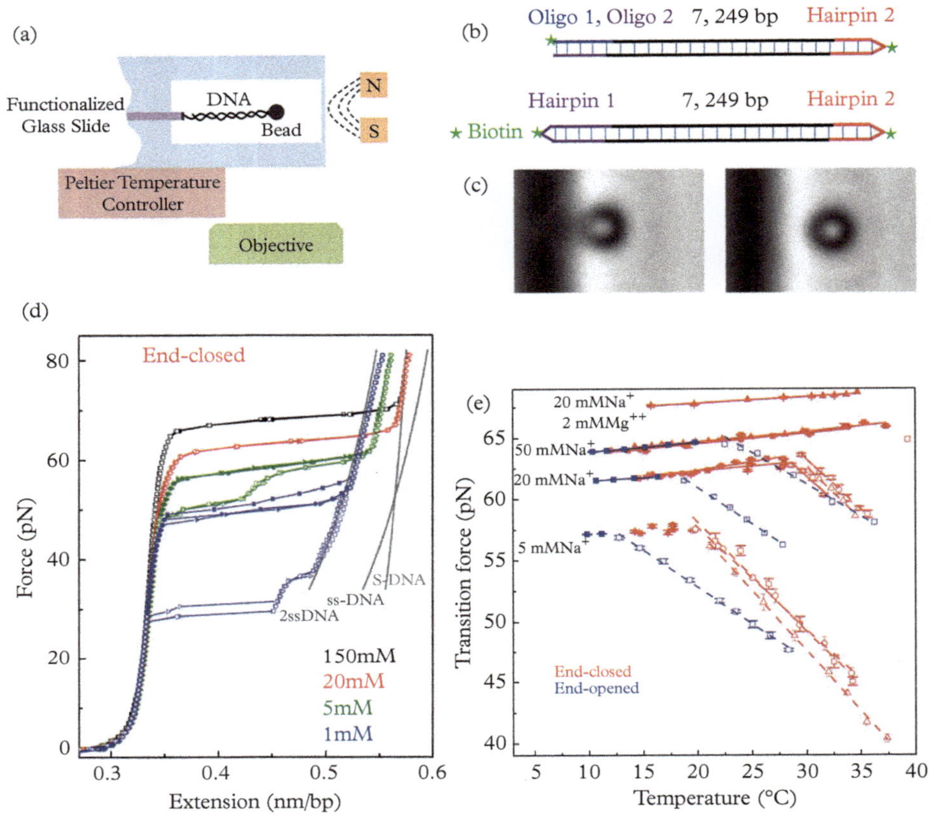

Fig. 5.5 Study of the stretch-induced transitions in DNA. *(a) The setup used in [271]: magnets are used to pull laterally on magnetic beads tethered to a surface by a single DNA molecule. (b) The DNA molecules are anchored to the bead and surface by a single bond, allowing for torsion-free experiments. Two types of DNA constructs were used: one whose end can peel (open end) and one which cannot (closed end). (c) At low force, the bead is close to the surface (left image), while at a force of about 60 pN the molecule extends significantly (right image). (d) When monitoring the extension at 24° C of a closed-end molecule versus force in different salt solutions, one notices a large change in extension at a force of about 60 pN. This change is hysteretic at low salt concentrations (< 10 mM, see blue and green curves) and non-hysteretic at high salt concentrations (> 10 mM, see red and black curves). The grey lines are extension versus force curves for S-DNA, ssDNA, or 2 ssDNA. (e) Variation of the force at the transition as a function of temperature. Notice the existence of two linear regimes: one at low temperatures with a small positive slope, and one at higher temperatures with a large negative slope. By the Clausius–Clapeyron relation, this slope is related to the entropy change ΔS at the transition. It is small and negative at low temperatures ($\Delta S \sim -2.5$ cal/K mol) and large and positive at higher temperatures ($\Delta S \sim 16$ cal/K mol). The transition associated with small negative entropy (at low temperatures) is to overstretched DNA (S-DNA) while the transition associated with a large positive entropy (at high temperatures) is to denatured DNA. Figures adapted from [271].*

there is no doubt that it is possible to denature DNA under a high enough tension. It was, however, argued that this denaturation is due to the presence of nicks or open ends in the anchoring DNA, from which ssDNA can be generated by peeling off the un-anchored strand. In support of that interpretation, one can cite experiments by the group of T. Perkins [268] on closed-end DNA molecules (see Fig. 5.5b), which displayed a non-hysteretic transition to a large extension as the molecule was stretched at about 65 pN.

A definitive understanding of these puzzling and seemingly contradictory results was achieved by the group of J. Yan [269, 270, 271] which has studied the phase diagram of overstretched DNA at different temperatures, salt conditions, and GC content. This group has shown that depending on the experimental conditions, both hysteretic and non-hysteretic transitions could be observed (with both closed and open ends), see Fig. 5.5d. This group also estimated the change in entropy ΔS associated with the transition at force F_t, using a generalization of the Clausius–Clapeyron relation [272, 273]:

$$\Delta S = -\frac{\partial F_t}{\partial T}\Delta b \tag{5.2}$$

where Δb is the change in extension per base pair at the transition.

The data, see Fig. 5.5e, displays two regimes: a low temperature or high salt regime where the entropy of transition is small and negative ($\Delta S \sim -2.5$ cal/K mol) and a high temperature or low salt regime where the entropy is large and positive ($\Delta S \sim 16$ cal/K mol), similar to the entropy of DNA melting. Hence, depending on salt, temperature, and GC content, overstretched DNA can adopt the structure of S-DNA or denatured DNA.

J.-F. Léger *et al.* [274, 275] have studied the B $\rightarrow$ S transition on torsionally constrained single molecules. In this case, the torsional constraint implies that the linking number is conserved through the transition. Therefore (unless S-DNA has exactly the same twist as B-DNA, which is not the case), the transition from B-DNA to S-DNA is coupled to the simultaneous formation of P-DNA regions. A complete analysis of the stretching behaviour of supercoiled DNA has been performed [274]. This analysis, which considers all force- and torque-induced phases of DNA, suggested that the S-DNA phase possesses a helical pitch of 22 nm with about 38 bases per right-handed turn, which would make it look like a slightly twisted ladder. The previously described results of Bryant *et al.* [229] support this interpretation.

5.3 Unzipping DNA

Single-molecule manipulation techniques have stimulated the imagination of many people. We and others [276, 277] have suggested that by pulling apart the two strands of a single DNA molecule (as in a zipper) it could be sequenced by monitoring the force required for breaking each base pair as one progresses along the molecule. Similarly, Thompson and Siggia [277] imagined following the 3D unfolding/folding of an RNA molecule (or a protein) stretched between a surface and the tip of an AFM cantilever. However, theoretical analysis [34, 276, 277] of these stimulating prospects revealed a number of conceptual difficulties, in particular a severe constraint on the stiffness of the measuring device.

This device consists typically of a macroscopic spring (e.g., cantilever, bead in an optical trap, microfibre) coupled to the stretched molecule via long polymers. Its effective stiffness ($k_{eff}^{-1} = \sum_i k_i^{-1}$) is dominated by its weakest, most flexible element. For example, when unzipping DNA, the double-stranded molecule is attached to the measuring device by flexible single-stranded DNA pieces, which get longer the more the molecule is unzipped. To observe the successive breaking of hydrogen bonds as the molecule unravels, the energy and spatial resolution should be respectively: $\delta\mathcal{E} \sim 10\ k_B T \sim 4\ 10^{-20}$ J and $\delta x \sim 3$ Å. Between two successive measurements, the energy accumulated in the measurement chain is, thus, $\delta\mathcal{E} = k_{eff}\delta x^2/2$, implying an effective stiffness $k_{eff} \sim 1$ N/m.

Although that value is not uncommon for AFM cantilevers, a polymer of reasonable size (> 10 nm) is much less stiff. For example, when unzipping 1,500 bp of DNA, typically at a force F of about 20 pN [25, 278], one ends up with two single-stranded DNA pieces (each about $l = 1\ \mu$m in length) connecting the molecule to an anchoring surface and a sensor. The elastic rigidity of these single-stranded DNA is $k_{mFJC} \sim 10^{-5}$ N/m [33] (see section 4.35), which limits the spatial resolution of unzipping at that stage to ~ 30 bp ($\delta x \sim 9$ nm). This resolution decreases as the length of the single-stranded DNA pieces increases.

Experimental unzipping of a λ-DNA molecule has been performed by Heslot and co-workers [25, 26] (see also Rief *et al.* [278]), first using a thin glass *fibre* as a sensor (stiffness $k = 1.7\ 10^{-6}$ N/m), attached to the DNA via a single-stranded piece, and later using an optical trap of higher stiffness. The unzipping signal displays a sawtooth pattern typical of a stick-slip process: as the molecule is stretched, energy accumulates in the measurement chain, see Fig. 5.6. When the force reaches the threshold to unpair two bases (about 15 pN in GC-rich regions and 12 pN in AT-rich ones), the bonds yield and a segment of n bases unbinds cooperatively. Because the loading rate in these experiments is much slower than the base-pair binding/unbinding rates, unzipping occurs at thermal equilibrium. The role of the force is to shift this equilibrium [279, 280] from a configuration where n bases are paired, to one where they are unpaired. Similar to the configurational changes described previously (e.g., B-DNA → S-DNA), the transition here is also a first order one, i.e., the behaviour is cooperative, hence the observed "molecular stick-slip" signal. The spatial resolution achieved in these experiments (about 500 bp) is in agreement with the theoretical limitations discussed here. To overcome those, one would have to stiffen the single-strand pieces of DNA as the molecule unravels, for example by coating them with RecA [281].

However, even if one could completely freeze the thermal fluctuations of the measurement chain (thus, stiffening it), zero temperature molecular modelling of DNA unzipping [34] reveals variations in the energy of the system which are not simply related to the sequence of the molecule. It is, therefore, not at all obvious that this ingenious technique could be used to read the sequence of a DNA molecule.

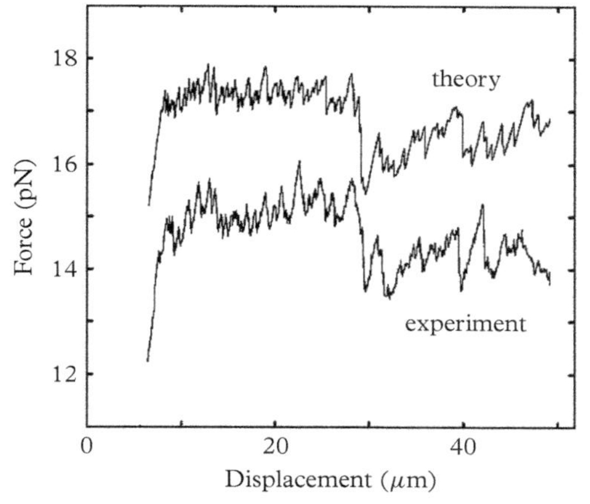

Fig. 5.6 Force–extension curves obtained when unzipping a λ-phage DNA. *The experimental curve has been shifted by 2 pN for clarity. The theoretical curve is derived from the GC content of the molecule. Figure adapted from [26].*

5.3.1 Mechanical sequencing of DNA by detection of blockages during rezipping

While DNA sequencing via a measurement of the unzipping force is unrealistic, a single-molecule mechanical sequencing method has been proposed, which uses the detection of blockages during the rehybridization phase of a stretched hairpin in order to determine the underlying sequence [282]. The principle of the method is very simple: start with a DNA hairpin (which can be engineered from an arbitrary dsDNA by end-ligating appropriate fragments); unzip it by pulling with a large enough force ($F > 15$ pN) and let it interact in the open state with short (8–10 bp) oligonucleotides of known sequence present in solution. Upon reducing the force, rezipping of the hairpin will be transiently blocked at positions where the oligos in solution have hybridized to their complementary sequence on the hairpin. These transient pauses in the rehybridization pathway can be easily detected in the signal measuring the distance of the bead to the surface (the end-to-end distance of the hairpin), see Fig. 5.7. Their positions indicate where on the hairpin there is a sequence complementary to the oligonucleotides in solution. The lifetime of the pauses depends on the temperature, the length of the oligonucleotides, and the presence of mismatches in their hybrids with the hairpin (a single mismatch reduces the blockage time by about an order of magnitude). This simple observation can be used to readily identify genomic DNAs (or cDNAs obtained from RNA) via their pattern of hybridization to a library of oligonucleotides (their "fingerprint").

The same method can also be used to implement various sequencing schemes: sequencing by hybridization, ligation, or polymerization[282], and by replacing the complementary oligonucleotide with an antibody against some specific DNA modification it can be used for the detection of epigenetic modifications on DNA. While the method

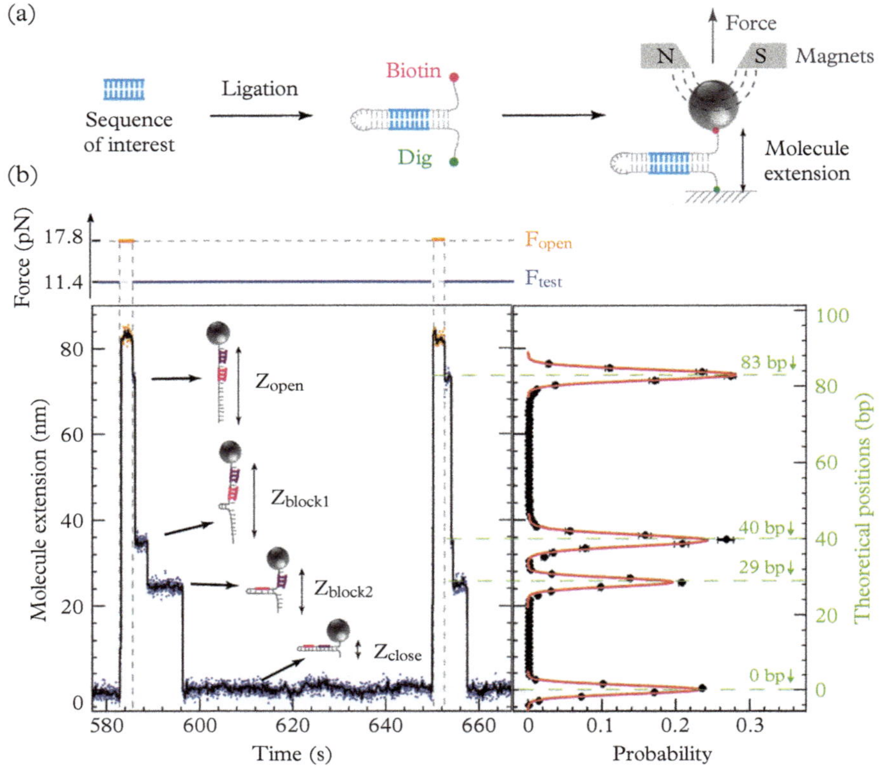

Fig. 5.7 Mechanical identification and sequencing of DNA. *(a) A DNA hairpin can be prepared from dsDNA fragments by ligating their ends with appropriate fragments: a DNA loop and a DNA stem labelled at its contiguous extremities with biotin and digoxigenin. (b) The hairpins can be used to tether magnetic beads to a surface. Upon pulling on the beads with small magnets, the hairpin can be unzipped. In this state, it can interact with oligonucleotides in solution. Upon reducing the force, rehybridization of the hairpin is transiently interrupted by the hybrids it forms with the oligonucleotides. The position of these pauses, determined with a ~ 1 nm resolution, provides sequence information on the DNA hairpin which can be used to identify or sequence it. Adapted from [282].*

is simple, easily parallelizable (the magnets exert the same force on many beads), and does not rely on PCR to amplify DNA or on fluorescence for detection (as most other methods do), only time will tell whether it is competitive with other more established approaches.

5.4 Conclusions and perspectives

As a result of the studies described in chapters 4 and 5, the mechanical properties of nucleic acids under both stretch and twist are well established. There is a very good agreement between the observations and the theoretical models of DNA under tension and torque (Worm-Like and Rod-Like Chain models). The behaviour of DNA at high tensions or high torques where new structures have been discovered is also well understood, as is the dynamics of free DNA molecules in solution. While a large part of this knowledge can be applied to RNA, the importance of specific base-pairing interactions in the determination of the RNA 3D structure complicates the understanding of the mechanical properties of RNA and the prediction of its structures. There is still much to be learned on the folding pathways of RNA and their perturbation by a tension on the molecule. This problem might be easier to understand than the equivalent one in the protein context. Nonetheless, the solid understanding of the mechanical properties of DNA sets the stage for the use of DNA extension as a proxy for the investigation of its interactions with many proteins, a programme we will outline in the remaining chapters.

6

Single-Molecule Studies of DNA-Associated Motors

6.1 Introduction to DNA–protein interactions

In the next chapters, we will present some case studies of DNA–protein interactions using single-molecule methods, either by manipulation or via fluorescence and FRET techniques. We have focused our attention on the cases we know best (polymerases, helicases, topoisomerases), but they clearly represent only a small fraction of the many proteins whose interactions with DNA have been studied (transcription factors, RecA, histones, DNA translocases, viral compaction motor, etc.). Our goal is not to present an exhaustive panorama of all systems studied (which would be impossible in this rapidly evolving field), but rather to present a few detailed examples which may inspire the readers in the study of their own systems.

The interaction of a protein with DNA can be investigated by single-molecule manipulation techniques if it affects the DNA's extension. This interaction may alter the variation of extension with force, degree of supercoiling, the stiffness of the molecule (i.e., its fluctuations in extension) or the response of the molecule's extension upon a rapid change in force (or degree of supercoiling). As we shall see in the following, from such mechanical measurements, much can be learned about the energetics and geometry of the enzyme–DNA complex (without ATP).

Before using single-molecule manipulation methods to study the interactions between DNA and proteins, one must consider a number of factors. First, to what extent is the supposed interaction going to alter the extension of DNA (either by bending or twisting)? Indeed, all DNA manipulation studies ultimately measure a change in DNA extension: either through the conversion of ssDNA into dsDNA (by DNA polymerases) or vice-versa (by helicases), through the bending or twisting of the molecule (e.g., histones), the coiling or uncoiling of the molecule (topoisomerases, gyrases), or its translocation (viral compaction motors, FtsK, chromatine remodelling factors, etc.). Proteins which interact with DNA without altering its extension, such as type II restriction enzymes, cannot be studied by single-molecule manipulation methods. Second, one must consider to what extent the interaction of the protein with DNA is cooperative or not. Proteins which interact with DNA as single isolated complexes are much better characterized by

Single-Molecule Studies of Nucleic Acids and Their Proteins. David Bensimon, Vincent Croquette, Jean-François Allemand, Xavier Michalet, Terence Strick 2019. Published in 2019 by Oxford University Press.
DOI: 10.1093/oso/9780198530923.001.0001

single-molecule manipulation methods, than proteins which form fibres or highly cooperative complexes along the DNA (such as Single-Strand Binding (SSB) proteins, RecA, certain structural proteins (IHF, HNS), or chromatine remodelling factors (ISWI)). In the latter cases, it is difficult to estimate the number of proteins interacting with DNA, and therefore characterize their individual interaction with the molecule. Third, the lifetime of the DNA–protein complex must be long enough (typically > 1 s) to observe and measure. Proteins which translocate along DNA with a low processivity (helicases, DNA or RNA polymerases) are difficult to study with manipulation methods. Fourth, what type of information does one look for? If one wants to measure average thermodynamic constants (e.g., Michaelis–Menten constants) or efficiency (average number of ATP consumed per cycle) one may consider bulk or quench flow measurements instead of single-molecule manipulation approaches (which will require many measurements (> 100) to get a precise estimate of the averages (better than 10%)).

On the other hand, single-molecule fluorescent or FRET measurements address many of these issues. The binding and unbinding kinetics of a DNA–protein complex can be studied by FRET between donor and acceptor dyes on the protein and DNA, even in the absence of any structural modification of the molecule (bending, twisting, etc.). Similarly, one can investigate the extent of cooperative interactions between protein and DNA by measuring the change in fluorescence of the growing complex or its bleaching by the illumination beam (RecA nucleation and growth on DNA have, thus, been studied [283]). Small structural changes, enzymatic step-sizes and low processivity enzymes are better studied by single-molecule FRET experiments. On the other hand, these fluorescent approaches are less adapted to investigate the mechanism of highly processive enzymes (for example, pausing during replication, transcription, or translocation) or the response of enzymes under load.

Obviously, the future of single-molecule studies involves the *in vitro* merging of manipulation and fluorescent/FRET approaches, as was already reported [284]. Another exciting prospect is the investigation of the dynamics [285] of single molecules in single cells *in vivo*, in particular using fluorescent approaches [286].

7

DNA and RNA Polymerases

7.1 Introduction to polymerases

During the course of its lifecycle, the cell must perform three fundamental manipulations of its genome: replication of one DNA molecule into two DNA molecules, DNA repair, and transcription of DNA into RNA. These tasks are accomplished by the polymerase family of enzymes. Nucleic acid polymerases were first identified in the mid to late 1950s as DNA template-dependent nucleotidyltransferases [287, 288]. These enzymes are remarkable molecular motors which either physically translocate along DNA or pull DNA through, reading the sequence of bases on one strand in a stepwise fashion and using it as a template for the stepwise polymerization of a new, perfectly complementary strand of DNA or RNA. Interest in the coupling between the chemistry of the nucleotide addition cycle and production of motion and force by polymerase molecular motors has made them a natural focus in single-molecule experimentation. In the process of replication, DNA polymerases (assembled into a complex known as the replisome) take a double-stranded DNA (dsDNA) molecule and make from it two identical dsDNA molecules. In the process of transcription, RNA polymerases localize to and transcribe the gene (DNA) encoding a protein: they polymerize a messenger RNA (mRNA) whose sequence is complementary to that of the original gene, and which serves as a template for the synthesis of proteins by the ribosome complex. Single-molecule experiments have provided insight into both the mechanism and the regulation of these enzymes.

7.2 An introduction to replication

The discovery of the double-helical structure of DNA immediately suggested a mechanism for its replication: the use of each strand as a template for the synthesis of the complementary one. This semi-conservative mechanism of DNA replication was confirmed in 1958 by Meselson and Stahl [7] in what is arguably "the most beautiful experiment in biology". At about the same time, the *E.coli* DNA polymerase I enzyme, also known as PolI, was discovered by Arthur Kornberg [289]. Using one DNA strand as a template, this enzyme could add complementary free nucleotides to the $3'$ end of

Single-Molecule Studies of Nucleic Acids and Their Proteins. David Bensimon, Vincent Croquette, Jean-François Allemand, Xavier Michalet, Terence Strick 2019. © Oxford University Press 2019. Published in 2019 by Oxford University Press.
DOI: 10.1093/oso/9780198530923.001.0001

the synthesized strand, resulting in its elongation in a 5′–3′ direction. DNA polymerase can add a nucleotide only to a pre-existing 3′-OH group: it therefore needs a double-stranded DNA (or DNA/RNA) nucleus (or primer) on which to initiate the synthesis of the complementary strand.

In DNA replication, DNA polymerase is part of a complex known as the replisome, which consists of various enzymes acting in concert with the polymerase. For a complete description, see the book by Kornberg and Baker [217]. One of these enzymes is a hexameric helicase which unwinds DNA at the replication fork to expose two ssDNA strands (see Chapter 8) that serve as templates for the synthesis of their complementary strands, consistent with the semi-conservative model of DNA replication. Another essential enzyme of this complex is the primase, which is synthesizing a small complementary RNA strand that serves as a primer for DNA polymerase. One of the peculiarities of DNA replication is the different manner in which the two strands are replicated. As they run in opposite directions, and as DNA polymerase elongates the newly synthesized DNA in the 5′–3′ direction only (there are no known polymerases working in the 3′–5′ direction), the 3′–5′ template strand (known as the leading strand) is copied continuously, usually by a single DNA polymerase, while the other (or lagging) template strand is copied discontinuously in small 5′–3′ segments (known as Okazaki fragments) by many different polymerases from numerous primers synthesized on the unwound lagging strand by the primase. These RNA primers are later degraded, the gaps they leave filled in by a DNA polymerase, and the various contiguous fragments ligated by a DNA ligase.

While DNA replication is the essential role of DNA polymerases, these enzymes (of which there are many variants in the cell) are also implicated in many other pathways, such as DNA repair and homologous recombination. In these other pathways, the role of DNA polymerase is to synthesize the complementary strand of an exposed ssDNA, starting from the 3′ end of a ds/ssDNA junction. An important property shared by some, though not all, DNA polymerases is the ability to correct for mistaken copies, i.e., insertion in the growing strand of a non-matching nucleotide. If such an incorporation occurs, the enzyme steps backwards and removes the inserted nucleotide using a domain possessing a 3′–5′ exonuclease activity. This proofreading ability of DNA polymerases results in an incorporation error that can be as low as 1 in 10^5 bp.

7.2.1 Structure of DNA polymerase

DNA polymerases can be divided into at least five different families [290], depending on their structure and function (replication, reverse transcription, or repair).The best studied of these families is the prokaryotic DNA PolI family, which includes the already mentioned *E.coli* PolI (and Klenow fragment), the Taq-Polymerase (widely used in PCR reactions and extracted from the thermophilic bacteria, *Thermus aquaticus*), and the phage T7 DNA polymerase. The second family of DNA polymerases is the Pol-α family. All eukaryotic replicating DNA polymerases and the polymerases from phage T4 belong to this family. The third family is composed of RNA-dependent DNA polymerases known as reverse transcriptases (RT), RNA-dependent RNA polymerases, and telomerases. Finally, DNA Pol-β, which is involved in base excision repair and

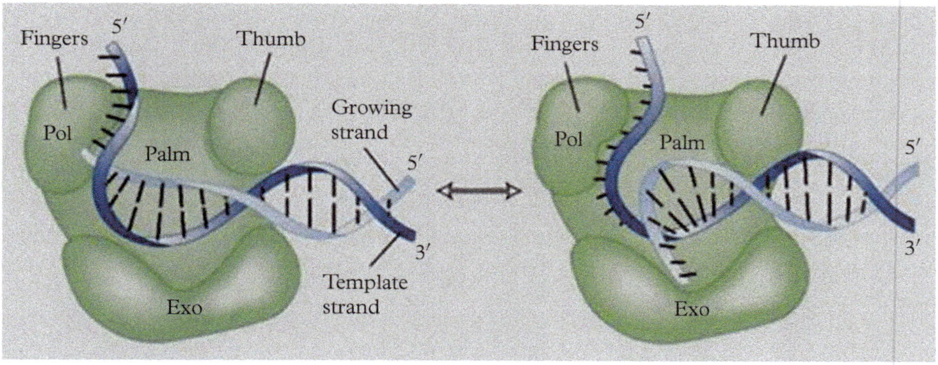

Fig. 7.1 Schematics of DNA and DNA polymerase interactions during synthesis and exonuclease activity. *Schematic view of a DNA polymerase displaying the three domains associated with polymerase activity (palm, thumb, and fingers) and the adjacent 3′–5′ exonuclease (proofreading) activity. Upon misincorporation, the 3′ end of the growing strand is shifted from the polymerase to the exonuclease domain. Adapted from [291] and [292].*

the bacterial DNA PolIII (the major polymerase implicated in the replisome), show no structural relation to any of these previous families and are classified as separate families.

In spite of these differences, all known polymerases appear to share a common overall architecture. They all possess a shape that has been compared to that of a right hand consisting of thumb, palm, and finger domains (15), see Fig. 7.1. The function of the palm domain is to catalyse the phosphoryl transfer reaction, whereby a dNTP is bound to the 3′ end of the growing strand with release of pyrophosphate. The role of the finger domain is to properly position the incoming dNTP onto its paired base on the template DNA. Finally, the thumb plays a role in positioning the duplex DNA, as well as in processivity and translocation.

The 3′–5′ exonuclease (proofreading) domain sits next to the palm domain, see Fig. 7.1. It competes with it for the 3′ end of the growing strand. The exonuclease site has high affinity for the 3′ end of ssDNA, whereas the polymerase active site has a ten-fold higher affinity for matched duplex DNA than for the mismatched kind. Moreover, polymerization is stalled following a misincorporation, presumably because of misorientation of the 3′-OH group, onto which the next dNTP is to be added. As a result, misincorporations destabilize the duplex DNA and shift the equilibrium towards binding of the 3′ end to the exonuclease active site, accounting for removal of the mismatched base and the proofreading ability of the enzyme.

7.3 Single-molecule *in vitro* DNAP experiments

Single-molecule *in vitro* experiments have been instrumental in revealing the various synthetic and proofreading activities of DNA polymerases, and in characterizing their rate, processivities, and coupling to other enzymes, such as the helicase implicated in DNA unwinding within the replisome complex. Two DNA configurations have been used to investigate the behaviour of this enzyme. The first [249, 250] consists of monitoring the change in extension of an ssDNA under tension, as a DNA polymerase synthesizes the complementary strand from an initial primer site. The second one consists of monitoring the opening of a DNA hairpin under tension, as a DNA polymerase displaces the fork while moving along the template strand in the 3′–5′ direction synthesizing the complementary strand [293]. In the following, we shall describe these two different experimental situations and dwell on their contribution to the understanding of the mechanism of DNA polymerases.

7.3.1 Polymerization on ssDNA under tension

Consider an ssDNA under tension, as in the magnetic trap or optical tweezer setups shown in Fig. 7.2a,b. Its elastic behaviour, namely its extension at a given applied force, differs from that of the dsDNA generated from it. At low forces ($F < 5$ pN), ssDNA is shorter than dsDNA, as it maximizes its entropy by forming a more compact coil than dsDNA (the persistence length of ssDNA is about twenty times smaller than dsDNA, see Chapter 4). At forces larger than about 5 pN, ssDNA, which is more flexible than dsDNA (hence its shorter persistence length), extends more than dsDNA. Thus, conversion of ssDNA under tension to dsDNA by the action of a DNA polymerase will cause the molecule to progressively lengthen at forces below 5 pN, while at forces above 5 pN, as the dsDNA portion grows, the molecule shortens.

Let $l_{ss}(F)$ and $l_{ds}(F)$ be the distance between neighbouring nucleotides at force F in ss and dsDNA respectively. The conversion within a time t of ssDNA into dsDNA at this force leads to a change in extension, $\delta l(F,t)$, which can be converted into the number of nucleotides synthesized by the polymerase:

$$n(F,t) = \frac{\delta l(F,t)}{l_{ds}(F) - l_{ss}(F)}$$

Because the change in extension resulting from a single nucleotide incorporation, $\sim$0.1 nm, is much smaller than the noise associated with the measurement of the molecule's extension (a few nm at best), this configuration has a very poor resolution along the DNA sequence.

Nonetheless, the average incorporation rate $\langle dn/dt \rangle$ can be estimated and studied as a function of the tension on the template ssDNA. It turns out that the mean enzymatic rate increases slightly up to $\sim$5 pN and decreases sharply above that force, stalling at about 30 pN, see Fig. 7.2c,d. This suggests that the work of the enzyme against the tension on the template is affecting a rate-limiting step of the polymerization reaction. Indeed, when $F < 5$ pN, the enzyme is working in conjunction with the force pulling on the template, and one may expect the tension to reduce the energy of the activated state. On the other hand, above 5 pN, the enzyme is working against the force, as it is shortening the molecule, and the energy of the activated state may actually be increased by the tension.

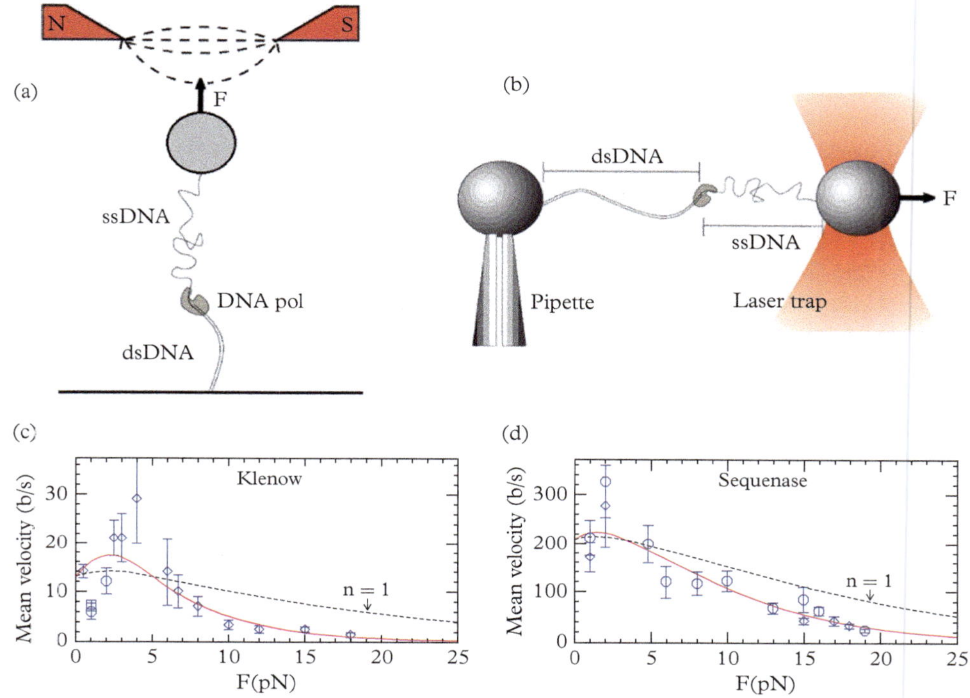

Fig. 7.2 Single-molecule studies of DNA polymerase on ssDNA. *Magnetic trap (a) and optical tweezer (b) setups were used to study the synthesis of a dsDNA on an ssDNA template by a single DNA polymerase, by monitoring the change in extension as ssDNA is converted into dsDNA. (c) The rate of incorporation in nucleotides/s (b/s) as a function of force for the Klenow fragment of DNA Pol I and for an exonuclease mutant of T7 DNA polymerase (sequenase). The continuous lines are fits to a model for which, in the rate-limiting step, n nucleotides on the ssDNA template ($n = 4$ for Klenow and $n = 2$ for sequenase) have to match the dsDNA structure. The dashed line is the theoretical prediction for $n = 1$. Adapted from [249] and [250].*

The question, then, is: what is the nature of this activated state (rate-limiting step)? From the crystallographic structures of DNA polymerases, one knows that closing of the fingers brings the incorporated dNTP and the template strand in phase with the growing dsDNA. Hence, at high force, closing of the fingers works against the tension, meaning at least one nucleotide in the stretched ssDNA will match the extension of dsDNA.

To bring a single nucleotide in phase with the dsDNA structure implies a change in free energy ΔG resulting from the work against the force $\Delta W = F(l_{ss}(F) - l_{ds}(F))$, and the change in entropy $T\Delta S$ going from ssDNA to dsDNA [250]:

$$\Delta G = \Delta W - T\Delta S = \Delta W - \left(\int_0^{l_{ss}(F)} F_{ss} dl - \int_0^{l_{ds}(F)} F_{ds} dl \right) \tag{7.1}$$

$$= \int_0^F [l_{ss}(F) - l_{ds}(F)]\, dF \tag{7.2}$$

where we used integration by part: $\int_0^F l(F')dF' = Fl - \int_0^l F(l')dl'$. That free energy change is, however, too small to account for the sharp decrease in the mean polymerization rate at high force, assuming an Arrhenius behaviour of the rate: $v = v_0 \exp -\Delta G/k_B T$. In fact, $n > 1$ nucleotides on the template strand have to be brought in phase with dsDNA to account for the force dependence of the mean polymerization rate. For T7 DNA-Pol [249, 250], n is about 2, while it is about 4 for the Klenow fragment of *E.coli* DNA PolI [249], see Fig. 7.2c,d. This result suggests that the $3'$ end of the growing dsDNA is frayed over a few bp when the finger domain is opened and the template is under tension. Closing of this domain brings this frayed DNA and the newly incorporated dNTP in phase with the fully matched dsDNA structure. This activated state associated with the closed configuration of the finger domain is consistent with the induced-fit mechanism for proofreading. According to this mechanism, the rate-limiting step in replication is due to a conformational change in the enzyme–DNA complex characterized by a transition between a loose and tightly fitted configuration of the two DNA strands. An error in the insertion of the last n bases skews the equilibrium to the loose (i.e., frayed) configuration, slowing down replication and shifting the $3'$ end to the exonuclease site. When pulling on the template strand, this equilibrium is similarly skewed to the frayed configuration, thus reducing the forward rate (even if in some of these experiments [249] the exonuclease site was inactivated by a mutation).

At forces higher than 35 pN, the tension on the template ssDNA is, in fact, large enough to prevent closing of the finger domain, thus stabilizing the frayed configuration. In this situation, the equilibrium is shifted from polymerization towards exonuclease activity, and the enzyme actually degrades the newly synthesized dsDNA [250].

7.3.2 Polymerization against a DNA fork under tension

The second configuration which has been used to study DNA polymerases, and more generally the complex it forms with other proteins during replication (the replisome), is that of a DNA fork under tension [293], see Chapter 4. In this configuration, a DNA hairpin is closed at one end with a DNA loop, while its other end is anchored on one strand to a surface and on the other to a small magnetic bead held in a magnetic trap, see Fig. 7.3a. By exerting a force larger than about 15 pN, the DNA hairpin can be completely unzipped. However, at forces smaller than about 12 pN, the junction between the dsDNA and the two ssDNA anchoring strands (the DNA fork) is stable, see Chapter 4. Hybridizing a small oligonucleotide to the $3'$ anchoring strand generates a primer from which a DNA polymerase can start polymerizing the complementary strand. However, a DNA polymerase is in general incapable of opening the fork, i.e., displacing the complementary strand. In the replisome, that role is played by a helicase, which unwinds the two strands and opens the way for the DNA polymerase to proceed. However, if the hairpin is under tension, the fork may be sufficiently destabilized such that a DNA polymerase can plough its way through, displacing one strand while synthesizing a new one on the complementary strand.

Clearly, for forces large enough to open the hairpin, the resulting configuration is similar to the one described previously, i.e., polymerization on an ssDNA under tension,

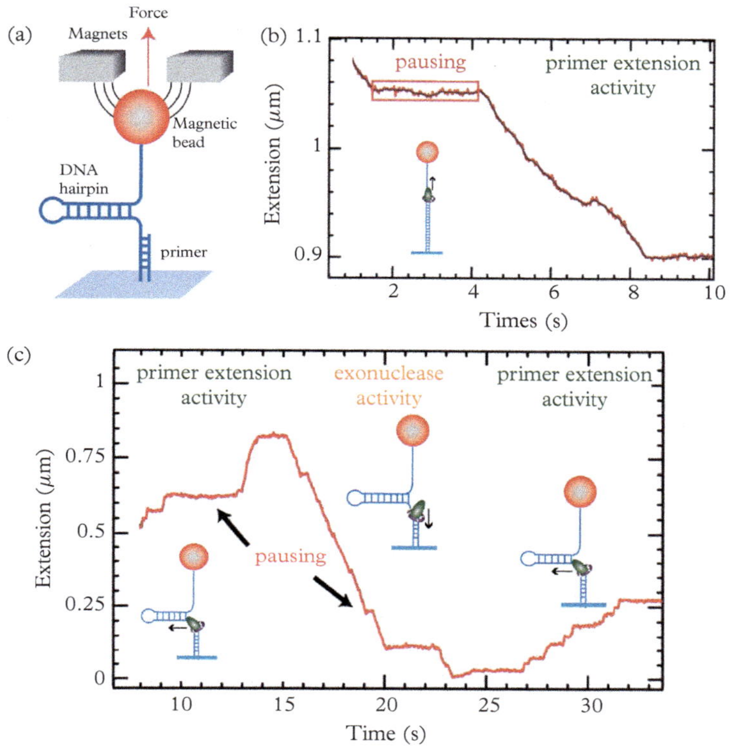

Fig. 7.3 Single-molecule study of a DNA polymerase on a DNA fork under tension. *(a) Schematic description of the magnetic trap used to apply a tension on a DNA hairpin and monitor the activity of a DNA polymerase (in its polymerization and exonuclease modes). (b) When the force is large enough to open the hairpin (here $F = 20$ pN), a DNA polymerase synthesizes the complementary strand on a stretched ssDNA. The use of small beads (1 μm in diameter) and short molecules (1.2 kbp) reduces the measurement noise to a level where pauses in the polymerase activity can be identified. (c) At forces below about 12 pN (here $F = 8$ pN), the hairpin is stable, but the fork is sufficiently destabilized for the enzyme to synthesize a complementary strand while displacing the existing one. The enzyme may also switch from a polymerization mode to an exonuclease mode, in which it degrades the complementary strand just synthesized. One notices the presence of pauses within and in between these activity modes. Adapted from [293].*

see Fig. 7.3b. By using small beads (1 μm in diameter) and short molecules (<1.2 kbp), the measurement noise can be reduced to such a level that pauses in the polymerase activity can be identified and the polymerization rate measured between pausing events. It turns out that, in contrast with the earlier measurement described in section 7.3.1, where pauses could not be identified, the variation of the polymerization rate with force can be accounted for by the free energy change due to incorporation of a single nucleotide. The mean incorporation rate (which takes into account pauses, as measured earlier) is more sensitive to tension on the substrate due to increased pause length at higher forces. Hence the effect of tension on the substrate (and presumably mismatches during incorporation) is to shift the enzyme to a polymerization-inactive state where the dsDNA is frayed by a few bp and from which it can either switch back into the polymerization state or into the exonuclease one.

This pause state is even more pronounced in the fork displacement assay, i.e., below a tension on the hairpin of about 12 pN. At low forces ($F < 7$ pN), pushed by the closing fork, the enzyme works mostly in the exonuclease mode, while at forces large enough to destabilize the fork ($9 < F < 12$ pN), it works mostly in the polymerization mode. At intermediate forces ($7 < F < 9$ pN), its activity is characterized by switches between these modes, interspersed with pauses. This observation suggests the existence of an inactive pause state between the polymerization and exonuclease ones, see Fig. 7.4a. While the transition from the polymerization to the exonuclease state always occurs through that intermediate inactive state (the transition time is characterized by a single exponential time scale), the reverse is not true: the transition between the exonuclease and polymerization states occurs only 10% of the time via this intermediate state, and mostly directly and quickly (<0.5 s). To explain this behaviour, the kinetic scheme shown in Fig. 7.4 has been proposed [293].

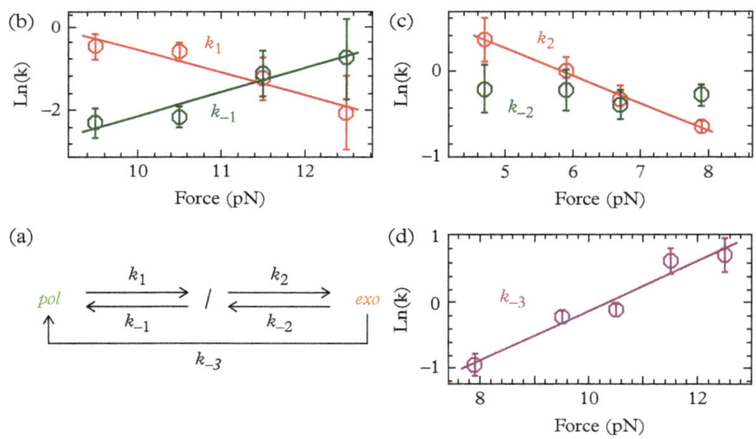

Fig. 7.4 Kinetic model of a DNA polymerase switching between synthesis and exonuclease activity. *(a) Kinetic scheme for the switch between polymerization (pol) and exonuclease (exo) modes of a DNA polymerase, via an intermediate inactive state (I). (b,c,d) variation with the force of the various kinetic rate constants shown in (a). These are deduced from the measurement of the probabilities of occupation of the various states:* p_{pol}, p_I, *and* p_{exo}, *and their mean lifetime:* $\langle\tau_{pol}\rangle = 1/k_1$; $\langle\tau_I\rangle = 1/(k_{-1}+k_2)$; $\langle\tau_{exo}\rangle = 1/(k_{-2}+k_{-3})$. *The various kinetic constants follow an Arrhenius dependence with the forces:* $k_i = k_i^0 \exp(d_iF/k_BT)$ $(i = 1,2)$ *and* $k_{-i} = k_{-i}^0 \exp(-d_{-i}F/k_BT)$ $(i = 1,2,3)$, *where the parameters* d_i *express the distance to the relevant transition state (for example,* $d_1 = 2.5$ *nucleotides is the distance to the transition state leading to state I, when in the pol mode). Adapted from [293].*

Extracting the kinetic constants of that scheme from the data [293] and studying their force dependence, see Fig. 7.4b–d, yields the following picture. The intermediate state is characterized by closure of the fork and displacement of the polymerase by about four nucleotides (with concomitant fraying of the newly synthesized dsDNA), while the exonuclease state is accompanied by a further displacement of the fork (and the polymerase) by about three nucleotides (with switching of the 3′-end of the synthesized strand into the exonuclease site). This experiment therefore yields a detailed and quantitative kinetic description of a DNA polymerase mechanism of polymerization and degradation.

The fork displacement assay has also been used to study the replication complex, and in particular the coupling between the helicase unwinding the fork ahead (see Chapter 8) and the DNA polymerase synthesizing the complement of the leading strand behind [294]. In stark contrast to the isolated enzymes, at low forces, the coupled system synthesized DNA at the maximum rate without exhibiting fork regression or pauses. It has been proposed [294] that the helicase releases the pressure of the fork on the DNA polymerase, allowing it to synthesize without pausing, and that, in return, the polymerase destabilizes the first few base pairs of the fork, thereby increasing the efficiency of helicase unwinding. This purely mechanical coupling between helicase and DNA polymerase was nicely demonstrated by replacing the phage T4 DNA polymerase of the T4 complex with a phage T7 DNA polymerase, resulting in an active heterologous complex, with a slight increase in polymerization activity consistent with the faster rate of synthesis by T7 DNA polymerase.

7.3.3 DNA sequencing by synthesis

The synthesis of a new strand by a single DNA polymerase can be observed using fluorescently labelled dNTPs. This approach has been used to read the sequence of 280 000 individual DNA templates sparsely immobilized on a surface by the hybrid they form with bound primers [295]. By sequentially supplying each type of labelled nucleotide (and a DNA polymerase), one can detect where the particular nucleotide was added by the presence of a fluorescent spot. Removing the fluorescent label, washing the solution and injecting a new labelled nucleotide allows one to read a new base. Using this method one can read the first ~30 nucleotides of the immobilized DNA templates, from which, using bio-informatics tools, the whole DNA sequence can be reconstructed [295].

Another approach allows one to observe the incorporation of labelled nucleotides in real time on a long (>10 kbp) DNA template bound to a surface by its complex with a DNA polymerase [296]. At the nucleotide concentrations required for replication (typically a few μM), the background fluorescence from the labelled dNTPs in solution is overwhelming for microscopic observation, even using a technique such as TIRF (see section 3.3) to illuminate the surface-bound enzyme–DNA complex. In order to observe the incorporation of individual dNTPs in the growing strand, Webb and co-workers [155] confined the light excitation in 3D by illuminating the backplane of a glass coverslip coated with a thin layer of aluminium in which ~100 nm diameter holes (wells) were made. The light is fully reflected by this aluminium mirror, except in the

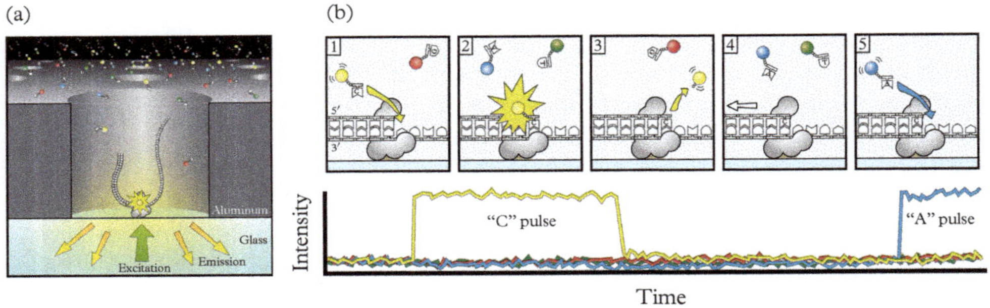

Fig. 7.5 Real-time sequencing of DNA using the detection of incorporation of labelled dNTPs. *(a) Schematics of the zero-mode waveguide (a small hole in an aluminium coating which creates a microscopic well above the glass surface) that confines the excitation light to a very small volume in which fluorophores labelling the various dNTPs can be excited and detected. (b) Cartoon of the incorporation of dCTP (labelled with a yellow fluorophore), which gives rise to a pulse of yellow fluorescence that lasts until the pyrophosphate (and the label) is excised by the DNA polymerase and diffuses away. Another nucleotide (here a blue-labelled dATP) can then be added and detected by the blue pulse that accompanies it. Adapted from [296].*

holes where an evanescent optical field is present and excites the fluorophores present in the well. The volume of the well (or zero-mode waveguide) is, however, so small ($\sim 10^{-18}$ l) that even at 10 μM concentrations, only a few labelled nucleotides are present within the excitation volume. A DNA polymerase forming a stalled complex with a DNA template is bound to the glass surface at the bottom of these holes, see Fig. 7.5. It is then supplied with dNTPs, each labelled on their last phosphate with a different fluorophore (so that, upon addition of the nucleotide to the growing DNA, the fluorophore is released with the pyrophosphate). Real-time incorporation of the labelled dNTP is thus detected as a transient pulse of fluorescence at a wavelength corresponding to the incorporated nucleotide. The time between incorporation events is random (with, occasionally, two successive incorporations mistaken as a single one) and dyes may bleach before they are detected. As a result, this single-molecule sequencing approach, while very elegant, is nonetheless affected by a large intrinsic error rate. However, this sequencing technique is characterized by long read-lengths (>10 kbp), and it allows for the detection of epigenetic DNA modifications of the template strand (e.g., methylations), as these affect the probability of incorporation (and thus the time between incorporation events) of the complementary nucleotides.

An alternative single-molecule approach for sequencing that uses neither fluorescent probes nor mechanical means (see section 5.3) to sequence DNA is based on the measurement of the current passing through a protein nanopore partially obstructed by an ssDNA threaded through the pore [298] (variations on α-hemolysine [299] or the MspA porin [300] are often used). As the molecule is reeled in by a DNA polymerase replicating it from its 3′ extremity, different combinations of n successive nucleotides passing through a constriction in the protein differentially affect the ionic current flowing through. For the MspA protein pore $n = 4$, and the current is thus set by the 256 possible quadromers [297]. By monitoring the current signal, the DNA sequence can be deduced in principle.

In practice, the signal/noise ratio is barely enough to distinguish between the 256 different current levels that sometimes differ by $\delta I < 0.01$ pA on a mean current $I_{mean} \sim 50$ pA. Indeed, for a DNA polymerase working at a rate of 20 nucleotides/s, i.e., with

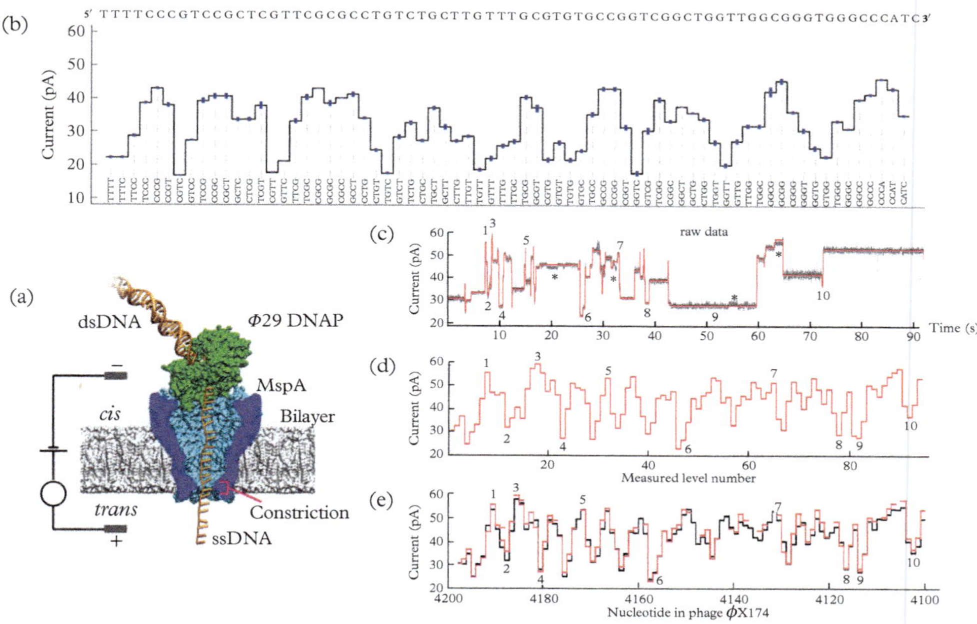

Fig. 7.6 Nanopore DNA sequencing. *(a) An ssDNA strand is threaded through an MspA pore. A highly processive ϕ29 DNA polymerase is synthesizing the complementary strand and reeling in the ssDNA. Partial obstruction of the pore constriction by four successive nucleotides affects the ionic current passing through the pore. (b) By measuring the current obtained with artificial ssDNAs (de Bruijn sequences) presenting all 256 combinations of four nucleotides (quadromers), one can match each combination with a measured current level. (c) The raw current signal (black trace) obtained when a phage $\phi X174$ DNA strand is threaded through an MspA pore, as shown in (a). This signal is fitted to a succession of current levels (red trace). Notice the large spread in the time interval between different levels (compare the time span of levels 8 and 9), a reflection of the fact that DNA synthesis is a stochastic process. (d) Representation of the signal shown in (c), in which DNAP backtrackings (indicated by a* in (c)) have been eliminated and the intervals between different current levels have been equalized to correspond to the reeling of ssDNA by single nucleotides. (e) Comparison between the measured current signal (red trace) and the one predicted (black trace) from the knowledge of the $\phi X174$ DNA sequence and the map between quadromers and current levels (shown in (b)). The excellent agreement between measurement and prediction can thus be used to identify a DNA sequence. Adapted from [297].*

an average cycle time $\tau = 50$ ms, the shot noise[1] associated with current fluctuations is $I_n = \sqrt{I_{mean}e/\tau} = 0.012$ pA. Nonetheless, by calibrating the 256 current levels on a synthesized DNA sequence containing all possible combinations of quadromers (a so-called de Bruijn sequence) and using the overlap between successive quadromers to correct for possible ambiguities, the predicted current trace of a known 5 kbp viral sequence was highly correlated with the observed signal, see Fig. 7.6.

Hence the current signal from a DNA reeled through a protein nanopore can be used to identify the DNA sequence [297]. Further slowing down of the polymerase or repeated cycles of synthesis followed by exonuclease activity, as shown in Fig. 7.3, may allow for an improvement in the signal/noise ratio sufficient for *de novo* DNA sequencing.

7.4 *In vivo* studies of single replication complexes

The development of novel single-molecule visualization methods (see Chapter 3: photo-activable fluorescent proteins, super-resolution microscopy, single-particle tracking, etc.) have opened a new vista on the study of processes at the single-molecule level in live cells. In particular, the dynamics, stoichiometry, and architecture of the replication complex in single *E.coli* bacteria has been studied by replacing the various components of the complex with fluorescently labelled analogues [78].

For example, the ϵ-subunit of the PolIII complex of *E.coli* (responsible for its exonuclease activity) has been labelled with the yellow fluorescent protein Ypet and expressed under its endogenous promoter. If the concentration of the labelled proteins

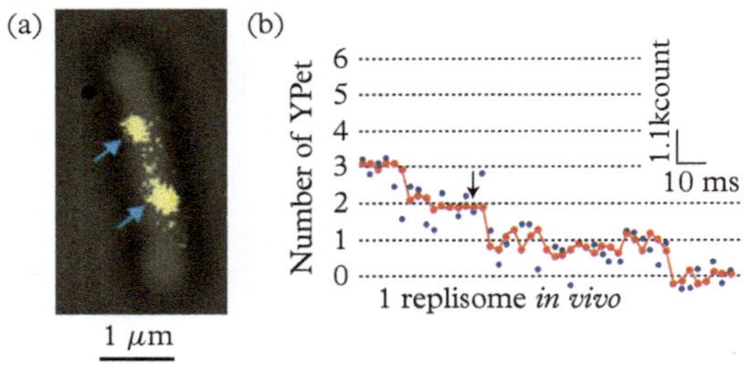

Fig. 7.7 ***In vivo*** **study of the stoichiometry of a single replisome.** *(a) The ϵ-subunit of the PolIII DNA polymerase was labelled with the yellow fluorescent protein Ypet and observed with epi-fluorescent microscopy in live E.coli. When the protein does not diffuse in the cell, it is localized at two bright fluorescent spots corresponding to the two replication forks (blue arrows). (b) The fluorescence of such a spot has an intensity corresponding to that expected from three Ypet molecules, and is observed to decay in three steps corresponding to the successive photobleaching of these three fluorophores. One thus deduces that there are three PolIII DNA polymerases per replication complex. Adapted from [78].*

[1] The shot noise is the ultimate noise associated with the fluctuations ($\alpha\sqrt{N}$) in the number N of charges passing through the pore in a time τ: $N = I\tau/q$ (for monovalent ions: $q \equiv e = 1.6\ 10^{-19}$C).

is not too high (i.e., having less than one fluorophore/pixel) they can be localized and tracked in the cell before they photobleach (within a few tens of ms in the conditions of the experiment reported in [78]), see Fig. 7.7. In the case of the ϵ-Ypet fusion, the protein was observed to be localized at one or two spots within the cell, corresponding to the two replication forks (that appear as one spot when the distance separating them is below the resolution of the microscope).

The number of ϵ-subunit, present in a single replication complex (their stoichiometry) could be determined by counting the number of fluorescent steps before photobleaching, see Fig. 7.7b, or by comparing the Ypet fluorescent intensity in the replisome to that of a single Ypet *in vitro*. Unexpectedly, the stoichiometry was determined to be three ϵ-subunits (i.e., three PolIII DNA polymerases) per replisome. From this observation (and others), one could deduce that two DNA polymerases are actively involved in replication of the leading and lagging strand, while the third one is waiting to be loaded onto the next lagging strand primer.

7.5 An introduction to transcription

DNA is the repository of the information required to make a cell. The first step in the translation of this information into functional molecules (RNA, proteins, enzymes, etc.) is known as transcription. It involves the synthesis from a single-stranded DNA template of a complementary strand of RNA by a specific enzyme, RNA polymerase (RNAP). Transcription by RNAP is a multi-step process involving binding of the enzyme to a specific site on the dsDNA (known as a promoter), unwinding (i.e., opening) of the DNA at that site, transcription initiation, promoter escape, transcription elongation, and transcription termination [301]. The main biochemical steps of this process, as well as its principal biomechanical components, are sketched in Fig. 7.8.

To perform transcription, RNAP must first locate a promoter site and bind to it. These promoter sites are located upstream of the gene which is being transcribed. *In vivo*, a major part of gene regulation is achieved by controlling the binding of RNAP to the relevant promoter using a variety of proteins known as transcription factors or regulators. Once bound to its promoter site, RNAP occupies roughly 60–70 bp of DNA, from position -55 to position $+20$ [302, 303] (position $+1$ is defined as the position of the first nucleotide transcribed into RNA or the transcription start site). Next, RNAP mechanically unwinds about one turn of the double helix, generating a $\sim$13 bp "transcription bubble" extending from about -11 to about $+2$ [304, 305, 306]. This "transcription bubble" is formed through interactions between the RNA polymerase σ subunit and the non-template strand of the bubble. Up to this point, the reaction is reversible: the transcription bubble can close and RNAP can dissociate. In the "open promoter" complex (RPo), the base pairs of the template (bottom) strand are available to the RNAP active centre for base-by-base, accurate, and processive transcription of the genetic information stored there.

Subsequent steps of transcription are less reversible. Indeed, upon addition of the full set of ribonucleotide triphosphates (ATP, UTP, GTP, and CTP), the RNA polymerase

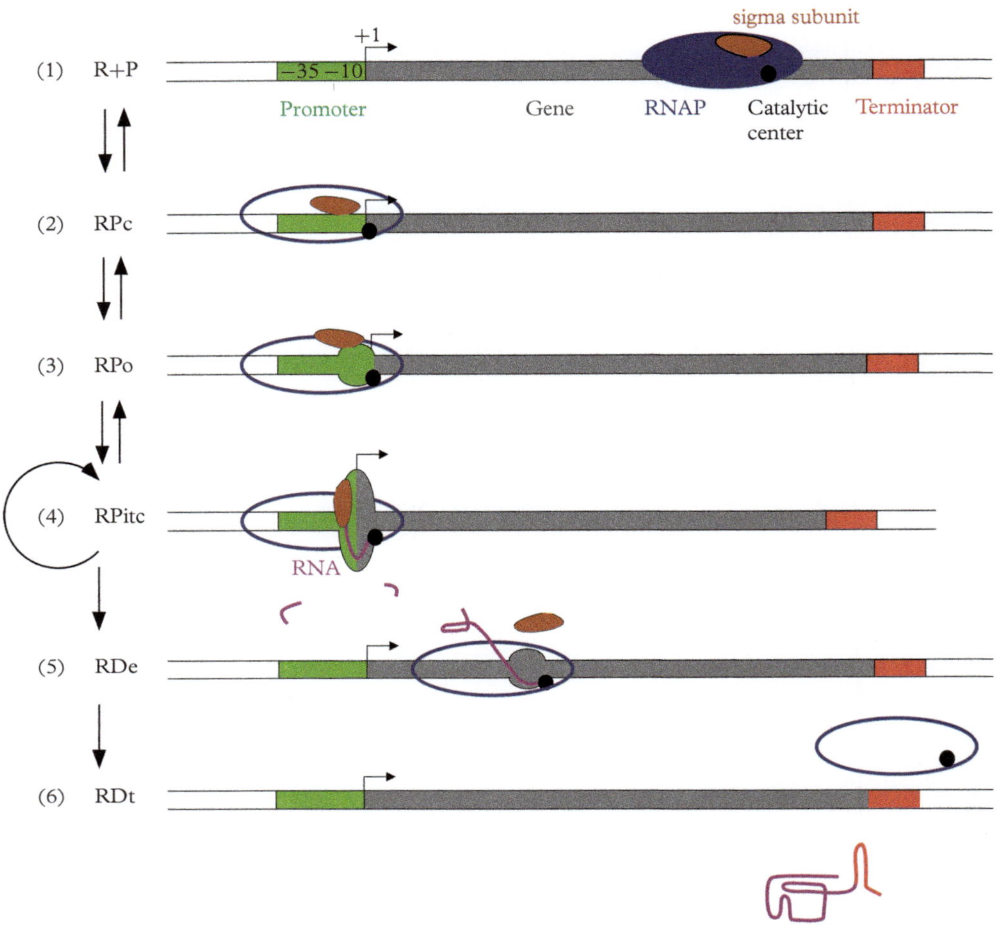

Fig. 7.8 Biochemical and biomechanical sketch of the different stages in transcription. *The RNA polymerase binds specifically, rapidly, and reversibly to the promoter (green) resulting in the RNA polymerase/promoter closed complex RPc. Upon promoter unwinding over ~12 bp, the open complex (RPo) is formed. Addition of ribonucleotides (NTP) leads initially to abortive transcription initiation (RPitc), during which DNA "scrunching" and formation of a so-called "stressed intermediate" allows for synthesis of a transcript of up to ~15 nucleotides prior to promoter escape. In this state the transcription bubble nearly doubles in size to ~25 bp. These short transcripts are often repeatedly synthesized and released before RNAP manages to escape from the promoter and fully transcribe the message. In this RNA polymerase/DNA elongation complex (RDe), the sigma subunit (brown) has unbound. RNAP processively transcribes RNA until it is forced to dissociate from the DNA template by the presence of a terminator sequence (red).*

in the open promoter complex begins to polymerize an RNA strand complementary to the sequence on the template DNA strand, beginning at the transcription start site. The initial stage of polymerization is often characterized by repetitive synthesis and release of incomplete RNA fragments containing only two to sixteen bases [307, 308, 309, 310, 311]. This process of "abortive initiation" is thought to involve the initial polymerization of a short RNA fragment, which is then released by the RNA polymerase as it simultaneously slides back on its DNA template (without dissociating from it) and returns to its initial position at the +1 transcription start site [308]. Typically, tens to hundreds of these short abortive transcripts are generated by the RNA polymerase before it is able to fully break free from the attractive interactions which tether it to the promoter site. The changes in RNAP/promoter structure which take place during this phase are of particular interest. They must eventually lead to sufficient weakening of the polymerase's interactions with the promoter, such that it becomes committed to transcription elongation, finally escaping from the promoter site [312]. These various structures of RNAP—open promoter, initially transcribing polymerase, abortive synthesis—have been observed in real time using a range of single-molecule methods, such as magnetic trapping and single-molecule fluorescence (see section 7.6), providing insight into the kinetics and energetics of transcription initiation, promoter unwinding, and promoter escape.

The subsequent phase, known as transcription elongation, constitutes the bulk of transcription. During elongation, RNA polymerase moves processively along the DNA at a rate of the order of tens of bp/s, synthesizing RNA [313, 314]. A single RNA polymerase is able to scan/read the entire gene and transcribe it into messenger RNA. Single-molecule experiments have shown that despite this very high processivity (several thousands of bp), RNAP can nevertheless pause and even backtrack on its template. Single base-pair-sized steps by individual RNA polymerase have been observed using optical trapping, by slowing down the enzyme so that the pauses between steps enabled resolving individual steps (see section 7.6.4). These experiments have provided a way of testing models of chemico-mechanical coupling in the mechanism of RNAP. Remarkably, the enzyme can sense certain common forms of DNA damage (such as those resulting from UV radiation), causing it to stall on the site of damage and recruit additional molecular motors to carry out DNA repair.

Finally, as the RNA polymerase reaches the end of the gene, it encounters a termination sequence characterized by a palindromic sequence (which leads to formation of a hairpin in the RNA transcript) followed by a stretch of adenine bases on the template strand [315, 316]. Folding of the hairpin at the back end of the enzyme, where the hairpin exits, is believed to favour irreversible pausing of the RNAP and leads to dissociation of the transcript from the RNA polymerase and presumably the RNA polymerase from its DNA template [317, 318].

7.5.1 Structure of RNA polymerases

RNA polymerases are remarkably conserved across all living species. Beyond the strong similarity in the primary sequence of genes encoding for RNA polymerases, there is a strong similarity in the actual structures of RNA polymerases [319]. For the sake of simplicity, we will concentrate on the RNA polymerase from *E.coli* which has been extensively studied using single-molecule techniques.

E.coli RNA polymerase is a multi-subunit protein with a total molecular weight of about 450 kDa, see Fig. 7.9. It is composed of two α subunits ($M_\alpha \sim 35$ kDa) which form an antiparallel dimer: α_2. This part of the protein is like the chassis of the RNA polymerase, holding together at its N-termini the β and β' subunits ($\sim$150 kDa each) which together form a pair of jaws reminiscent of a crab claw. The C-termini of the α subunits are responsible for contacting protein or DNA signals upstream of the promoter. The active centre is formed towards the junction of the jaws by domains provided by both β' and β and includes a catalytically essential magnesium ion. A fifth subunit, ω ($M_w \sim 6$ kDa), assembles onto the β' subunit and is thought to have a role in assembly of RNAP and its stabilization [320]. These five subunits constitute the "core" RNA polymerase, which is capable of non-specifically transcribing RNA from a single-stranded DNA template, but which cannot unwind the promoter DNA. Finally, a sixth subunit, known as σ, is essential only for promoter recognition in transcription initiation: sigma "crowns" the core polymerase by sitting atop the crab claw structure and also in part integrating into the RNAP core structure itself; σ directly contacts the key DNA promoter elements denoted -10 and -35 and confers on RNA polymerase a specificity for a given type of promoter sequence [321]. For the most common σ subunit, σ^{70}, which controls transcription of many important ("housekeeping") genes, the consensus -10 sequence is TATAAT, and the consensus -35 sequence is TTGACA (as read off the non-template top strand). Other σ subunits are involved in transcribing genes required for sensing environmental stress, such as heat shock (σ^{32} [322, 323]) or nitrogen metabolism (σ^{54} [324]).

7.5.2 Initial abortive transcription

Following binding to and unwinding of the promoter DNA, RNA polymerase must break free from it in order to move down the length of the gene and transcribe it. Achieving this promoter escape is not a trivial task, and RNAP typically undergoes multiple rounds of "abortive initiation", during which it repeatedly transcribes and releases short 3–15 nt long RNA fragments before stochastically managing to break free from the promoter and transcribing a full-length RNA [308]. Although the active centre of RNAP must, therefore, move relative to DNA in order to transcribe these short abortive RNA fragments, footprinting experiments[2] show that the upstream boundary of the RNAP–promoter complex (i.e., the back end of the enzyme) remains fixed during this abortive cycling [326]. This apparent contradiction—that the back of the enzyme appears not to move, whereas the active centre appears to move—can be resolved by several models. The first, dubbed "inchworming", proposes that the active centre of the enzyme belongs

[2] Footprinting experiments are DNA digestions in the presence of DNA-binding proteins, which reveal how much DNA was protected from digestion due to its binding to the protein.

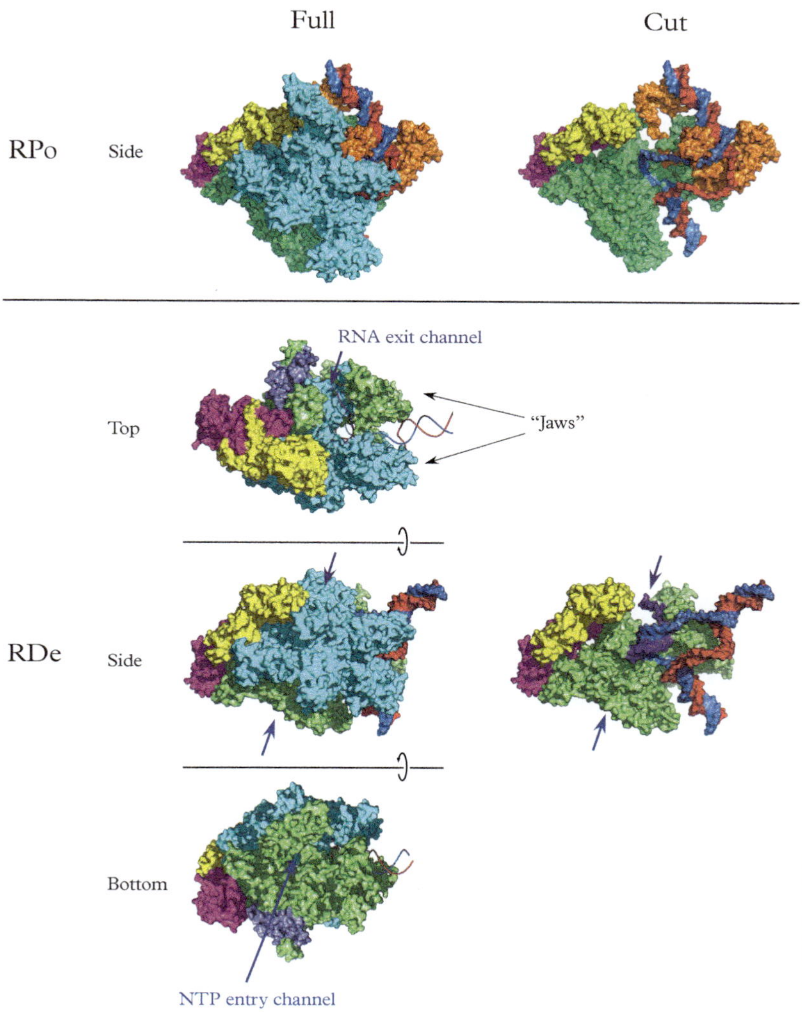

Fig. 7.9 Views of RNA polymerase and DNA in the open-promoter state (RPo) and in the actively transcribing state (RDe). *Side views are shown facing the β subunit (cyan), and with the polymerase transcribing towards the bottom of the image. Cut views lack the β subunit. α subunits are in yellow and pink; β' in green; σ is coloured orange and ω grey; template DNA strand is blue; non-template DNA strand is red; and RNA is in magenta. Top view of the elongating complex shows how the β and β' "jaws" hold the DNA in place. The top view also has the so-called "β-flap" portion of the β subunit removed so as to better highlight the RNA exit channel. Bottom view of the elongating complex shows the nucleotide entry channel with RNA just visible at the end of the channel. Model courtesy of S. A. Darst, adapted from [325].*

to a flexible element of the RNAP, which can advance along the DNA relatively independently of the back end of the enzyme. The scrunching model suggests instead that it is the DNA which gets reeled into the protein, causing the transcription bubble to grow, and thereby allowing the active site to see the succession of bases in the initial transcript. This latter model correctly predicts a transient increase in the size of the transcription bubble prior to promoter escape, as seen in single-molecule experiments [327, 328], see section 7.6.3. Structural data and modelling [329] indicate that the transition from initiation to elongation begins when the nascent RNA has grown by about fifteen nucleotides, at which point it is sufficiently long to begin displacing the σ subunit's domain.

7.5.3 The transcribed region: elongation and pausing

As mentioned earlier, a single RNAP is capable of synthesizing mRNAs thousands of bases long. To add a base to the 3′-OH end of the growing RNA chain, the active centre coordinates nucleophilic attack by the 3′-OH group on the incoming nucleotide triphosphate. The incoming nucleotide is selected as complementary to the "bottom" DNA strand. In near-physiological conditions (pH $\sim 8, 37\,^{\circ}$C, $\sim$50 mM monovalent ions and 5–10 mM magnesium), elongation is characterized by a Michaelis–Menten constant for ribonucleotide binding of the order of 200 μM, and a maximal rate of elongation of the order of 30 nt/s [330, 331]. Unlike the earlier stages of transcription, regulatory signals are less dominant here: after accomplishing promoter unwinding and promoter escape, RNAP has to a large extent been "licensed" to transcribe and will typically complete this task and then release the mRNA. The large-scale macroscopic motion of RNAP along DNA has long made it a prime topic of interest for biophysicists using single-molecule approaches in order to watch living matter working at the level of its constituent units.

7.5.4 Transcription termination

Transcription termination concludes the process; it encompasses release of RNA from the RNAP–DNA complex, followed by dissociation of RNAP from DNA [332]. Two major kinds of transcription termination signals have been described.

"Intrinsic" terminator sequences direct release of the full-length RNA, and are characterized by the presence of palindromic sequences in the DNA, typically followed by a stretch of adenines [315, 316]. This palindromic sequence allows formation of a stable hairpin in the RNA, perhaps helping to yank the 3′ end of the RNA out of register with the active centre of the polymerase. The subsequent stretch of As, transcribed into poly-U, indeed provides the RNA polymerase with an isoenergetic nucleic acid profile, perhaps facilitating slippage of the RNA polymerase off the DNA. Pausing of RNAP at the termination site is an important first step of this process, which is believed to lead to irreversible trapping of a stalled complex and ends up releasing RNA [333].

"Extrinsic" transcription termination is assisted by proteins: in *E.coli*, in particular, one talks of "Rho-dependent" termination, mediated by the Rho-protein [334].

7.6 Single-molecule *in vitro* RNAP experiments

From this introduction, it is apparent that RNA polymerase is a remarkable molecular motor, capable of a range of large-scale mechanical interactions with its DNA substrate. The diversity of movements of RNA polymerase relative to its DNA template makes single-molecule analysis a powerful tool for studying the coupling between biochemistry and biomechanics during the different phases of transcription. During transcription initiation, the RNA polymerase physically unwinds the DNA, perhaps by the application of a mechanical torque to the DNA. During transcription elongation, the RNA polymerase must physically translocate along the entire length of the DNA, an activity typical of molecular motors. Finally, at transcription terminators, RNAP must stop. Thus, the RNA polymerase is capable of first rotating (or torquing) the promoter DNA at transcription initiation, and then moving (translating) along the DNA during elongation—but also of stopping at the appropriate sequence. These properties make RNA polymerase an excellent candidate for single-molecule experiments designed to understand how mechanical constraints such as force and torque affect the conformational changes an enzyme must undergo to complete its catalytic cycle.

Our discussion of single-molecule experiments involving RNA polymerase will proceed according to the enzyme's catalytic cycle: beginning with transcription initiation and promoter escape, continuing onto transcription elongation, and ending with transcription termination.

7.6.1 DNA binding and promoter search

This aspect of transcription has been studied using single-molecule fluorescence experiments to measure in real time the association and dissociation kinetics of individual RNAP molecules from single molecules of DNA containing or lacking a promoter [335, 336, 337]. In these experiments, DNA is attached to a surface and illuminated with an evanescent field. Binding/unbinding of RNAP labelled with a fluorophore or a quantum dot is imaged in real time. Overall, there is agreement on the value of the binding rate constant of RNAP to promoter DNA: $k_{on} = 10^7$ M^{-1} s^{-1} [335, 336], close to the diffusion limit of $\sim 10^8 - 10^9$ M^{-1} s^{-1}. The time RNAP spends on non-specific DNA appears to be short: $\tau_{diss} \sim 10 - 100$ ms [335, 336, 337]. The 1D diffusion constant of RNAP along DNA during this time is not well determined, with rough estimates of the order of 10^{-10}cm^2/s [335], suggesting that RNAP could diffuse typically nearly 100 nm (or approximately 300 bp) before dissociating. Whether non-specific diffusion consists in only hopping/sliding along DNA, or also contains a rotational component, is unclear; indeed, some experiments have suggested that during non-specific 1D diffusion on DNA, the RNAP should track the helical groove of the DNA, which is reasonable given the requirement for RNAP to detect promoter DNA sequence [338]. Nevertheless, diffusion along DNA does appear to be limited in scope (both temporally and spatially), raising the question of its role in promoter search *in vivo* [337].

7.6.2 Formation of the RNAP-promoter open complex

Detection scheme. Single-molecule detection of promoter unwinding by RNAP was first reported using a magnetic trap system. The system was used to supercoil DNA molecules and monitor their topology in real time [236], as RNAP reversibly formed the open complex, *RPo*. This approach was used to explore the mechano-chemistry of promoter unwinding. It showed that the rate of formation of the open promoter is limited by the torque acting on the DNA.

Figure 7.10 shows the principle behind the topological detection of promoter unwinding by RNA polymerase. Conservation of linking number (see Chapter 4) implies that unwinding of the promoter, which leads to a decrease in the DNA twist $\Delta Tw \sim -1$,

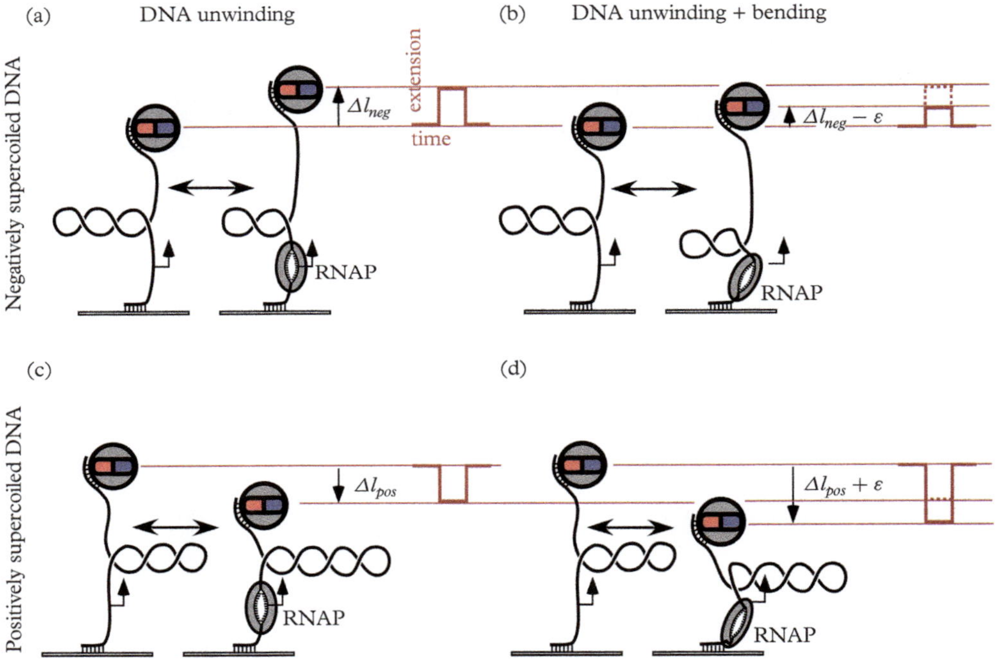

Fig. 7.10 Principle of detection of RNAP-induced promoter unwinding using a single supercoiled DNA molecule. *For negatively supercoiled DNA, unwinding of the promoter site leads to the absorption of a negative plectonemic supercoil and an increase in DNA end-to-end extension. For positively supercoiled DNA, unwinding of the promoter site leads to the addition of a compensatory positive plectonemic supercoil and a decrease in DNA end-to-end extension. If pure DNA unwinding was generated by its interaction with the RNAP, with no accompanying DNA bending or compaction, then the amplitudes of extension changes should be the same for positively and negatively supercoiled DNA (left column). However, the presence of DNA bending leads to a differential signal when comparing positive and negative supercoiling (right column); this differential signal can be analysed to extract the extent of bending/compaction and the extent of pure unwinding. Adapted from [236].*

must be accompanied by a unit increase in the DNA's writhe: $\Delta Wr = +1$. While the unwinding of 10.5 bp of DNA (one helical turn) is associated with an undetectable small change in the molecule's extension, the associated change in the DNA writhe is easily detected in real time by monitoring the end-to-end distance of a mechanically stretched and supercoiled DNA molecule. At low forces ($F \sim 0.3$ pN), a change in writhe of one unit is accompanied by a change in extension of about 60 nm, see Chapter 4. This "topological amplification" of changes in DNA twist is the basis for the investigations of promoter unwinding and promoter escape by RNAP.

As shown in Fig. 7.10a,c, the change in the DNA's extension observed upon unwinding of the promoter site by RNAP is expected to have the same amplitude but opposite sign for positively or negatively supercoiled DNA. In both cases, unwinding of the promoter site leads to a unit increase in the molecule's writhe. For negatively supercoiled DNA, increasing the writhe by one is equivalent to removing one negative supercoil, allowing the DNA end-to-end extension to increase by an amount noted as $\Delta l_{Wr} \sim 60$ nm. For positively supercoiled DNA, increasing the writhe by one leads to the addition of one positive supercoil, causing the molecule's end-to-end distance to decrease by an amount Δl_{Wr}. If, in addition to promoter unwinding, the RNAP also bends the molecule, the decrease in extension due to this bending Δl_{bend} has to be taken into account. For negatively supercoiled DNA, the absolute change in extension is thus: $\Delta l_{neg} = \Delta l_{Wr} - \Delta l_{bend}$, while for positively supercoiled DNA it is: $\Delta l_{pos} = \Delta l_{Wr} + \Delta l_{bend}$.

Experimental data. Overall, these predictions are borne out by experimental data obtained using negatively and positively supercoiled 4 kbp DNA containing a single lactose promoter, Fig. 7.11a–c.

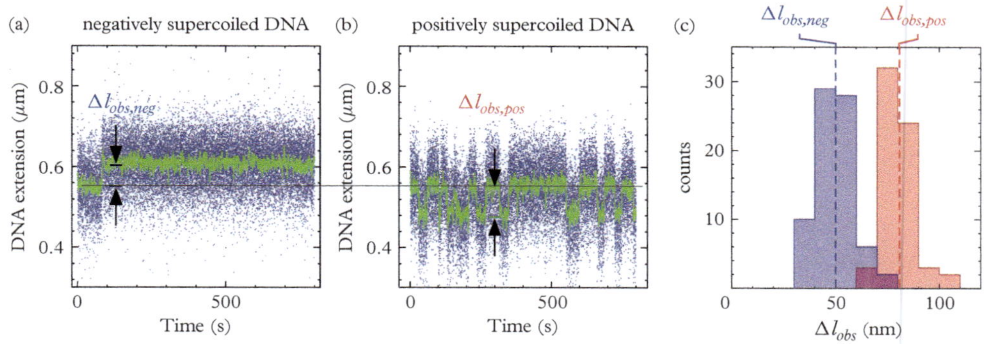

Fig. 7.11 Real-time detection of *lac***-promoter unwinding** *observed using a magnetic trap at a tension $F = 0.3$ pN. (a) Time dependence on negatively supercoiled DNA ($\sigma = -0.018$) of the molecule's extension upon stochastic switching of the RNAP from closed (low extension) to open (higher extension) complex (green: data acquired at video rate; red: same data filtered at 1 Hz). (b) The same on positively supercoiled DNA. The closed complex is now the higher extension state. (c) Histogram of the absolute change in extension Δl_{neg}, Δl_{pos} going from closed to open complex for negatively and positively supercoiled DNA. The difference in the mean value is due to DNA bending by the RNAP, see text in section 7.6.2. Adapted from [236].*

Reversible unwinding of the promoter site by an RNA polymerase gives rise to a random "telegraphic" extension signal, which is due to the promoter site switching between a "closed" and an "open" (unwound) state characterized by different DNA end-to-end extensions. The amplitude of the change in DNA extension upon promoter unwinding/rewinding is related to the size of the unwound segment of the promoter and to its possible bending by RNAP. The time interval between successive unwinding events characterizes the rate of formation of the closed promoter/RNAP complex. It is influenced by a wide range of basic experimental parameters, including temperature, ionicity, and buffer formulation. As we will discuss later on, the rate of formation of this complex responds to protein concentration according to classical Michaelis–Menten kinetics. Finally, the lifetime of an unwinding event is a measure of the stability of the open complex.

From the absolute change in extension on positively and negatively supercoiled DNA molecules, one can deduce the change in extension due to bending of the promoter: $\Delta l_{bend} = (\Delta l_{pos} - \Delta l_{neg})/2 \sim 14$ nm. Correcting for the stretching of the molecule (at a force of 0.3 pN, it extends only to about two thirds of its full crystallographic length, see Chapter 4), one deduces that the change in DNA contour length due to binding of the RNAP to the promoter is about 22 nm. This corresponds to a bending of the DNA by about 80°, in qualitative agreement with models and prior measurements [339, 340].

From the change in extension when positively and negatively coiling the DNA, one can similarly deduce the amount due to unwinding: $\Delta l_{Wr} = (\Delta l_{pos} + \Delta l_{neg})/2 \sim 65$ nm. From a measurement of the decrease in extension per added supercoil at the same force (56 nm/turn) and the DNA pitch (10.5 bp/turn), one deduces that about 12 bp of the promoter have been unwound in the open complex.

Kinetics of formation and lifetime of the open RNAP/DNA complex. The probability distribution of waiting times T_{wait} for formation of an open promoter and its lifetime T_{open} display single-exponential behaviour [236], suggesting that the kinetics of formation and rewinding of the complex are characterized by a single time scale. For the formation of the complex, this time scale is set by the time for localization of the promoter by the RNAP. Indeed, the on-rate for formation of the complex k^{on}_{RPo} obeys a typical Michaelis–Menten kinetics, see Fig. 7.12a:

$$k^{on}_{RPo} = \frac{k_{on}[RNAP]}{[RNAP] + K_M} \tag{7.3}$$

with $k_{on} = 0.3$ s^{-1} and $K_M \sim 100$ nM. The dependence of the on-rate of open complex formation on the enzymatic concentration in solution suggests that dissociation of RNAP from the DNA in its closed complex form is faster than switching to the open complex ($\tau_{diss} k_{on} \ll 1$), consistent with fluorescence observations [336]. Hence every open complex is essentially dependent on novel associations of RNAP with the promoter. On the other hand, the rate of conversion of the open complex back into the closed complex, $k^{off}_{RPo} = 0.03\ s^{-1}$, is essentially independent of the enzyme's concentration, see Fig. 7.12.

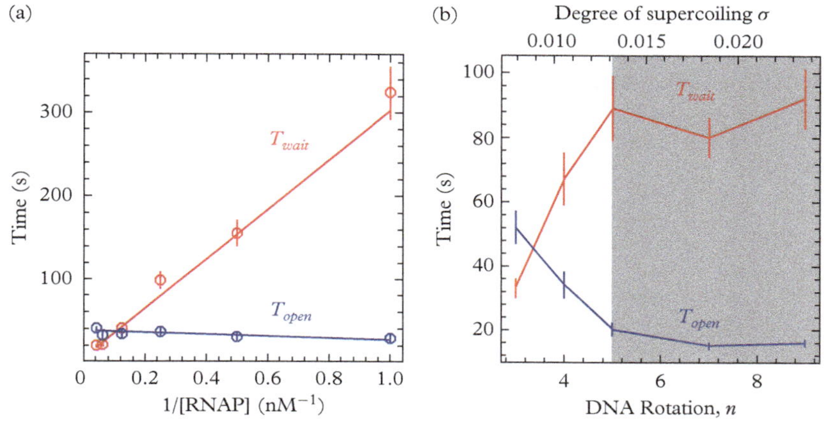

Fig. 7.12 Kinetics of promoter unwinding and rewinding by RNAP. *(a) The mean time $T_{wait} = 1/k^{on}_{RPo}$ (red) for open-complex formation and its lifetime $T_{open} = 1/k^{off}_{RPo}$ (blue) are plotted as a function of inverse RNAP concentration (nM^{-1}). The linear fit to T_{wait} corresponds to a Michaelis–Menten kinetics (see section 7.6.2). (b) Increasing the positive torque on the DNA (i.e., preventing unwinding) increases the waiting time (red) for open complex formation and decreases its lifetime (blue). The torque varies non-linearly with supercoiling (see Chapter 4): although it initially increases linearly with supercoiling, it saturates once the molecule buckles. The supercoiling dependence of the kinetics of promoter unwinding, thus, mirror the torque dependence on DNA supercoiling. Adapted from [236].*

Effect of supercoiling on the kinetics of formation and the lifetime of the open RNAP/DNA complex. The response of the system to supercoiling was also studied using a magnetic trap, see Fig. 7.12. By varying the degree of positive or negative supercoiling, one could confirm that positive (resp. negative) supercoiling reduced (resp. increased) the rate of formation and the lifetime of the open complex, see Fig. 7.12a. The on- and off-rates are in fact determined by the torque on the molecule. The dependence of the observed rate constants on the torque can be accounted for by a simple Arrhenius law:

$$k_{on}(\Gamma) = k_{on}(0)\exp[-\theta_{on}\Gamma/k_B T] \tag{7.4}$$

$$k^{off}_{RPo}(\Gamma) = k^{off}_{RPo}(0)\exp[\theta_{off}\Gamma/k_B T] \tag{7.5}$$

where Γ is the torque estimated at that force ($F = 0.3$ pN; see Chapter 4 and Fig. 4.10), $\theta_{on} \sim 1.8$ rad and $\theta_{off} \sim 2.2$ rad are the angular distance to the transition state of the relevant RNAP/DNA complexes.

7.6.3 Abortive initial transcription

The challenge for RNAP in the open-promoter complex is to initiate RNA synthesis and break the very favourable energetic ties that bind it to the promoter. Recall that typically,

RNAP must first transcribe ~15 nucleotides of RNA before being able to break free from the promoter. The magnetic trap technique has proved useful in detecting and characterizing that transition, from formation of the open promoter complex to promoter escape. These experiments (using the same "topological amplification" approach as for promoter unwinding) demonstrated that when RNAP is undergoing abortive cycles of transcription, the associated transcription bubble is growing (as evidenced by the compensatory generated supercoils), in a manner consistent with the scrunching model for abortive initiation [327]. The mean lifetime of this "scrunched" state was measured to be about 5 s.

Independent but complementary single-molecule FRET experiments [328] further showed that during this transient increase of the transcription bubble, it is downstream DNA that is pulled into the RNAP, while the enzyme remains fixed on the promoter site during these abortive cycles. In these experiments, the RNAP or DNA was labelled at various positions with a donor dye, and the DNA with an acceptor. The shift in the FRET signal, following initiation of abortive transcription (ensured by supplying only three out of the required four NTPs), informs on the relative movement of the RNAP on the DNA and on the relative translocation of DNA itself, see Fig. 7.13a,b. By appropriate labelling, this experiment did confirm that the mechanism of abortive initial transcription is by "scrunching" of the downstream DNA and not by transient excursion of the RNAP or translocation of its frontal subunits.

Together, these experiments demonstrate that during initial abortive transcription the RNAP remains fixed on the promoter, while reeling in downstream DNA and increasing the size of the transcription bubble. The "scrunching" events represent the majority (80%) of the events eventually leading to promoter escape and transcription (since some of the events may be too short to detect, the proportion might even be larger). The "scrunched" state can thus be viewed as a stressed intermediate, in which mechanical strain is accumulated to assist the system in overcoming the energy barrier preventing it from escaping the promoter.

7.6.4 Transcription elongation

Transcription elongation is highly processive: it can involve translocation of RNAP on DNA over tens of thousands of base pairs (microns), which makes the detection of that translocation rather easy to detect at the single-molecule level. The first such study was reported in 1991 by Schafer *et al.* [342].

In these experiments, see Fig. 7.14a, a DNA molecule was prepared with a promoter site close to a biotin-labelled extremity. Transcription elongation (towards the other end) was initiated by adding RNA polymerase and three of the four ribonucleotides. Transcription stalled when the missing nucleotide was required after the first 15 bp of transcript (needed for promoter escape). The stalled complexes of *E.coli* RNAP have lifetimes of a few hours, which allows their manipulation prior to restart (by supplying the missing nucleotide). The DNA was tethered to streptavidin-coated colloidal gold particles and deposited on a glass surface, where the RNAP adsorbed while remaining catalytically active. Restarting the stalled complex caused the DNA tether between the

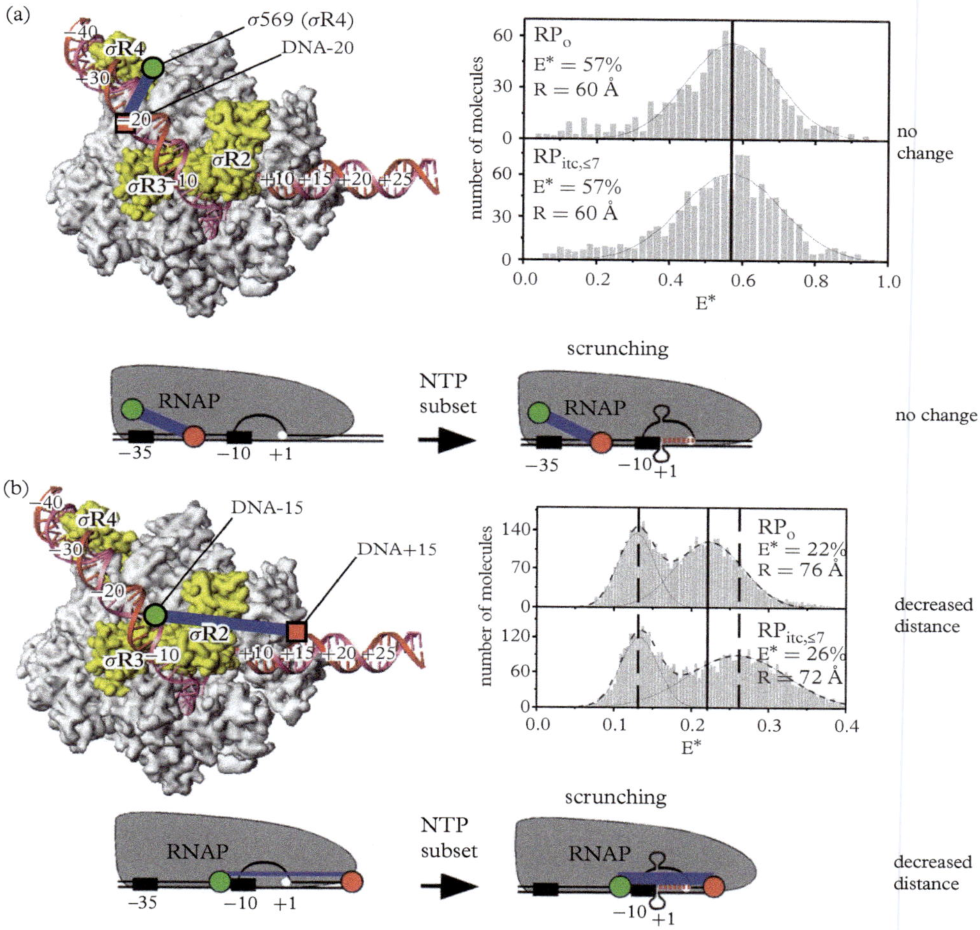

Fig. 7.13 FRET study of abortive initial transcription in RNA polymerase. *(a) Labelling sigma factor with a donor dye (green) and the −20 position on the promoter with an acceptor dye (red). One notices that upon abortive transcription initiation (ensured by supplying only three out of the required four nucleotides), the FRET signal (right) does not change, indicating that the enzyme remains locked on the promoter, as expected from the scrunching model of abortive transcription. (b) Labelling the −15 position on the promoter with a donor dye (green) and the downstream +15 position on the DNA with an acceptor dye (red). One notices that upon abortive transcription initiation, the highest FRET peak is shifted to higher values, indicating that the two dyes got closer as expected from the scrunching model of abortive initiation. The two FRET peaks observed in this experiment correspond to FRET between the DNA dyes in the absence of RNAP, and in the open complex with RNAP, when DNA is unwound and bent, thus bringing the two dyes closer to one another. Adapted from [328].*

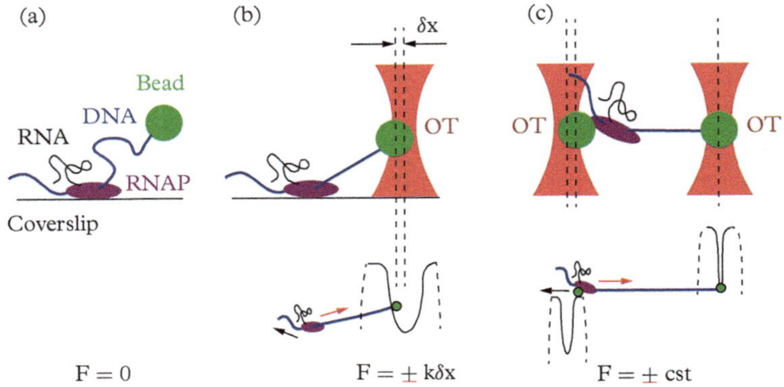

Fig. 7.14 Experimental designs for studying transcription elongation. *(a) Tethered-particle motion: a bead is anchored by a DNA molecule to a stalled transcription complex bound to a surface. Upon restart of transcription, by supplying the missing nucleotide the DNA molecule is translocated by the RNAP. The change in the tether length is detected by a change in the bead's amplitude of fluctuations (~DNA's radius of gyration). (b) Single-beam optical trapping: a transparent bead tethered to RNAP, as described in (a), is held in optical tweezers. DNA translocation by the RNAP can thus be studied in the presence of a hindering force (enzyme working against the tension) or an assisting one. (c) Dual-beam optical trapping with passive force-clamp: similar construct as in (b), but the RNAP is bound to a second bead also held in optical tweezers, whose strength is chosen (by tuning the light intensity) such that its stiffness is very small ($dF/dx \sim 0$), so that the force applied on the bead is constant [341].*

RNA polymerase and the gold particle to lengthen as the enzyme translocated along the DNA, transcribing it into mRNA. The extension of this DNA tether, l, was deduced from the increased mean fluctuations $\langle \delta x^2 \rangle$ of the gold particle (a technique known as Tethered Particle Motion (TPM)): from polymer theory, see Chapter 4, one expects $\langle \delta x^2 \rangle \sim l$. The transcription rate dl/dt was then deduced and found to be of the order of ten to twenty ribonucleotides per second [343, 344], in agreement with results obtained from bulk biochemical experiments.

This work was soon followed by experiments of a similar conception, where, in addition, optical tweezers were used to trap the tethered bead [346], see Fig. 7.14b. Translocation of the RNAP towards the bead reduced the DNA tether and caused the tension in the molecule to increase (as the bead was pulled away from its trapped position). Beyond the possibility thus offered to study translocation of RNAP under load, this technique has the advantage over the TPM approach of allowing a much more precise measurement of the DNA change in extension (via the displacement of the bead in the trap). At a force of about 10 pN, one can achieve a typical extension noise of a few nanometres at 1 kHz, see Chapter 2. Optimization of the optical tweezers with respect to environmental, thermal, and optical noise, and reduction of the turnover of RNAP by limiting the NTP concentrations, has enabled [345] the resolution in real time of RNAP individual steps of ~3.4 Å(averaged over ~1 s), see Fig. 7.15.

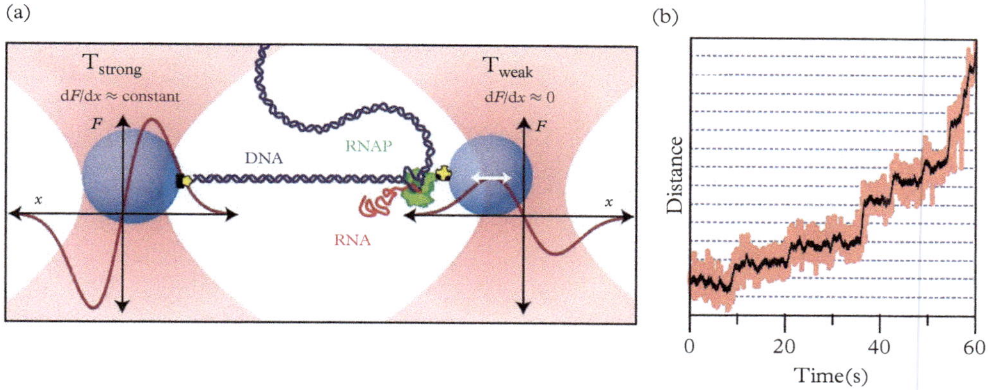

Fig. 7.15 Observation of RNA polymerase progression by single base-pair steps. *(a) The dual-beam optical tweezers setup used to study single-nucleotide stepping by RNAP. One optical tweezers (T_{strong}) had high stiffness and set the force, while the second had low stiffness ($dF/dx \approx 0$), so that the bead in that trap moved at fixed force. (b) In limiting NTP concentrations, the RNA polymerase advanced slowly enough that individual steps ~3.4 Å apart (the distance between nearby base pairs in DNA) could be resolved. Adapted from [345].*

Force–dependence of the rate of transcription. The RNAP transcription rate v_{trans} was measured (at high NTP concentrations) to be about $v_{trans} \sim 10-12$ nt/s over a wide range of forces assisting (≤ 37 pN) or hindering (≤ 20 pN) RNAP translocation [347, 348]. However, above an opposing force of about 20 pN, the translocation rate of the enzyme decreased rapidly and stalled at a force of about 25–30 pN. Various scenarios have been proposed to explain this observation [346].

One possibility is that above 20 pN, a force-dependent step in the enzymatic translocation becomes rate limiting, so that the total translocation cycle time is the sum of a force-independent time scale τ_1 and a force-dependent time scale τ_2, with an Arrhenius dependence:

$$1/v_{trans} = \tau_{trans} = \tau_1 + \tau_2 e^{F\Delta/k_B T} \quad (7.6)$$

with $\tau_1 \gg \tau_2$ and where Δ is a typical length (about 10 bp) associated with the force-dependent step. This is a rather large conformation change to be associated with RNAP translocation along DNA.

Another possibility is that one (or a few) step(s) along the kinetic pathway consists of a reversible force-dependent transition between two on-pathway states (a forward-active state with probability p_+ and backward-inactive state with probability p_-). The relative occupancy of these states obey a Boltzmann relation: $p_-/p_+ = \epsilon \exp[F\Delta/k_B T]$ (where ϵ is the zero force equilibrium), so that the backward state is favoured by a hindering force. In that case, the translocation rate being proportional to the probability of being in the active state, one obtains:

$$v_{trans} = \frac{v^0_{trans}(1+\epsilon)}{1+\epsilon e^{F\Delta/k_BT}} \tag{7.7}$$

backward translocations of the RNAP with associated long pauses (~10 s) have indeed been observed [349, 350, 351, 352], see Fig. 7.15, and were shown to account for this observation [347].

Finally, a third possibility is that the RNAP pauses in an off-pathway state which is favoured by the hindering force [346, 353]. Transition to this off-pathway state causes a delay which increases with the force. This delay, when combined with other force-independent translocation times, leads to an equation for τ_{trans} similar to Eq. 7.6. Sequence-dependent short pauses (~1 s) have indeed been identified [353], and they could also be responsible for the observed force dependence of the RNAP translocation rate.

Drug-induced pausing of RNAP during elongation. Because optical tweezer experiments can resolve pauses of various duration during the elongation phase of transcription, they have been used to investigate the mechanism of transcription inhibitors, notably the antibacterial drug microcin J25 [348]. This antibiotic was known to inhibit transcription, possibly by obstructing the nucleotide entry channel. The reduction in RNAP elongation rate could be explained either as a decrease in its mean transcription rate or as an increase in its frequency of pausing. Single-molecule experiments, by allowing one to differentiate between these possible explanations, showed that in the presence of the inhibitor, while the rate of elongation was unchanged, the probability and duration of pauses were significantly increased. The rate constant associated with pausing was rather small: $k_{on}^{J25} \sim 10^4$ M^{-1} s^{-1}, while the duration of pausing: $1/k_{off}^{J25} \sim$ 4 s, yielding a constant of inihibition $K_i = k_{off}^{J25}/k_{on}^{J25} \sim 25\ \mu$M. The observed inhibition phenotype—sudden, yet reversible cessation of elongation with no gradual change in the elongation velocity—is consistent with a model whereby microcin J25 inhibits elongation by occluding the nucleotide entry channel for RNAP.

It is remarkable how such single-molecule experiments can improve our understanding of the mechanisms of action of various drugs, see also Chapter 9. First of all, direct observation of the interaction allows one to discriminate between different modes of action of the drug. Secondly, single-molecule kinetic analysis provides, in a set of unified experimental conditions, not just the binding affinity of the drug (its constant of inhibition K_i), but also a more complete kinetic picture of its association and dissociation rate constants. Because the dissociation rate is related to the binding energy of the drug to its target, such experiments may provide a means of better characterizing rationally designed drugs, for instance by comparing the binding energetics of different iterations of a compound.

Observation of RNAP tracking of the DNA helix during elongation. It is expected that, as RNA polymerase transcribes DNA into RNA, it will track along the helical path of DNA, leading either to rotation of the RNAP around the DNA or rotation of the DNA as it is threaded through the RNA polymerase. Elegant experiments by Kinosita *et al.* [354] convincingly demonstrated this point. Elongation complexes stalled at one end of the DNA were bound via the RNAP to a glass surface (as in a TPM assay). The other end of the DNA was tethered by multiple points to a magnetic bead to ensure that the molecule was torsionally constrained. Small annular magnets generated an azimuthally symmetric field, which while exerting no torque on the bead, could pull on it with a force of about 0.1 pN. Using fluorescence imaging to visualize the motion of the magnetic bead during transcription elongation (initiated by supplying the required nucleotides), Kinosita and co-workers could directly observe its expected clockwise rotation (as seen from above) as the DNA was being threaded through the immobilized RNAP and translocated. By comparing the bead's rotation rate to the transcription rate, the authors concluded that the bead rotated by about one turn for every ten transcribed base pairs (as expected for an RNAP that tracks the helical pitch of dsDNA).

In more recent experiments, Wang and collaborators [355] have used the optical trapping of a cylindrical object to apply a torque on DNA during transcription. By positively supercoiling the molecule ahead of the polymerase, they showed that it can transcribe against a positive torque of about 12 pN·nm. RNAP can thus plough its way through DNA by applying a torque which is large enough to melt the downstream segments, see Chapter 4.

7.6.5 Transcription termination

Once a gene has been transcribed by an RNA polymerase, the mRNA as well as the RNAP must be released. This process is not always 100% efficient (RNAP may continue to transcribe past the polyA tail). To ensure termination, in prokaryotes about 50% of transcripts are terminated by specific DNA sequences (intrinsic terminators) that encode for RNA consisting of a GC-rich hairpin followed by a 7- to 9-nt U-rich tract. One possible model for transcription termination posits that closing of the hairpin pulls the slippery U-rich tract out of register with the underlying A-rich DNA segment, thereby stalling the enzyme and inducing its unbinding, with concomitant release of the mRNA.

To test that model, Block and co-workers [356], used double-well optical tweezers to pull on an elongated transcript, see Fig. 7.16a,b, and investigated the probability of termination under load with various termination sequences (with a U-tract and with or without various hairpin sequences). They observed that in the presence of a U-tract only, the probability of termination was an increasing function of the tension on the RNA, see Fig. 7.16c. This increase in termination efficiency results from a decrease ΔG_{hybrid} of the activation energy for termination that can be accounted for, see Eq. 7.2, by the slippage of the mRNA by an amount δ_{hybrid} corresponding to about one nucleotide:

$$\Delta G_{hybrid} = F\delta_{hybrid} - T\Delta S_{ssRNA} = \int_0^{\delta_{hybrid}} x_{ssRNA} dF \tag{7.8}$$

Slippage of the mRNA by one nucleotide would be enough to stall transcription by removing the $3'$ end of the RNA transcript from the RNAP active site.

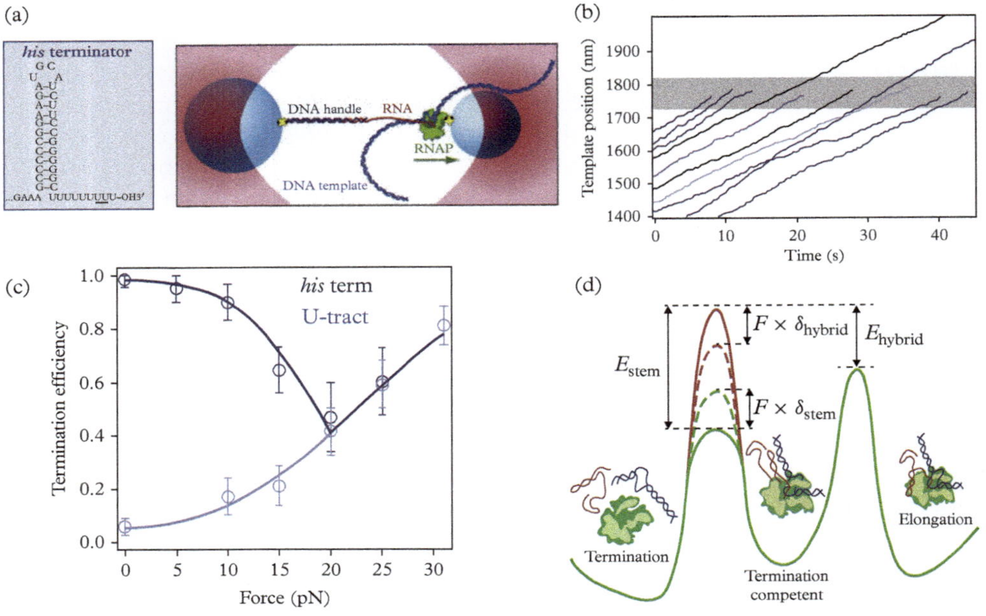

Fig. 7.16 Single-molecule study of transcription termination. *(a) A stalled RNAP complex was anchored to a bead by the biotin-labelled RNAP and to another by hybridization of a bead-anchored ssDNA to the complementary first twenty-five nucleotides of the RNA transcript. The use of a dual-beam optical trap allowed pulling on the RNA transcript with a preset force. Particular termination sequences were tested such as the his-terminator shown on the left, which consists of a GC-rich hairpin followed by a poly-U tract. (b) The extension vs time trace of RNA transcripts demonstrates that transcription does sometimes proceed past the termination sequence (region shown in grey). (c) The efficiency (probability) of termination as a function of force for two different termination sequences: a poly-U tract and the his-terminator shown in (a). Notice the continuous increase with force of the termination efficiency when a poly-U tract is used. As shown in (d), this can be accounted for by a decrease of the activation energy for termination (in red) by about $F\delta_{hybrid}$ (with δ_{hybrid} about one nucleotide) due to slippage of the poly-U tract on its complementary poly-A tract on the DNA template. In the presence of the his-terminator, the termination efficiency is increased due to a lowering of the activation barrier by an energy E_{stem} due to the presence of the hairpin. However, when pulling on the RNA transcript, the hairpin is frayed by an amount δ_{stem} (of about four nucleotides) which reduces the efficiency of the hairpin (i.e., increases the barrier to termination) by about $F\delta_{stem}$. Beyond a critical force of about 15 pN, the hairpin is fully open and the termination efficiency of a his-terminator sequence is the same as that of a poly-U tract. Adapted from [356].*

In the presence of both a U-tract and terminating hairpin, the termination probability displayed bi-phasic behaviour. Termination efficiency decreased with increasing force, see Fig. 7.16c, until the hairpin was fully opened. From that point on, it increased as in the absence of a hairpin (with a U-tract only). This bi-phasic behaviour can be explained by a decrease of the termination barrier by an amount E_{stem} due to hairpin formation, see Fig. 7.16d. When the hairpin is under tension, its base can fray by around four nucleotides, which reduces E_{stem} and increases the termination activation energy. Above the critical force for hairpin unwinding (about 15 pN, see Chapter 4), the tension on the RNA transcript acts only on the U-tract, causing its slippage, as described above, and increasing termination efficiency.

Single-molecule *in vitro* experiments have thus illuminated many aspects of transcription. As shown in this section, they have provided a much better and more accurate description of the various stages of transcription, from formation of the open complex to abortive initiation, via transcription elongation, and ending with transcription termination.

7.7 *In vivo* studies of transcription

The study of transcription *in vivo* has been conducted by the group of R. Singer in a series of pioneering experiments [185, 357, 358, 356]. In order to monitor single mRNA in live cells and overcome the strong auto-fluorescence of eukaryotic cells, this group introduced in the studied gene a cassette of twenty-four repeats of the same motif which upon transcription yields a 3D mRNA structure that is recognized by an MS2 protein fused to a fluorescent GFP protein. Thus, single mRNA in the cell can be detected and tracked, as they are labelled by as many as twenty-four fluorescent molecules [185, 357]. To localize a given gene in the nucleus, a similar cassette of 256 lacO repeats was inserted upstream of the gene of interest. These DNA segments are recognized by a LacI fused to a CFP (or RFP) fluorescent protein, creating a strong blue (or red) fluorescent signal at the site of insertion, i.e., next to the gene of interest [358]. A stable cell line was engineered containing at a single genomic locus 200 repeats of a gene cassette, each containing 256 lacO repeats followed by a transcription unit, see Fig. 7.17j. That unit comprised an inducible promoter driving the expression of a functional gene having at its 3′ untranslated region (3′-UTR) a segment of twenty-four MS2 repeats. Finally, to monitor transcription, single RNA PolII proteins were labelled with a YFP fluorescent protein. By counting the number of mRNA following induction, it was determined that there were on average two RNA polymerases per transcription unit. Hence *in-vivo* transcription of the induced locus involved about 400 RNA PolII transcribing 200 identical genes in tandem. The level of fluorescence associated with these transcribing RNA polymerases was high enough above the background to allow for the monitoring of transcription at this single engineered locus.

In vivo transcription was studied through Fluorescence Recovery after Photobleaching (FRAP). The engineered locus was identified by the fluorescence from Red Fluorescent Protein (RFP) fused to the lactose repressor (lacI) that binds to a lacO site. Upon induction of the promoter, transcription was detected by a spot in the YFP channel co-localized with the RFP spot, see Fig. 7.17b,c. The YFP spot was bleached, see Fig. 7.17d, and the recovery of fluorescence at that spot was monitored. This increase in fluorescence is due to the recruitment of new YFP labelled PolIIs to the transcription locus. Additional experiments were conducted that measured the fluorescence recovery

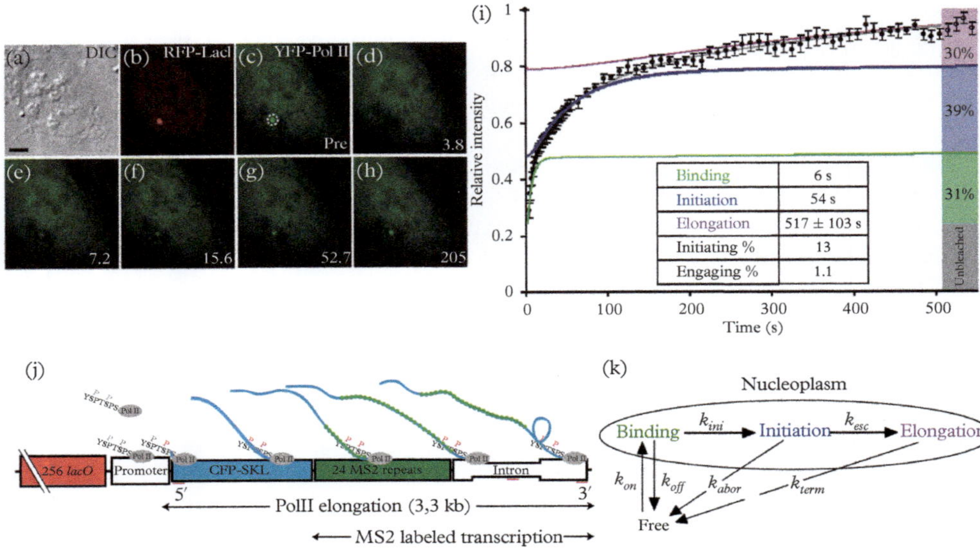

Fig. 7.17 ***In vivo*** **study of transcription at a single locus.** *This locus consists of about 200 repeats of a cassette (j) comprising 256 lacO sites and an inducible promoter driving the transcription of a gene having in its 3′-UTR region a tandem of twenty-four MS2 binding motifs. That locus is stably integrated in the genome of a cell (a) which also expresses a PolII RNA polymerase labelled with YFP. It is identified by binding of a LacI-RFP protein (b). Upon induction, transcription is visible in the YFP channel by the recruitment of about two RNA-pol per transcription unit (i.e., a total of about 400 enzymes per locus): the green spot in (c) co-localized with the LacO sites shown in (b). That fluorescence is bleached (c), and its recovery is monitored as new labelled RNA-PolII are recruited to the transcriptions sites (d–h). An analysis of the fluorescence signal (i) recovered after photobleaching (FRAP), allows one to deduce the various constants of the transcription model shown in (k), see section 7.7. Adapted from [358].*

following bleaching of the GFP fluorescence signal, which measures the insertion of fluorescent MS2-GFP proteins on the newly synthesized mRNAs.

Alternatively, the fading of GFP as mRNA transcripts are released was monitored using a photo-activated variant of GFP fused to MS2-proteins [358]. These data (and others) were fitted to a kinetic model of transcription, see Fig. 7.17k, that involved three successive states—binding, abortive initiation, and elongation—with possible dissociation from each one of these states:

$$\begin{aligned}
\frac{d[Binding]}{dt} &= k_{on}[Free] - k_{off}[Binding] - k_{init}[Binding] \\
\frac{d[Initiation]}{dt} &= k_{init}[Binding] - k_{abor}[Initiation] - k_{esc}[Initiation] \\
\frac{d[Elongation]}{dt} &= k_{esc}[Initiation] - k_{term}[Elongation]
\end{aligned} \tag{7.9}$$

The best fit rate constants were: $k_{off} = 0.145\ \text{s}^{-1}$, $k_{init} = 0.0216\ \text{s}^{-1}$, $k_{abor} = 0.0170\ \text{s}^{-1}$, $k_{esc} = 0.00159\ \text{s}^{-1}$, $k_{term} = 0.0193\ \text{s}^{-1}$, from which one deduces that PolII transcribes the 3.3 kbp gene in $1/k_{term} = 517$ s, i.e., at an average rate of 6.4 bp/s. This is an unexpectedly slow transcription rate. Further analysis of the dynamics of mRNA release following elongation (by measuring the MS2-GFP signal) revealed the existence of two time scales characterizing that process: a short one associated with active elongation at a rate of 72 bp/s, and a longer one associated with pausing during elongation with an average pause time of about 4–5 min. Another surprising outcome of these investigations was the revelation that transcription is rather inefficient. At steady state, only 13% $(= k_{init}/(k_{init} + k_{off}))$ of bound PolII go into abortive initiation and of those only 8.6% $(= k_{esc}/(k_{esc} + k_{abor}))$ go into elongation, i.e., only 1% of bound enzymes actually proceed to elongation!

These *in vivo* studies exemplify the power of fluorescent methods for the characterization of dynamic processes in live cells.

7.8 Conclusions and perspectives

In this chapter we have described the use of single-molecule manipulation and visualization methods to study the mechanism of some DNA and RNA polymerases. Much has been learned on the behaviour of an isolated polymerase on a DNA template under various tensions and torques. The stage is now set for the investigation of more complex systems such as the replisome [359], transcription regulation, and DNA repair (such as DNA transcription-coupled repair [360]). The coupling between manipulation (which best addresses and characterizes the function of the enzymes) and fluorescence visualization methods (to understand the stoichiometry of the complexes or to reveal structural changes during the enzymatic cycle) will shed new light on these systems which are very difficult to study using bulk enzymatic methods. Finally, the development of more sensitive fluorescent probes may allow the study of transcription *in vivo* at native transcription loci in prokaryotes and possibly even eukaryotes.

8

DNA Helicases

8.1 Introduction to helicases

Helicases are enzymes (proteins or subdomains of proteins) whose role is to unwind the double helix of dsDNA, dsRNA, or DNA/RNA heteroduplexes (hence, their name), in order to provide other enzymes with an ssDNA or ssRNA substrate on which to perform their respective function. DNA helicases are involved in DNA replication, recombination, and repair [361, 362, 363, 364, 365], whereas RNA helicases are involved in transcription termination, translation initiation, and RNA splicing, among other roles [366]. As part of this unwinding activity, helicases translocate along their substrate using NTP hydrolysis, making them true molecular motors. This unwinding of duplex nucleic acids associated with so many different roles in the cell cycle, makes helicases ubiquitous in the tree of life [363]. The term "helicase" covers a wide variety of molecules whose genetic sequences are hypothesized to represent about 1% of each individual genome [367]. We will focus in this chapter on DNA helicases, for they are the main ones to have been studied with single-molecule approaches (see, however, [368]). Complete descriptions of DNA and RNA helicase families can be found in the article of M. Spies [369], and the citations therein.

DNA helicases are ATP-dependent enzymes capable of unwinding double-stranded DNA (dsDNA). They can be classified according to diverse parameters among which is their stoichiometry in the active state: helicases can be monomeric (e.g., PcrA), dimeric (e.g., Rep), trimeric (e.g., RecBCD), or hexameric entities (e.g., DnaB), but *in vivo* they usually interact with a number of other factors in complex molecular assemblies, limiting the pertinence of this classification.

Another classification is based on the polarity of translocation along the DNA strand, which has classically been deduced from the overhang required to load the enzyme at the ss–dsDNA junction, where unwinding is initiated. Helicases that translocate along DNAs from the $3'$ to the $5'$ end are designated as $3'$–$5'$ helicases, and often need a $3'$ overhang to load onto the junction. Similarly, enzymes that translocate and unwind in the $5'$ to $3'$ direction are designated as $5'$–$3'$ helicases, and often require a $5'$ overhang loading site.

Structure and sequence data suggest another classification of helicases into six superfamilies (SF1–SF6), with toroidal ring-forming enzymes sorted into SF3 to SF6, while

Single-Molecule Studies of Nucleic Acids and Their Proteins. David Bensimon, Vincent Croquette, Jean-François Allemand, Xavier Michalet, Terence Strick 2019. Published in 2019 by Oxford University Press.
DOI: 10.1093/oso/9780198530923.001.0001

non-ring-forming ones are found in SF1 and SF2. Most 3′–5′ helicases are members of superfamilies 1 and 2. These two families have similar sets of seven conserved structural motifs, and the proteins are usually considered to be dimeric. Superfamily I (SF1) enzymes, with members such as UvrD or Rep, exhibit a large contact area with the ssDNA. For this reason, they are classified as ssDNA translocases. They melt downstream dsDNA as they proceed, see sections 8.4 and 8.5 below.

Superfamily II (SF2) enzymes, with members such as RecG, RecQ, NS3, PriA, interact nonspecifically with the phosphodiester backbone and are assumed to be dsDNA translocases, see section 8.3.

Superfamily III (SF3) comprises helicases from small DNA and RNA viruses, while superfamilies IV (SF4) and V (SF5) contain 5′–3′ hexameric helicases related either to the *E.coli* replicative helicase DnaB (SF4) or the bacterial termination factor Rho (SF5).

Beyond such classifications, a full understanding of the activity of these motors in a living cell requires an understanding of their assembly, structure, and the conformation(s) of their nucleic acid target, their sequence specificity if any, and their binding modality. Equally important is the mechanism used during their translocation, their directionality, the number of bases they can unwind before dissociating, their velocity, their step size and stoichiometry of NTP hydrolysis (how many base pairs are unwound per NTP consumed), and, of course, their coupling with other proteins. Many of these elements are related to the mechanical aspects of their activity, and thus single-molecule micromanipulations are very useful tools to investigate these issues, together with classical bulk assays.

A helicase operating in isolation is difficult to assay, since the ssDNA intermediates of the unwinding reaction are transient and may re-anneal in the wake of the enzyme. To monitor a helicase, one has to use cofactors: proteins (e.g., Single-Strand Binding (SSB) proteins) or enzymes (e.g., Single-Strand Nucleases or the cell replication machinery) which trap or process the ssDNA generated by the helicase activity [370, 361, 362]. The characterization of a helicase therefore depends on the experimental conditions where it is assayed [370]. In fact, most helicases function poorly (inefficiently and with reduced processivity) when isolated from the macromolecular machinery and coupling factors with which they are intended to operate [370]. One of the challenges of studying helicases as molecular machines is thus to understand not only the mechanisms of the helicase itself, but also how these are altered in different *in vitro* and *in vivo* contexts.

From a mechanistic point of view, the most important issue concerning the function of helicases is the coupling between ATP hydrolysis and DNA unwinding. The simplest model assumes that the helicase is "passive": ATP hydrolysis is required only for translocation of the helicase which traps unbinding fluctuations of the DNA junction, see Fig. 8.1. DNA fraying may-be thermally induced, or the junction may be destabilized by its interactions with the helicase or other enzymes (e.g., polymerases active with the helicase at the replication fork [371]).

In other models, the helicase is more "active": it uses part of the energy of ATP hydrolysis to melt the dsDNA and translocate. One such model assumes that ATP hydrolysis is coupled with a switch in the affinity of the binding sites of a dimeric enzyme, from single-strand to double-strand DNA. In this "rolling model" of DNA

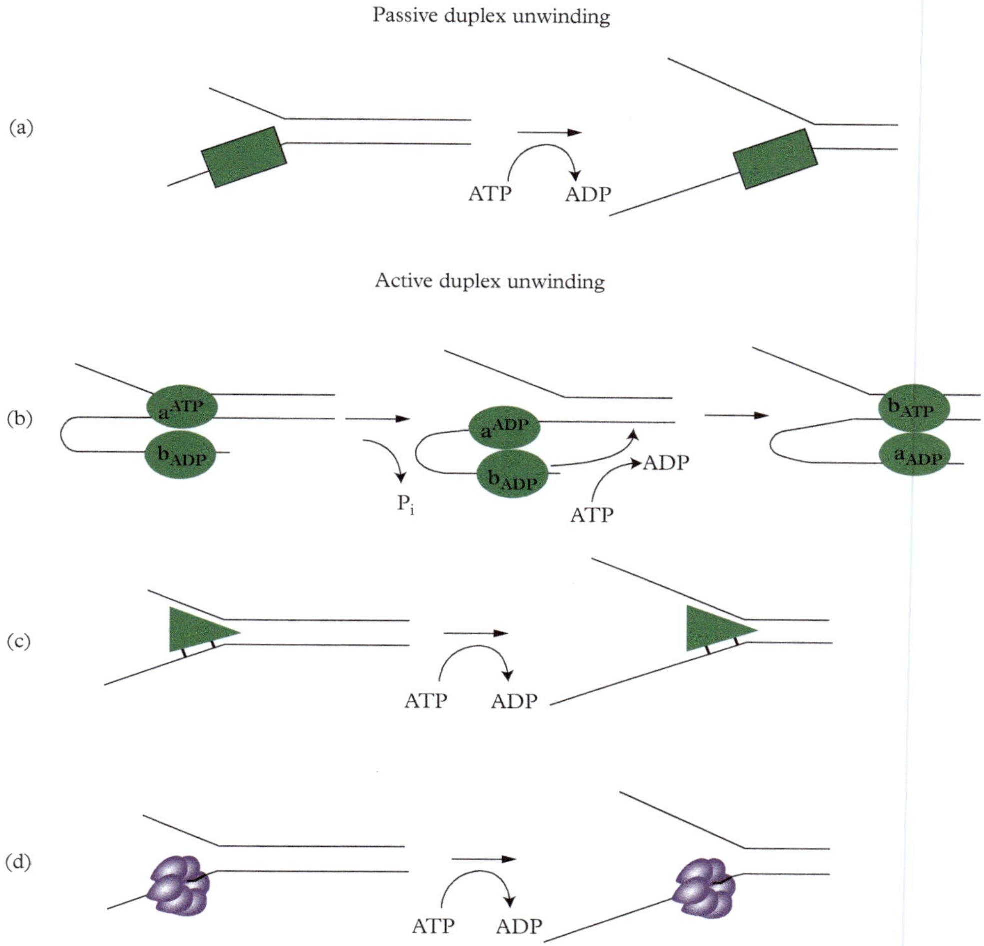

Fig. 8.1 Possible translocation mechanisms of helicases. *(a) The helicase can be passive, trapping fork fluctuations through an ATP-driven progression. Alternatively, helicases can actively unwind DNA as they progress along by a "rolling model" (b) where each monomer takes turns at unwinding the molecule by driving a "plough" through the DNA fork (c), or by pulling the DNA through the helicase as in some hexameric helicases (d). So far all these mechanisms are found in helicases, with the exception of the rolling model, which has not yet been observed in experiments. Adapted from [372].*

unwinding, see Fig. 8.1, in the absence of ATP, both subunits have a high affinity for ssDNA. Binding of ATP to the subunit upstream of the junction causes it to detach from ssDNA and rebind to the adjacent (downstream) dsDNA, which melts apart when ATP is hydrolysed. Product release reverts the system to its initial state, with the subunits having switched order. In this model, the enzymatic step size is at least as large as the number of DNA bases contacted by each monomer.

A second active model, the "inchworm model", assumes that ATP hydrolysis is used to translocate the enzyme along single- or double-stranded DNA, with the opening of the junction a consequence of that translocation. For example, if the helicase translocates along ssDNA (SF1 enzymes), it will unwind the junction like a snowplough, see Fig. 8.1. If it translocates on dsDNA (SF2 enzymes), it may use a wedge domain positioned at the junction to open the DNA like a wire-stripper.

In spite of these precise descriptions of the enzymatic mechanism, exactly how active or passive an enzyme is has rested on problematic estimations of the free energy of interaction of the enzyme with the DNA fork. To cut this Gordian knot, we will define a passive enzyme as one whose rate is very sensitive to the DNA sequence and increases significantly when a force is acting to unzip the two DNA strands. By contrast, the unwinding rate of an active enzyme will be insensitive to the DNA sequence and unchanged (or even possibly decreased) when a force is acting to separate the two strands.

Hexameric helicases raise yet another problem: how is their action coordinated to induce DNA unwinding? An analogy with the F1-ATP-ase rotary motor [373] has been proposed, in which three of the units have no ATP-ase activity, whereas the three other units are successively engaged in ATP binding, hydrolysis, and product release. The affinity of the subunits for the various substrates is coupled to their interaction with the DNA molecule, one strand of which is being threaded through the centre of the hexameric unit.

In this chapter, our goal is to illustrate the use of single-molecule techniques in elucidating the mechanisms of various helicases. While we will deal with specific helicases, the methods used to study and analyse them may have applications for a broader array of enzymes.

8.2 Bulk versus single-molecule assays: the case of RecBCD

To address helicase mechanisms in bulk experiments is very difficult, and hardly convincing, since the presence of coupling factors and their own binding dynamics to ssDNA may interfere with the analysis of enzymatic activity. For instance, the interaction between DNA polymerase and helicase at the replication fork increases the unwinding rate by between 15- and 20-fold [371].

One helicase, RecBCD, is, however, free of these problems. RecBCD consists of two helicases, RecB and RecD. The RecB subunit is a helicase with $3'$–$5'$ directionality. Its first two domains define it as a member of the SF1 family. Its third domain is a nuclease domain, connected to the remainder of the protein by a long linker region of about seventy amino acids. The RecD subunit is also a SF1 helicase, but with a rather uncommon $5'$–$3'$ unwinding activity. The RecC protein contacts both strands of DNA and splits them (like the shovel of a snowplough) before feeding the $3'$ strand to the RecB protein and the $5'$ tail to the RecD subunit. The two helicase motors apparently work independently of each other, and because of their opposite polarity they both pull the anti-parallel DNA strands in. A detailed picture of the mechanism of coupling between these units has emerged from single-molecule experiments and structural data.

Bulk measurements of the activity of RecBCD (estimated by the amount of DNA hydrolysed by the RecB nuclease domain) yield a very large unwinding rate [374] ($v_m \sim 930$ bp/s), a huge processivity [375] (about $P_m = 27{,}000 \pm 3{,}000$ bp), and a rate of ATP hydrolysis [376] (with $k_{cat} \sim 740\ \text{s}^{-1}$) all characterized by the same Michaelis–Menten coefficient [374, 376] ($K_M \sim 100\ \mu$M). These results imply that the unwinding rate and processivity are controlled by ATP hydrolysis, and that about one ATP is required to unwind a single base pair. However, in these bulk experiments an unknown portion of the enzymatic population may be inactive, which may affect these estimates.

To address this issue, Bianco *et al.* [377] have directly visualized the motion of RecBCD on single DNA molecules. In their experiment, a DNA molecule was tethered at one end to a small polystyrene bead held in optical tweezers. The DNA molecule was stained with a fluorescent dye (YOYO1), stretched by the flow of the enzyme-containing buffer, and visualized on an intensified CCD camera. As it is unwound and degraded by RecBCD, its extension shortens, see Fig. 8.2. By monitoring the change in extension as a function of time, one can directly measure the rate and processivity of RecBCD. A surprising result of these single-molecule investigations was the extreme variability in the unwinding rates measured for various enzymes on the same substrate. Although the rate of DNA unwinding by any individual RecBCD was uniform (within experimental error) on any given DNA molecule, the rate for different enzymes deviated by up to 5-fold. Subtle changes in the conformation of the complex as it is loaded on DNA may result in this large difference in unwinding rates. In spite of these large enzyme-to-enzyme fluctuations, the average rate ($v_m \sim 970$ bp/s at 37°C), average processivity ($P = 38{,}000 \pm 5{,}700$ bp), and Michaelis–Menten coefficient ($K_M \sim 150\ \mu$M) were all compatible with bulk measurements. This is a comforting result, since for this enzyme no coupling factors are required to observe its unwinding activity, i.e., to prevent reannealing of the two DNA strands. Hence, direct observations of DNA unwinding and degradation by a single enzyme and bulk measurements of DNA unwinding are expected to yield similar results (as we shall see, that need not always be the case).

Another result from these studies concerns the activity of RecBCD after it encounters a χ sequence (5′-GCTGGTGG-3′). Recognition of χ was known to reduce both nuclease activity and translocation speed of RecBCD, and to activate RecA loading. What the single-molecule experiments have shown is that upon encountering a χ sequence, the enzyme paused for a few seconds and then continued to translocate as a full trimeric complex [378] and unwind DNA, but *on average* at about half the previous rate (for some enzymes, the rate after χ could be one tenth of the prior rate). While before χ both single strands were degraded, after χ the 3′ strand is not degraded: its end is sequestered by the enzyme, and as RecBCD translocates a ssDNA loop is extruded (visible as an increasing bright spot translocating along the fluorescently labelled DNA) [379].

A full mechanistic picture of the working of RecBCD emerges when these single-molecule results are combined with the crystallographic data of RecBCD. The location of the nuclease active site (the third domain of RecB) allows processive hydrolysis of the 3′ tail, as it emerges from the RecC subunit. However, proximity of the 5′ tail would enable it to compete with the 3′ tail for binding at the nuclease. Thus, before χ, both strands can be degraded by the nuclease activity of RecB. Upon encountering χ, the

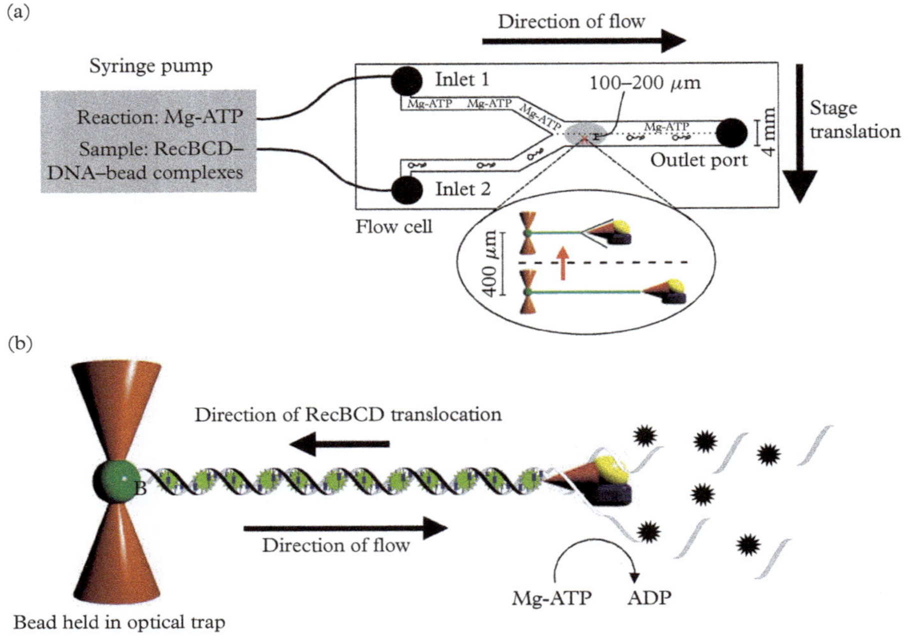

Fig. 8.2 Single-molecule study of RecBCD. *(a) The flow cell introduced by the Kowalczykowski group to study the interaction between DNA and RecBCD. It consists of a microfluidic cell with two (or more) inlets. Through one inlet a solution of ATP, dye, Mg^{++}, etc., can be introduced, and through the other, DNA bound to transparent beads and possibly proteins (such as RecBCD). In this laminar flow regime, the two solutions do not mix. One bead is captured by optical tweezers. The cell is then translated, and the trapped DNA molecule brought into contact with ATP (to launch the unwinding reaction) and/or a dye (to stain the molecule). (b) As the molecule is unwound by the progression of RecBCD, it is degraded by the RecB exonuclease activity. Its length can be deduced from the fluorescence of the stained and stretched remaining dsDNA. Hence, the progression of the helicase complex on the molecule can be monitored in real time. Adapted from [377].*

RecC subunit binds tightly to the 3′ tail, preventing its further digestion. This induces a conformational change in RecD that slows it down. The 5′ tail is now able to access the nuclease site more frequently, and is degraded more fully. The enzyme continues to advance along the DNA, extruding a loop out from the RecB subunit that can be loaded with RecA proteins.

8.3 Active and passive unwinding mechanisms: RecQ versus gp41

We have suggested above to define active helicases as those whose rate is force- and sequence-independent, and passive ones as those whose rate is force- and sequence-dependent. In the following, we shall exemplify these features on two helicases which are

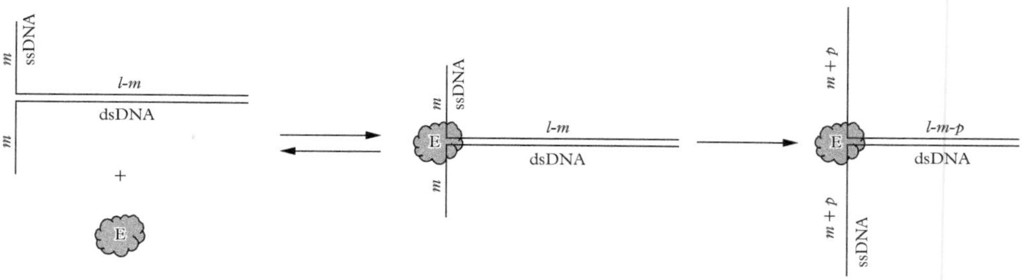

Fig. 8.3 Michaelis–Menten scheme for the action of a helicase on a DNA fork *during one enzymatic cycle. The enzyme may bind or unbind at the fork between a dsDNA of $l-m$ bp and two ssDNA of m nucleotides. The progression of the enzyme by one step opens the fork by p bp.*

typical of each class: RecQ, an active helicase implicated in recombination, and the T4 replication helicase, gp41, a passive one.

First, let us make the trivial remark that helicases are enzymes, and as such they act by lowering the activation barrier of the reaction they catalyse. This reaction consists in unwinding by p bp a DNA molecule consisting of two ssDNA of m nucleotides and a dsDNA of $l-m$ bp, see Fig. 8.3:

$$(ss_m)_2 \cdot E \cdot ds_{l-m}DNA + ATP \underset{k_{-1}}{\overset{k_1}{\rightleftharpoons}} (ss_m)_2 \cdot E^* \cdot ds_{l-m}DNA \tag{8.1}$$

$$\overset{k_2}{\longrightarrow} (ss_{m+p})_2 \cdot E \cdot ds_{l-m-p}DNA + ADP + P \tag{8.2}$$

where k_1, k_{-1} are binding/unbinding rate constants of ATP and k_2 is the rate constant for ATP hydrolysis/translocation, which it is reasonable to assume is force-dependent [380].

$$k_2(F) = \nu_m e^{-\Delta G(F)/k_B T} \tag{8.3}$$

where ν_m is the maximal value of the rate constant reached when the associated activation barrier vanishes: $\Delta G(F) = 0$. From that point of view, the difference between an active and a passive helicase rests on the amplitude of $\Delta G(0)$. If $\Delta G(0) \ll k_B T = 4$ pN·nm (where T is the temperature and k_B is Boltzmann's constant), then the helicase is active (the barrier is negligible). On the other hand, if $\Delta G(0)$ is of the order of a few $k_B T$, then hopping over the activation barrier may be the rate-limiting step in the reaction and the enzyme is passive. In such a case, a variation of $\Delta G(0)$ with the DNA sequence will affect the unwinding rate of the enzyme. Second, notice that application of an unzipping force on a DNA fork also lowers the activation barrier to DNA melting to a point where, at $F \sim 14$ pN, the dsDNA may spontaneously unwind within the experimental time scale [26]. In general, therefore, we may write the various contributions to $\Delta G(F)$ as:

$$\Delta G(F) = n(\Delta G_0 - \Delta G_{heli} - \Delta G_F) \tag{8.4}$$

where n is the number of unwound bp to reach the transition state, ΔG_0 their binding mean free energy, and ΔG_{heli}, ΔG_F the reduction in that binding energy due to the helicase and the unzipping force, respectively. When the rate-limiting step is set by hopping over the activation barrier to melting, as is the case for a passive enzyme, one expects that application of a force on the DNA fork will reduce this barrier and result in an increase in the enzyme's unwinding rate, see Fig. 8.4a. When the enzyme is active, the effect of the force may be negligible (as expected for an "inchworm" active model [363]) or counterproductive, i.e., slowing down the enzyme (as expected for an active "rolling" model [363]). As we shall see, these qualitative considerations, coupled to single-molecule measurements of the rate of a single helicase unwinding a DNA fork under a given tension, are enough to establish the mechanism of the studied helicase.

To monitor the activity of a single helicase, one can use a magnetic trap [17, 249, 380], see section 2.4.5. Briefly, a DNA hairpin (with duplex length of up to 7,000 bp) is tethered by its extremities to a glass surface and a magnetic bead. Small magnets positioned above the sample apply a constant force on the molecule. Unwinding of the hairpin by a single helicase results in an increase of the end-to-end distance of the DNA

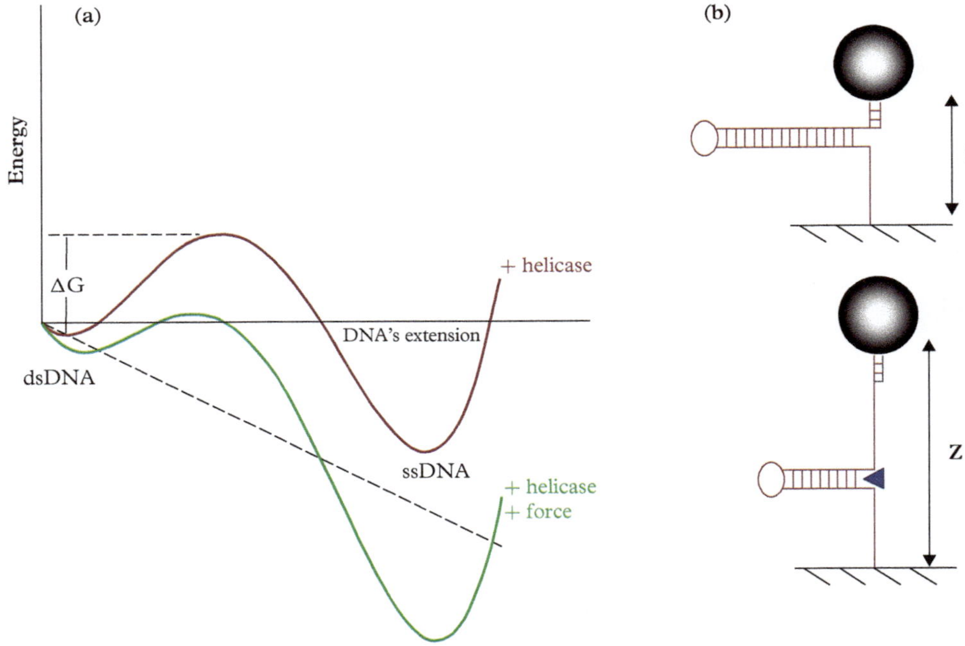

Fig. 8.4 Effect of the tension on a DNA fork during helicase unwinding. *(a) Reduction in the activation barrier of unwinding upon the application of a force on the DNA fork. (b) Schematics of the experiments investigating the translocation of a helicase on a hairpin substrate submitted to an unzipping force with a magnetic trap.*

molecule measured as a change of the distance between the bead and the surface, see Fig. 8.4b. Since we can measure the force–extension behaviour of stretched ssDNA, we can convert the observed increase in the end-to-end distance into the number of unwound base pairs. The mechanical unfolding of the hairpin construct in the absence of helicase can be characterized first: above a critical force of $F_u = 14 \pm 1$ pN the hairpin unfolds, while it remains otherwise stably folded at forces $F_c < 11 \pm 1$ pN. Thus, at forces $F < F_c$, any unfolding observed in the presence of a helicase is a result of its activity. Indeed, in its absence, the extension of the DNA molecule remains constant and equal to that of the folded hairpin.

RecQ, a 3′–5′ helicase belonging (as NS3, PriA, and RecG) to the SF2 superfamily of helicases, has been studied. Its eukaryotic homologues, BLM and WRN, are implicated in some autosomal diseases (Werner and Bloom syndrome) characterized by genomic instabilities, probably resulting from illegitimate recombination events. Adding RecQ in a solution with hairpin molecules under moderate tension ($F < F_c$) results in processive opening of the hairpin at a constant unwinding rate $v_U \sim 68$ bp/s, see Fig. 8.5a. By mechanically and transiently increasing the force (to a value $F > F_u$), the hairpin is completely opened during an unwinding event, allowing one to deduce the enzyme translocation rate on ssDNA v_{ss}, see Fig. 8.5b. As the force is reduced (down to its initial value $F < F_c$), the hairpin refolds but is shorter by an amount δx corresponding to the distance travelled by the helicase on ssDNA. Notice that this rate is the same as the unwinding rate of RecQ. In fact, as shown in Fig. 8.6, the unwinding rate of RecQ varies little with force.

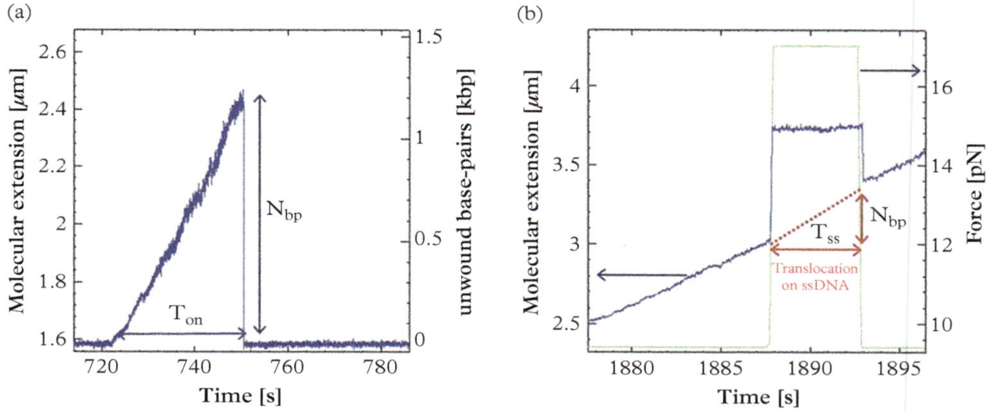

Fig. 8.5 RecQ translocation on ds- and ssDNA. *(a) Observing RecQ unwinding of the DNA fork through increase of the extension of the tethering ssDNA portions. The enzyme processivity is given by the number of base pairs unwound N_{bp}, the unwinding rate v_U of the enzyme by: $v_U = N_{bp}/T_{on}$. (b) Investigating RecQ translocation on ssDNA when the hairpin is transiently opened by an increase in the unzipping force. The velocity of RecQ translation is obtained as: $v_{ss} = N_{bp}/T_{ss}$. Adapted from [380] and [381].*

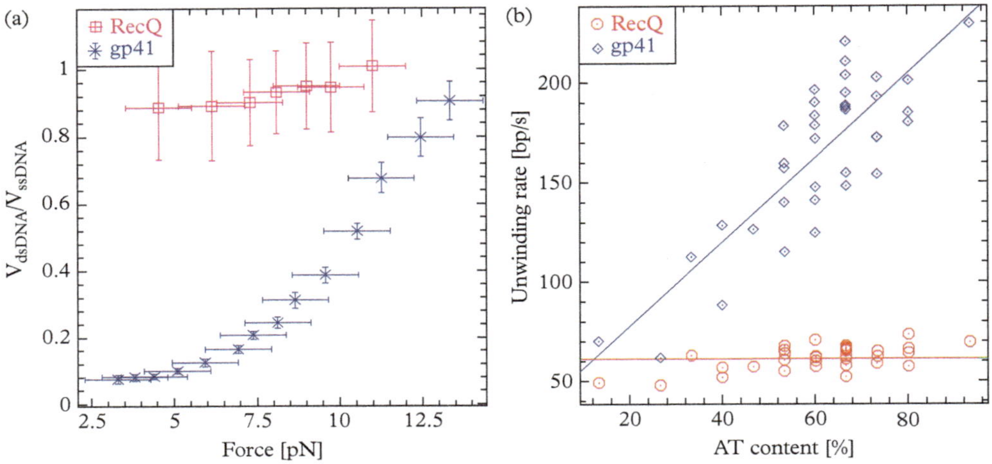

Fig. 8.6 Active (RecQ) versus passive (gp41) helicase. *(a) Variation with the unzipping force F acting on the DNA fork of the mean unwinding rate of RecQ (red points) and gp41 (blue points). (b) Variation with the local AT content of the dsDNA of the unwinding rate of RecQ (red points) and gp41 (blue points) at $F = 9$ pN and linear fits. Adapted from [381].*

This is in stark contrast with the hexameric helicase of the T4 replication complex, gp41 [380, 381]. In this case, the unwinding rate varied by almost an order of magnitude between $F = 0$ and $F = F_c$, see Fig. 8.6a. As argued previously, gp41 can be fitted by a purely passive model of helicase translocation [380]. One thus concludes that RecQ is an active helicase, processing along DNA by an inchworm-like mechanism. This conclusion is reinforced when considering the variation of helicase activity with DNA sequence (or AT content). As shown in Fig. 8.6b, while the rate of unwinding by RecQ is independent of the local AT content of the hairpin (averaged over 500 bp) the rate of unwinding of the same hairpin by gp41 is highly correlated with the local percentage of AT bp.

We conclude that these two helicases, RecQ and gp41, provide ideal examples of two different unwinding mechanisms: an active one for RecQ and a passive one for gp41. The mechanisms of other helicases can be studied by similar means and assessed against these pure cases.

8.4 The UvrD helicase studied by single-molecule manipulations

UvrD, a member of the SF1 superfamily, plays a crucial role in nucleotide excision repair and methyl-directed mismatch repair [363, 364, 365]. It has been shown to initiate unwinding from a 3′-end ssDNA tail, or from a gap or a nick, and to translocate along ssDNA in a 3′–5′ direction [382, 383]. Most bulk assays of its activity require the presence of co-factors, such as SSB proteins or a high concentration of UvrD (which also has a high affinity for ssDNA) to prevent reannealing of the separated strands in the wake of the advancing helicase complex. Thus, without SSB proteins, full unwinding of DNA is observed with near stoichiometric ratios of UvrD monomers/nucleotides [383] (between 4% and 60%). Clearly, in these conditions it is difficult to assess the intrinsic unwinding rate, processivity, or efficiency of an active enzymatic complex.

8.4.1 Quench flow studies

To overcome some of these difficulties, rapid quench flow techniques have been used to study the unwinding of relatively short oligonucleotides, with duplex regions ranging between 10 and 40 bp, preloaded with UvrD complexes [384]. The unwinding reaction is initiated by rapid mixing with ATP and a large excess of free ssDNA to trap unbound enzymes. It is quenched at various times T by rapid mixing with Na_3EDTA, and the fraction $F(T)$ of unwound DNA is then quantified, see Fig. 8.7. In these experiments, it is assumed that the helicase complex itself acts as a kinetic barrier against quick rehybridization of the single strands in its wake. Although not fully justified (less than 20% of 40 bp oligonucleotides are eventually unwound), this assumption allows the rate and step size of the helicase complex to be extracted quite easily.

From the fraction $F(T)$ of unwound DNA, one can calculate the probability $p(T)$ of unwinding an oligonucleotide of length l in a time T: $p(T) = dF(T)/dT$. From this probability distribution, one can compute the average time for unwinding $\langle T \rangle = \int_0^T tp(t)dt$ and its second moment $\langle T^2 \rangle = \int_0^T t^2 p(t)dt$. These moments yield an estimate of the mean enzymatic rate, $\langle v \rangle = l/\langle T \rangle$, of the mean time between steps, τ, and of the number of steps required for full unwinding, N. Indeed, assume that the time t_i between steps $i-1$ and i is characteristic of a Poisson process:

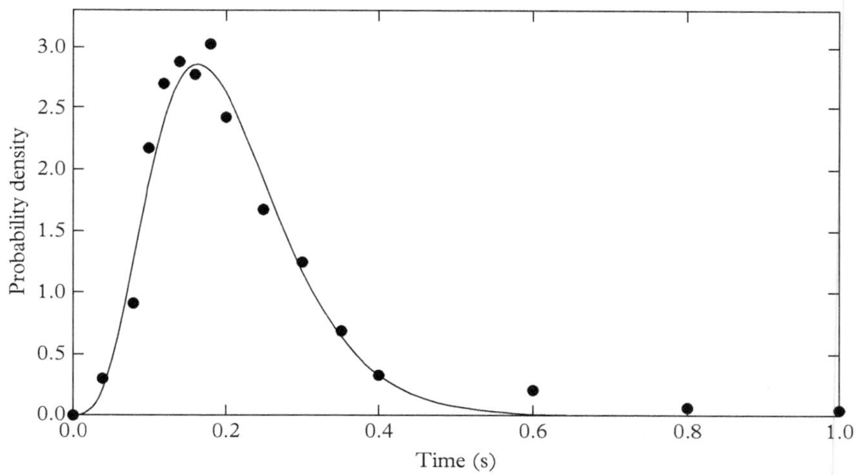

Fig. 8.7 DNA unwinding from quench flow studies. *Probability density $p(t)$ of unwinding a DNA molecule of length $l = 18$ bp. The data points are best fitted with a distribution $\gamma(t) \sim t^\alpha \exp(-t/\tau)$ (with $\alpha \sim 4$ and $\tau \sim 41$ ms), which implies that the step size is about 3.6 bp $(= l/(\alpha+1))$. Data extracted from [384].*

$$p_1(t_i) = \exp(-t_i/\tau)/\tau$$

If N steps are required to unwind a given oligonucleotide of length l in a time $T = \sum_1^N t_i$ (assuming no rehybridization), the average time for unwinding is:

$$\langle T \rangle = N\tau$$

and its variance is:

$$\langle T^2 \rangle - \langle T \rangle^2 = N\tau^2 = \frac{\langle T \rangle^2}{N}$$

The number of steps necessary to fully unwind an oligonucleotide of length l is then:

$$N = (\langle T^2 \rangle / \langle T \rangle^2 - 1)^{-1}$$

The step size is $d = l/N$ and the mean time between steps is $\tau = \langle T \rangle / N$. More generally, the probability distribution $p(t)$ of unwinding an oligonucleotide of length l in N steps (taken at times: t_i) within a total time span $t = \sum_1^N t_i$ is a convolution of N Poisson processes:

$$p(t) = \int_0^t dt_1 \int_{t_1}^t dt_2 \cdots \int_{t_{N-1}}^t dt_N P_1(t_1) P_1(t_2) \ldots P_1(t_N) \tag{8.5}$$

$$= \frac{t^{N-1} e^{-t/\tau}}{\tau^N (N-1)!} \equiv \gamma(t) \tag{8.6}$$

From the data of Fig. 8.7, one thus deduces that the step size of UvrD is $d = 3.6 \pm 0.2$ bp [384], the mean time between steps is $\tau \sim 41$ ms, and the mean unwinding rate is $\langle v_U \rangle \sim 87$ bp/s [385]. This analysis is, of course, invalid if reannealing in the wake of the advancing complex occurs in these experiments. In that case, the observed fraction of unwound DNA is not simply related to the probability of unwinding anymore, but depends on the joint probability that the DNA has been fully unwound at time T and did not rehybridize until that time. Single-molecule methods, to be presented next, are useful in overcoming such limitations.

8.4.2 Single-molecule estimation of the UvrD step size

The magnetic trap system previously described (see section 2.4.5) can be used to pull on a single nicked DNA molecule anchored at one end to a surface and at the other to a small magnetic bead [386]. The change in length accompanying DNA unwinding (single- and double-stranded DNA have different extensions) is then used as a proxy for monitoring, in real time, the progression of a single helicase complex, see Fig. 8.8a (just as the opening of a hairpin can be used to monitor the progression of a helicase, see section 8.3). At a high enough tension ($F > 25$ pN), the mismatch between the stretched and unstretched single strands creates a kinetic barrier that prevents their reannealing in the wake of the enzyme as the DNA is unwound. This force-induced kinetic barrier allows one to study the activity of a single enzymatic complex at very low concentration (well below the enzyme's dissociation constant $K_d \sim 10$ nM) and without co-factors. Because at this force ssDNA is longer than dsDNA, unwinding (U) results in a continuous (ATP-dependent) increase in time of the length of the molecule, see Fig. 8.8b. The slope of this signal yields the unwinding rate v_U, its spatial extent—the number of base pairs unwound N_U and its duration—the on-time τ_U of the enzymatic complex. Upon dissociation, the DNA fork serves as a nucleation point for the quick rehybridization (H) of the two unwound strands, resulting in a rapid (ATP-independent) shortening of the molecular extension, see Fig. 8.8b.

UH bursts of unwinding activity are thus seen as bursts of increase (Unwinding) in the molecule's extension, immediately followed by rapid shortening (re-hybridization). The large temporal separation between those bursts (relative to their duration) makes it very unlikely that they are due to the simultaneous action of many enzymatic complexes. These bursts of unwinding result from successive opening of the dsDNA by d base pairs per enzymatic cycle.[1] Often the noise due to thermal fluctuations of the sensor (e.g., the magnetic bead) prevents the observation of individual cycles of unwinding. However, due to the different statistics and spectral properties of unwinding (shot noise) and thermal fluctuations ("white" noise) one may deduce the enzymatic step size (i.e., number of base pairs opened per cycle) from a spectral analysis of the unwinding bursts.

Let $x(t)$ be the time-dependent extension of DNA following unwinding within a time t by N enzymatic cycles (at various time points t_i $(i = 1, \ldots, N)$), each cycle resulting in unwinding of d base pairs:

$$x(t) = d\sum_{i=1}^{N}\theta(t - t_i) + \eta(t)$$

where $\theta(t)$ is the Heaviside function, which is unity for positive t and zero otherwise, and $\eta(t)$ is the Brownian noise of the sensor (the bead). The Fourier transform of the extension signal is given by:

$$\tilde{x}(\omega) = \frac{d}{i\omega}\sum_{i=1}^{N} e^{i\omega t_i} + \tilde{\eta}(\omega)$$

[1] Strictly speaking, d is the number of open bp between microscopic pauses, which might be linked with ATP binding, hydrolysis, or release, but may possibly be due to other causes. So one should view d more as an upper bound on the enzymatic step size, than as a definite measure of it. That would require the observation of a coupling between the pause time and the ATP concentration.

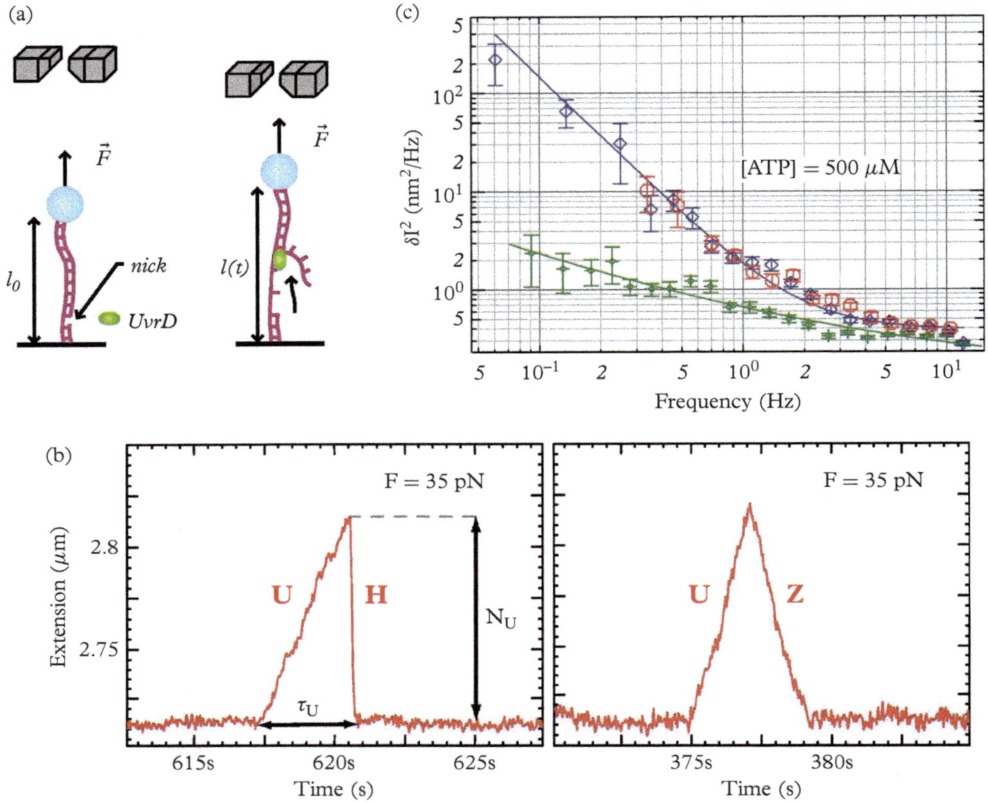

Fig. 8.8 Estimating the enzymatic step size from the noise in extension during unwinding events. *(a) Schematic description of the unwinding of a dsDNA under tension by a UvrD helicase from a nick in the molecule. At high enough force, the unwound free strand does not rehybridize behind the complex and the extension of the molecule increases. (b) Typical signals observed with UvrD. On the left: Unwinding (U) of the dsDNA results in a continuous increase of the extension from which the rate, the processivity, and the step size can be determined. Unbinding of the enzyme results in rehybridization (H) of the two strands. On the right: Unwinding (U) of the dsDNA is followed by strand switching of the enzyme, which results in an ATP-dependent re-zipping (Z) of the two strands in the wake of the helicase. (c) Fourier spectrum of the signal (unwinding data (U): blue and red points) and noise (data between bursts: green points). From the dependence of the signal spectrum at low frequencies, one can deduce the enzymatic step size. Adapted from [386].*

For a stochastic enzyme, the step times are uncorrelated; hence, the sum is over unit vectors pointing at random in the complex plane. As for a random walk, the mean of the sum is zero, while its variance is N. This allows one to compute the average power spectrum of the extension measurements, $\langle S_x(\omega)\rangle$:

$$\langle S_x(\omega)\rangle \equiv \frac{2|\tilde{x}(\omega)|^2}{t} = \frac{2d^2N}{\omega^2 t} + \langle S_\eta(\omega)\rangle \tag{8.7}$$

$$= \frac{2dv_U}{\omega^2} + \langle S_\eta(\omega)\rangle \tag{8.8}$$

where $v_U = Nd/t$ is the mean unwinding rate and $\langle S_\eta(\omega)\rangle$ is the noise power spectrum of a particle (a tethered bead) undergoing Brownian motion in a harmonic potential with stiffness k (the stiffness of the optical tweezers or of DNA in the case of a magnetic trap):

$$\langle S_\eta(\omega)\rangle = \frac{k_B T}{\pi k \omega_0} \frac{1}{1 + (\omega/\omega_0)^2}$$

where ω_0 is the roll-off frequency of the noise, which decreases with increasing drag on the system, see section 2.5. Importantly, at low frequencies ($\omega < \omega_0$), the power spectrum of the noise is flat (frequency-independent), whereas the contribution of the stepping process to the fluctuations in extension (the first term on the right of Eq. 8.8) increases as $1/\omega^2$. For sufficiently long traces, i.e., sufficiently low frequencies in the power spectrum, the noise power arising from random stepping will dominate the Brownian noise. The step size d can then be derived from a fit of the low-frequency part of $\langle S_x(\omega)\rangle$ to the functional form: $2dv_U/\omega^2$, see Fig. 8.8c. In practice, the extension fluctuations are obtained by subtracting a line of slope v_U, the average velocity, from the experimentally measured extension trace. The average power spectrum of several such traces can be fitted to obtain a step size: $d = 6 \pm 1.5$ bp, in rough agreement with quench flow estimates.

8.4.3 DNA reannealing in the wake of a helicase and strand-switching

We have seen how to extract unwinding rates, processivity and step size by analysing the unwinding events of UvrD. Surprisingly and unexpectedly, this helicase exhibits another type of unwinding burst. In these events, unwinding (U) is followed by a slow, ATP-dependent re-zipping (Z) of the two separated strands, see Fig. 8.8b. The reannealing rate v_Z obeys the same Michaelis–Menten kinetics as the unwinding rate v_U, suggesting that these (UZ) events are due to the enzyme-switching strands, moving 3′–5′ on the complementary ssDNA and limiting, thereby, the reannealing rate of the two separated strands in its back. This behaviour is missed in bulk (and quench flow) studies as they are averaged out by looking only at complete unwinding of the two strands. The similarity of the unwinding and reannealing rates suggests that the rate of UvrD on DNA is only slightly affected by the enzyme having to open the strands or by the fork closing in its wake, and points to an active mechanism of unwinding, as for RecQ (see section 8.3).

Single-molecule experiments yield the probability distribution of unwinding rates, which is Gaussian with a mean $v_U \sim 248$ bp/s and standard deviation $\sigma_U \sim 100$ bp/s.

It is interesting to notice that the spread of the velocity histogram is larger than the experimental noise, but consistent with the statistical variance expected from a local estimate of the velocity obtained from a small number of steps ($\sim$10). In other words, the intrinsic stochasticity of stepping can account for the variance in the velocity histogram. By introducing a nick in the DNA molecule at a known position, one can study the dependence of the mean unwinding rate on the local DNA sequence, namely check how $\langle v_U(t) \rangle$ varies with the number of steps $N(t)$. One can also investigate enzyme-to-enzyme variability by studying the histogram of the time-averaged unwinding rate: $\bar{v}_u = \int_0^{T_0} v_U(t)dt/T_0$ in different unwinding bursts of duration $T > T_0$. Breaking of ergodicity, i.e., enzymatic individuality, could happen if the ratio of the variance of $\bar{v}_U$ to its mean is significantly larger than the statistically expected $1/\sqrt{N(T_0)}$.

8.4.4 UvrD activity at various ATP concentrations

By varying the ATP concentration, the various parameters characterizing enzymatic activity (rate, processivity, on time, etc.) may vary. For some of these parameters, single-molecule experiments may take more time and be less accurate than bulk measurements. Nevertheless, they may be less affected by some systematic errors (for example, the proportion of inactive enzymes or the reannealing in the back of the complex) than bulk data. On the other hand, experience with the study of single helicases is very recent, and might be plagued by unknown systematic errors of its own. It is therefore useful to compare the variation of the average enzymatic rate with ATP concentration in both bulk and single-molecule experiments.

For UvrD helicases, the average rate follows a Michaelis–Menten (MM) kinetics:

$$\langle v_U \rangle = v_U^{max}[ATP]/(K_M + [ATP])$$

with $v_U^{max} = 275$ bp/s and $K_M = 53\ \mu$M. The value of v_U^{max} is about three times larger than quench flow estimates [384], a significant but not unusual difference when comparing single-molecule and bulk assays. Considering the caveats accompanying bulk measurements and the fact that the SM data were obtained at a rather large stretching force ($F = 35$ pN), this discrepancy is reasonable. More to the point, the value of K_M (which should be less sensitive to systematic experimental errors) is consistent with the value deduced from bulk ATP-ase assays:

$$\langle V_{ATP} \rangle = k_{cat}[ATP]/(K_M + [ATP])$$

with $k_{cat} = 95 \pm 3\ \text{s}^{-1}$ and $K_M = 53 \pm 4\ \mu$M. The ratio $v_U^{max}/k_{cat} \sim 3$ implies that one ATP could be used to unwind as many as three base pairs. This result implies that helicases could be much more efficient than presumed from bulk data, where that ratio has often been reported to be much smaller than one. Though the value deduced from single-molecule experiments is an upper bound, reannealing of the separated strands in bulk assays contributes to reducing the efficiency of unwinding.

8.5 Single-molecule FRET studies of the Rep helicase

Rep is an *E.coli* DNA helicase from superfamily I (SF1), involved in DNA replication, DNA repair, and transcription [387]. Ensemble studies have shown that it is a 3′–5′ DNA helicase [388], and that it forms stable dimers on binding either ss- or dsDNA [387]. It was unclear whether the unwinding activity is thus due to a Rep monomer (similarly to the PcrA helicase [389], another member of SF1) or to a Rep dimer, which opens the possibility for the two monomers to alternate their roles in the dimeric complex [363]. For instance, it was proposed that an isolated monomer would bind ssDNA, whereas a second bound monomer could bind dsDNA. The determination of a high-resolution structure of a Rep dimer on dsDNA exhibited two monomeric conformations, which were taken as evidence for a "domain swivelling" mechanism involved in the enzyme function [390]. Accordingly, two models were proposed: an "inchworm" model (attributing the Rep activity to a monomer) and the "active rolling" model, postulating the need for a second monomer [363], see Fig. 8.1. However, as emphasized by Velankar *et al.* [389], "experiments to distinguish conclusively between these models have been difficult to design, largely because the two models actually share a number of similarities in terms of the biochemical events taking place, and most experimental observations are in fact consistent with either mechanism". This type of remark is a calling sign for single-molecule experimentalists.

8.5.1 FRET studies

Two single-molecule experiments [391, 392] were conducted, using FRET measurement on surface-immobilized DNA–enzyme complexes and illumination in Total Internal Reflection (TIR) mode (see section 3.3). This choice allows for a high-resolution study (sub-nm) of the Rep translocation and conformational changes on a small lengthscale, in contrast to the experiments with RecBCD, UvrD, or RecQ described in the previous sections, which studied long-range translocation and unwinding activity of highly processive enzymes.

The experiments with Rep indicate unambiguously that the monomer is not capable of unwinding dsDNA, but binds to the trailing ssDNA, translocates along it in a 3′–5′ direction, and sits at the ssDNA/dsDNA junction in an inactive (closed) conformation. Unwinding is triggered by the binding of a second monomer of Rep, and stops when it is released. In these experiments [391], the separation of the two strands of immobilized, short (18 or 40 bp) dsDNA molecules was monitored by measuring the evolution of the FRET signal between two probes attached to the two separate ssDNA strands, as depicted in Fig. 8.9. The DNA constructs are immobilized on a surface and observed in parallel after rapid mixing (< 0.2 s) with the solution of Rep enzyme with or without ATP. The time-varying FRET signal is interpreted as a change in the distance between the two fluorophores. In the presence of ATP, two regimes can be observed, depending on the Rep enzyme concentration.

Low concentrations ([Rep] < 2 nM). DNA unwinding rarely occurs, but using a ssDNA overhang with a donor at the 3′ end (rather than at the ss–dsDNA junction),

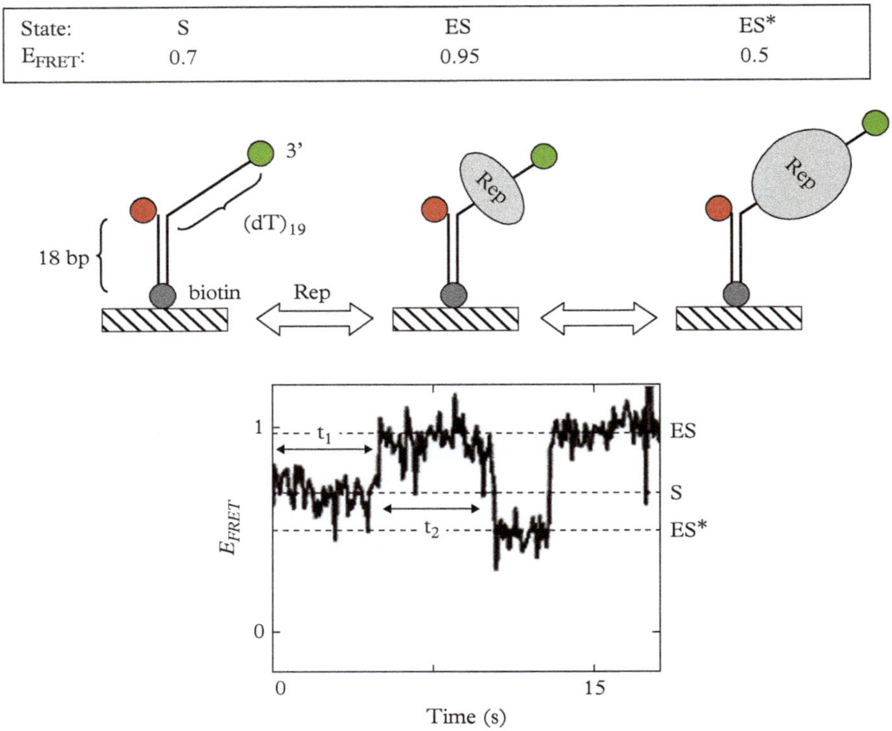

Fig. 8.9 Single-molecule FRET studies of the Rep helicase. *In the absence of ATP and at low Rep concentrations ([Rep] < 2 nM), observed FRET efficiencies are distributed around three values, corresponding to three different states: S corresponds to the isolated DNA, ES is a Rep–DNA complex with a shorter donor (D)–acceptor (A)—distance, ES* is a later state, most likely corresponding to an extended conformation of the Rep monomer. The analysis of distributions of lifetimes in the S, ES, and ES* states obtained from time traces as the one shown here, lead to the mechanism described in section 8.5.1. Adapted from [391].*

bursts of FRET signals are observed with a rate slightly smaller than the binding rate of Rep to ssDNA, and with durations comparable to the inverse of the dissociation rate of Rep from ssDNA. This observation is thus interpreted as the signature of Rep binding to the ssDNA overhang, rapidly translocating to the ss–dsDNA junction and eventually dissociating from it, to leave room for another binding and translocation event.

Large concentrations ([Rep] > 20 nM). DNA unwinding is observed, with an initiation rate proportional to the enzyme's concentration and an unwinding rate comparable to the values measured by ensemble techniques. Unwinding interruptions followed by either resuming of unwinding or rewinding are observed at intermediate concentrations ([Rep] = 100 nM) and for long dsDNA stems (40 bp), indicating that the presence of two Rep monomers might be necessary to ensure processivity of the enzyme.

Interestingly, in the absence of ATP, FRET fluctuations between three distinct values (E_S, E_{ES}, E_{ES*}) are also observed in the construct just described (see Fig. 8.9). The

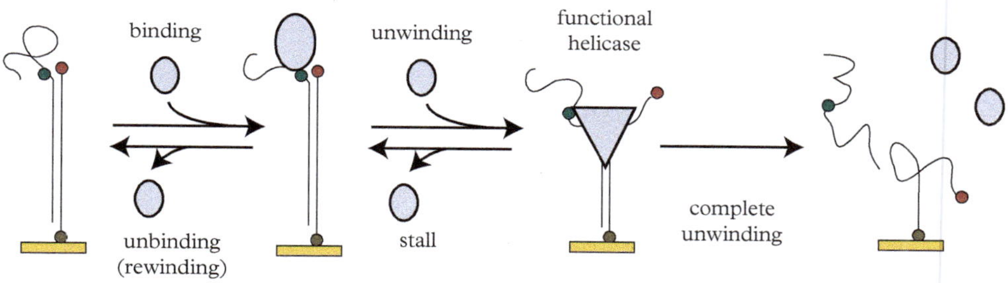

Fig. 8.10 Suggested mechanism of a Rep-helicase on DNA. *A single enzyme binds to an ssDNA and progresses to the DNA fork, where it stalls. Binding of a second monomer is required for switching the complex into unwinding mode. Dissociation of this monomer stalls the complex, which may lead to the re-initiation of unwinding (upon binding of a second enzyme) or to reannealing of the strands (i.e., spontaneous rehybridization upon dissociation of the remaining monomer) (see section 8.5.1 and [391] for details).*

first FRET efficiency value ($E_S \sim 0.7$) is that of an isolated ssDNA overhang and corresponds to the absence of Rep. The two other values are observed on either side of E_S (higher and lower FRET efficiencies) and seem to correspond to two distinct conformations of the Rep monomer on ssDNA (ES and ES*). Although the observed duration of the S state scales inversely with [Rep], indicating that the subsequent state corresponds to Rep binding, the duration of the ES state seems independent of [Rep], suggesting that the transition to the other state ES* corresponds to a thermally activated conformational change. Interestingly, the lifetime of the ES* state is variable, implying that it could depend on the binding site on the ssDNA tail. At higher Rep concentrations, these states disappear, replaced by a fourth one corresponding to the binding of two Rep monomers to the ssDNA.

Figure 8.10 summarizes the likely scenario of the Rep mechanism emerging from these studies. Rep binds as a monomer to the 3′ end of the ssDNA overhang and rapidly progresses to the DNA fork, where it stalls in the absence of a second monomer. Upon interaction with a second unit, the enzyme switches into an active state and proceeds to unwind dsDNA. Unbinding of one unit results in a stall (which can be re-initiated upon rebinding of a Rep monomer), while a detachment of both units results in rapid reannealing of the unwound strands.

8.5.2 Conclusions and perspectives

The examples described above demonstrate the power of single-molecule approaches in the study of helicases. They have permitted the discovery of their unexpected strand-switching behaviour. They have allowed extracting important unknown parameters, such as the helicase processivity and step size. Finally, they have provided details on the enzymatic mechanisms and the kinetic parameters of the enzymatic cycle. Since these methods solve the problem of the reannealing of the unwound strands which plague bulk studies, they are rapidly adopted as the standard for the study of helicases.

8.6 Problems

Monitoring helicase activity with evanescent field illumination. Consider monitoring the activity of a helicase unwinding a hairpin DNA molecule using a magnetic trap setup, see Fig. 8.4b, and monitoring the distance of the bead to the surface via the diffusion by the bead of the EM field generated by evanescent field illumination, see Fig. 3.7c.

1. For a typical helicase (e.g., RecQ), what spatial and temporal resolutions are required to monitor the enzymatic activity with single-bp resolution?
2. From an estimate of the trap stiffness and drag, which bead size and which tension on the molecule should be used to achieve the required spatio-temporal resolution?
3. Explain how the distance of the bead to the surface can be deduced from the intensity of its image on a CCD camera.
4. The bead is imaged on a CCD camera with 100 Hz frame rate and a pixel saturation of 20,000 e^-. For an evanescent illumination with a 200 nm penetration length, how many pixels should the image of the bead occupy to ensure the required spatio-temporal resolution during the full unwinding of a 1 kbp DNA hairpin?

Extracting the kinetic parameters for a model of UvrD activity. Occasionally, more complex types of events are observed during UvrD activity, such as **U**nwinding, followed by slow re-**Z**ipping, then **U**nwinding and finally re-**H**ybridization (UZUH events), see Fig. 8.11b. The purpose of this exercise is to show that a statistical analysis of these events allows extracting the rate constants associated with the enzymatic pathway defined in Fig. 8.11c.
First, find the rate constant for unwinding k_u from the mean rate of unwinding $\langle v_U \rangle / = 248$ bp/s using the step size estimate: $\delta d = 6$ bp.
Then let p_u^s or p_u^d be the probabilities per cycle that the enzyme switches strands or dissociates in the course of an unwinding event. The processivity p_u, defined as the probability per cycle to perform one enzymatic step forward (i.e., neither switching strand nor dissociating) is related to p_u^s and p_u^d via $p_u = 1 - p_u^s - p_u^d$. From the mean number of enzymatic cycles performed during a burst of activity: $\langle N \rangle = \langle N_U \rangle / \delta d = 240/6 = 40$ cycles, show that $p_u = 0.975$.
From this result and the ratio of UZ to UH events (288/155), deduce that $p_u^s = 0.016$ and $p_u^d = 0.009$. Relate these probabilities to the rate constants for unwinding, switching, or dissociating: k_u; k_u^s; k_u^d. Deduce that $k_u^d = 0.36\ \mathrm{s}^{-1}$ and $k_u^s = 0.67\ \mathrm{s}^{-1}$.

A back-tracking helicase. Consider a helicase that has a probability p_+ of unwinding DNA by one bp and a probability $p_- < p_+$ of back-tracking by one bp within

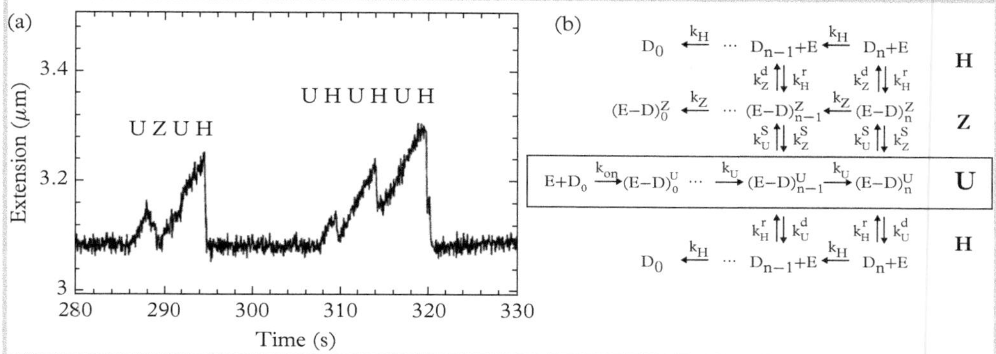

Fig. 8.11 Deducing kinetic constants from single-molecule data. *(a) More complex signals resulting from a combination of the elementary processes described in the text. (b) A model of UvrD progression along a dsDNA alternating between binding, unwinding, rezipping, and unbinding. The various rate constants of this complex scheme can be determined directly from the probability of observing the appropriate transitions in single-molecule experiments, see section 8.4.2 and text of the question. Adapted from [386].*

a mean time $\tau = 10$ ms. The helicase starts to unwind a dsDNA at a nick n_0 nucleotides from one end, and only falls off the molecule when it has fully unwound the two strands, i.e., when reaching $n = 0$.

1. Write the master equation describing the probability $P(n, t+\tau)$ of finding the helicase at position n at time $t+\tau$.
2. Let $x = nb$ be the position of the enzyme along the molecule, where $b = 0.32$ nm is the base-to-base distance along the DNA. By going into the continuum limit, show that the probability $P(x,t)$ obeys a biased diffusion equation: $\partial_t P = D\partial_x^2 P + v\partial_x P$. Express D and v in terms of p_+, p_-, b, and τ. If $p+ = 0.5$ and $p_- = 0.4$, what are D and v? How do these values compared with the velocity of helicases and the diffusion constant of proteins (in solution)?
3. The condition that the helicase falls off the fully unwound molecule implies: $P(0,t) = 0$. Use the method of images to solve for $P(x,t)$, i.e., introduce an "image helicase" at position $-x_0 = -n_0 b$ with "strength": $-\exp(-x_0 v/D)$.
4. The probability that the enzyme is still on DNA at time t is $P_s(t) = \int_0^\infty dx P(x,t)$. Hence, the probability of fully unwinding DNA at time t is: $P_{off}(t) = -dP_s(t)/dt$. Show that in the limit $t \to \infty$: $P_{off}(t) \sim \exp(-v^2 t/4D)$. Compare that behaviour with the behaviour of an enzyme that does not back-track.

9

Topoisomerases

9.1 Introduction to topoisomerases

Soon after a double-helical structure was proposed for DNA [1], the replication of the molecule raised an intriguing issue: if the chromosomes are long DNA molecules, how do they manage to disentangle in a few minutes, as they have to, prior to cell division? Indeed, the reptation theory for polymers, see section 4.2, Eq. 4.21, predicts that the typical time for disentanglement, t_{rep} of polymers of length L, is:

$$t_{rep} \sim \frac{1.2\eta}{k_B T} L^3 \sim 3\ 10^{17} L^3 \quad \text{s}$$

For many years, it was thus believed that chromosomes consisted of small DNA molecules ($L \sim 10$ μm) which could disentangle within a reasonable time interval ($t_{rep} \sim 300$ s). That picture was supported by the fact that for about ten years, DNA extracted from cells had a similar size. However, the observation in the fly—following careful manipulation to prevent hydrodynamic shear—of very long chromosomal DNA ($L \sim 10$ cm), see Fig. 9.1, rekindled the issue with a vengeance: DNA molecules that long would take ten million years to disentangle passively by pure reptational diffusion! The question remained unsolved until the discovery of the topoisomerase (topo) family of enzymes by J. Wang [393]: these are ubiquitous enzymes, capable of changing the topology of DNA molecules by transiently cleaving one or both strands of a dsDNA (type I and II topoisomerases, respectively) [394, 395, 396, 397, 398]. Topoisomerases regulate the degree of supercoiling in DNA molecules, but they also affect their degree of catenation and unknot entangled structures, thus allowing chromosomes to disentangle in a very short time.

Most Type I topoisomerases relax DNA supercoiling and intertwining without the need for an energy cofactor (such as ATP). A notable exception are the reverse gyrases. The canonical type I topoisomerases are divided into two sub classes: topo IA and topoIB, which differ in their structure and functional mechanisms.

For the canonical prokaryotic topoIA, it has been proposed that the enzyme binds to a single strand in a locally melted dsDNA (possibly induced by binding of the enzyme). It transiently breaks that strand and transfers the second strand through it,

Single-Molecule Studies of Nucleic Acids and Their Proteins. David Bensimon, Vincent Croquette, Jean-François Allemand, Xavier Michalet, Terence Strick 2019. Published in 2019 by Oxford University Press.
DOI: 10.1093/oso/9780198530923.001.0001

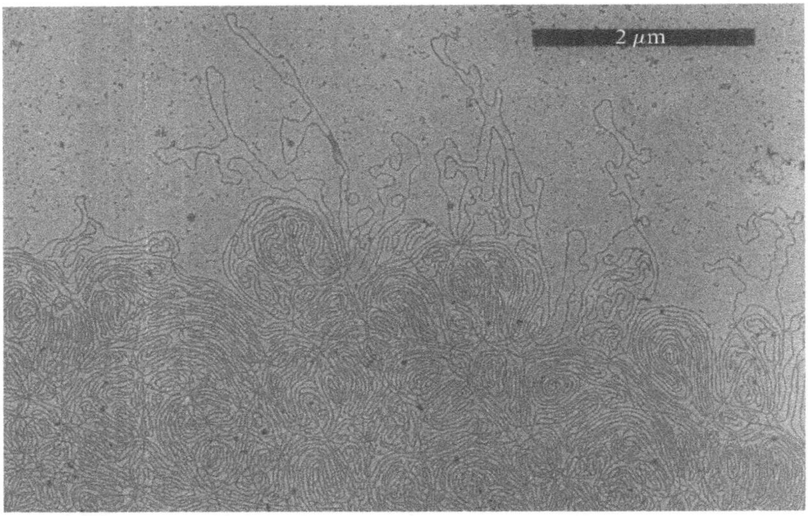

Fig. 9.1 Electron microscopy image of histone-depleted DNA *from HeLa cells spread on an EM grid. Notice the long continuous DNA loops. Adapted from [8].*

before religating the broken strand, see Fig. 9.2. It therefore changes the dsDNA linking number (*Lk*, which stands for the number of times one has to cut and pass one single strand through the other to unlink them) in well-defined quanta of $\Delta Lk = +1$ per enzymatic cycle. This enzyme can relax the twist present in negatively supercoiled DNA [399, 400, 401] and can change the degree of catenation of two molecules, provided there is a nick in one of them. Because it has to bind to ssDNA, which usually arises in denaturation bubbles (see Chapter 5), it is inactive on positively supercoiled DNA.

For the canonical eukaryotic topoIB, a different mechanism has been proposed: the enzyme binds to dsDNA and creates a transient break in one of the strands. By allowing limited rotation about the intact strand, it changes the linking number of positively or negatively supercoiled DNA, so as to relax the torsional constraint. Consequently, ΔLk is modified in stochastic steps, see Fig. 9.2. In contrast with topoIA, topoIB cannot alter the degree of catenation of nicked DNA molecules.

As we shall see in the following, single-molecule manipulations have provided the first direct evidence for the proposed mechanisms of action of these two different enzymes.

While type I enzymes do not need energetic cofactors, type II topoisomerases require consumption of two ATP molecules per enzymatic cycle. Based on biochemical studies [396, 402, 397] and the crystal structure of the yeast type II topoisomerase [403], the following functional mechanism was proposed: the enzyme creates a double-strand break in one DNA segment (the "gate" or G-segment), uses ATP to transport another DNA segment through it (the "transport" or T-segment) [404], and religates the gate segment, see Fig. 9.2. That model is akin to the one just described for the type IA enzymes, with which the type II topoisomerases share a similar structure. Using this mechanism, type II topoisomerases change the linking number of supercoiled DNA by exactly two

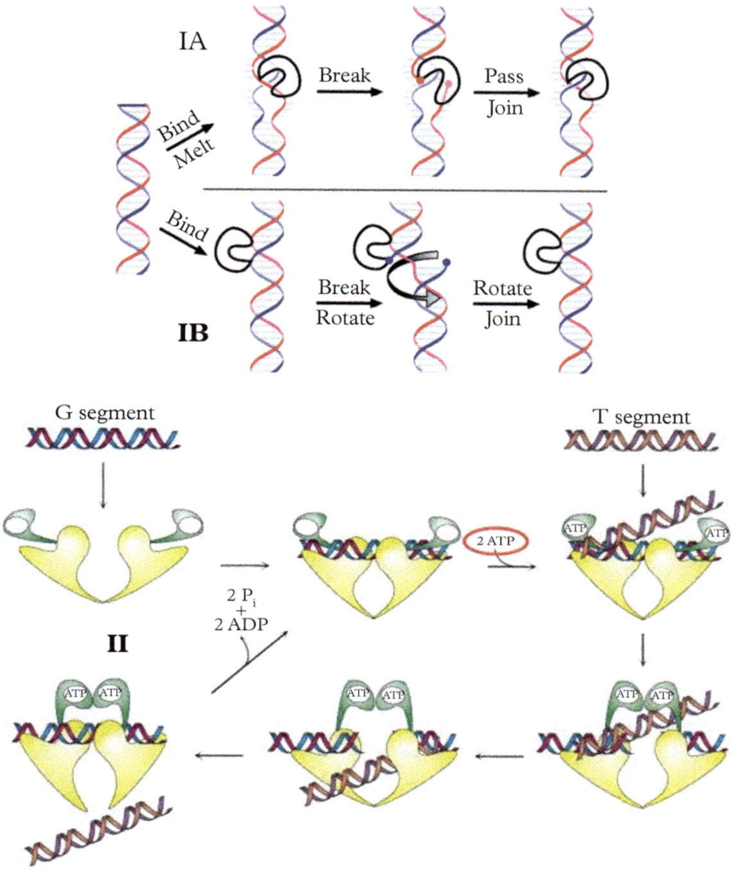

Fig. 9.2 Mechanistic models of topoisomerases. *Type IA topoisomerases induce the local melting of DNA to bind a single-stranded DNA. They transfer the other single strand through a transient break in the first. Consequently, they increase the linking number by one unit: $\Delta Lk = +1$. Type IB topoisomerases create a break in one of the strands and relax the torsion in the DNA by permitting its free rotation about the intact strand. Therefore, the change in linking number is a random integer n: $\Delta Lk = \pm n$. Adapted from [413]. Type II topoisomerases bind to a dsDNA segment (the gate, G-segment) and transport a second (T-)segment through a transient break in the first. In that process, they consume two ATP molecules and change the linking number of supercoiled DNA by two units: $\Delta Lk = \pm 2$. Adapted from [414].*

units ($\Delta Lk = \pm 2$). *In vivo*, they are known to regulate the level of supercoiling during replication, transcription, or recombination. They also disentangle chromosomes during DNA replication: their malfunction during mitosis or meiosis ultimately causes cell death. As a consequence, type II topoisomerases are a major target of antibiotics and anti-cancer drugs.

A particularly puzzling characteristic of the type II enzymes is their ability to decatenate DNA molecules well below the thermodynamic entanglement equilibrium

[398]: type II topoisomerases systematically shift the partition between knotted and unknotted DNAs, towards unknotting. How can an enzyme that "senses" locally, act globally? How can its action always result in unknotting, rather than knotting? This puzzle is particularly striking in bacteria where replication of the circular chromosome could end up in two highly interlinked DNAs. These two molecules have to be precisely decatenated in order to be dispatched to each of the two daughter cells. However, the entanglement of DNA molecules (their catenation and knotting) is a global topological property, whereas the enzyme is catalysing locally the transfer of DNA segments through each other. From pure thermodynamic considerations, these characteristics must be related to the fact that these enzymes use the energy of ATP hydrolysis [405] to unlink the molecules. Indeed, although type II topoisomerases resemble the famed "Maxwell demon" of thermodynamics (which is capable of altering the distribution of particles in a box), they do not break the second law of thermodynamics, since to perform their feat, they consume energy. The way DNA transport and ATP hydrolysis are coupled to allow for that remarkable property has been investigated for years [396, 402, 406, 407], yet the mechano-chemistry of these enzymes is still mysterious.

9.2 Bulk assays

The migration in an electrophoretic gel of a circular DNA (plasmid) of given length varies with its compactness, which in turn depends on its topological state (topoisomers), i.e., its degree of supercoiling, knotting, or catenation, see Fig. 9.3. Bulk assays of the activity of topoisomerases use that property to test the distribution on an electrophoretic gel of the various topoisomers of an initially supercoiled (or catenated) plasmid, after incubation with increasing amounts of enzyme or longer times. These assays allow for a quantification of the activity of topoisomerases in terms of the amount (μg) of supercoiled (or catenated) plasmid relaxed (or decatenated) per s, per μg of enzyme. Knowing the initial degree of supercoiling of plasmids (typically $\sigma = -0.06$), one can deduce the rate of supercoil removal per enzyme. The typical rates of supercoil removal for topoIA deduced from these bulk data are $v_{topo1A} \sim 5$ links/min [408]; for topoIB: $v_{topo1B} \sim 90$ links/min [409]; and the decatenation rate for an efficient type II topoisomerase is $v_{topoIV} \sim 8$ links/min [410]. These numbers are surprisingly low, since the whole *E.coli* circular genome (which has a linking number $L_k \sim 450{,}000$) is replicated within ~45 min. This requires the genome to be decatenated at a rate of 10,000 links/min! With the rates deduced from bulk data, thousands of enzymes would be needed to unlink the *E.coli* chromosome during replication, i.e., topoisomerase concentrations >1 μM would be required, which appears excessive [411].

The interaction of topoisomerases with DNA is also often studied by nitrocellulose filter retention assays. These filters immobilize the proteins and their complexes with DNA, but let free DNA pass through. Release of the DNA complex is then achieved by washing the filter with an appropriate buffer (e.g., one that denatures the adsorbed proteins). With these assays the affinity of proteins for various forms of DNA (supercoiled, nicked, catenated, etc.) can be studied in different conditions.

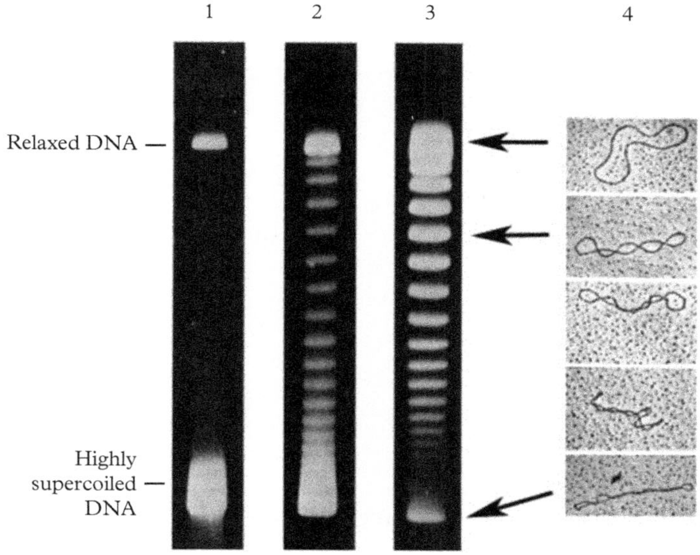

Fig. 9.3 Migration in an electrophoresis gel of various topoisomers. *Lane 1: the native plasmid consists of a majority of highly supercoiled loops and a minority of relaxed, nicked ones. Upon relaxation with increasing amounts of topoisomerase, a population of loops with decreasing degrees of supercoiling is generated (lanes 2 and 3). Electron microscopy images of some of these topoisomers are shown in lane 4. Adapted from [217].*

9.3 Single-molecule *in vitro* assays: type I topoisomerases

To measure the uncoiling activity of a single topoisomerase, a magnetic trap is used. A DNA molecule is coiled beyond the buckling threshold, where it forms supercoils by spinning the magnets (typically by more than ten turns for a 10 kbp DNA). As a result, the bead to surface distance z (i.e., the molecule's effective extension) shortens. Introduced in a proper buffer, type I or II topoisomerases will relax the supercoils leading to a progressive increase in z, see Fig. 9.4. A great advantage of single-molecule experiments on topoisomerases as compared to bulk assays is that they are reversible: by mechanically rewinding the molecule, the supercoils can be regenerated and the experiment repeated.

9.3.1 Topoisomerase 1A

To observe a single enzymatic cycle of nicking and religation by a type I topoisomerase, one must ensure that a single enzyme is actually nicking the molecule, and one must slow down its activity such that individual cycles can be resolved within the time resolution of the experiment (usually about >0.1 s). To fulfil the first requirement, one has to reduce the enzyme's concentration to such an extent (< 1 nM) that the simultaneous binding

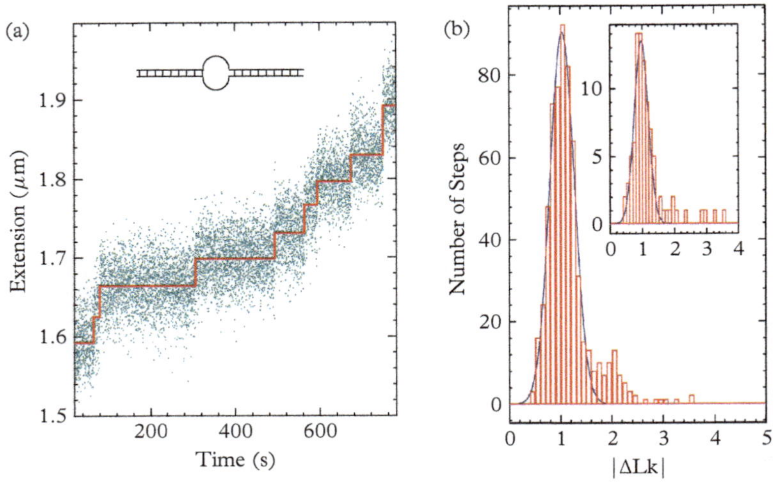

Fig. 9.4 **(+) supercoil removal by *E.coli* topoIA** *in a DNA construct containing a mismatched region of 12 bp, shown schematically at the top of (a). (a) Single relaxation events observed at* $F = 2$ *pN: blue points (raw data), red line (best fitted steps). (b) Histogram of topoIA step sizes (N = 682) in six experiments with forces* $1.65 < F < 2.1$ *pN (inset: results for a single experiment at 2 pN). The average step size is* $\Delta Lk = 1.04 \pm 0.09$. *Adapted from [413].*

and activity of two enzymes on a single DNA molecule is unlikely. To slow down the enzyme, one can often reduce the amount of fuel (i.e., ATP) which is powering it. That procedure is unfortunately not an option for type I topoisomerases that do not require ATP to function.

For the type IA enzymes, which have to bind to a ssDNA to be active, a strategy introduced by Kirkegaard and Wang [412] was adopted: a DNA molecule was constructed with a central 12 bp mismatch where an oligo (dC_{12}) is opposite to an oligo (dT_{12}). Although topoIA cannot relax regular (+)scDNA (which cannot denature to generate the required single DNA strands), the engineered molecule can be relaxed by topoIA when overwound. The size of the mismatch region can accommodate only a single enzyme, whose activity is greatly reduced when the DNA is under tension [413]. This construct thus solves the two problems mentioned above: ensuring single-molecule activity and slowing it down. In the present case, it permitted the direct monitoring of topoIA activity through the increase in the molecule's extension. The data, obtained using a magnetic trap setup, showed that the enzyme was increasing the molecule's span by steps of 35 nm, which corresponds to the removal of one (+)supercoil ($\Delta Lk = -1$) at a time, see Fig. 9.4.

Of course, one might argue that by slowing down the enzyme, one reduces its step size. A means is thus needed to assess the enzymatic step size at low forces, when the enzymatic cycles cannot be resolved by the time trace of the molecule's elongation. Such a means exists: it involves an analysis of the noise during a burst of enzymatic activity, see section 8.4.2. This noise is due to two sources: a frequency-independent

Brownian noise coming from the random fluctuations of the tethered bead and a frequency-dependent shot noise coming from the stochastic activity of the enzyme. The latter source is the dominant one at low frequencies. Its amplitude is directly proportional to the enzymatic "step size": the change in DNA extension per enzymatic cycle. For topoIA, the value of $\Delta Lk = -1$ thus deduced coincided with the value determined at high forces, when the enzyme was slowed down to allow for observation of individual cycles [413]. The force alters the frequency of relaxation events, not the number of supercoils relaxed per cycle. This experiment did confirm that the bacterial type I topoisomerase removes DNA supercoils by a strand passage mechanism.

9.3.2 Topoisomerase 1B

For topoIB, it was found that increasing the concentration of Mg^{++} ions to 10 mM decreased the activity of the enzyme to a level where single relaxation events could be observed at nM concentrations of topoisomerase [415]. These relaxation events were observed as a succession of variable increases in the extension of a (+)scDNA molecule, see Fig. 9.5b. By measuring the decrease in the molecule's extension per turn as it is coiled in a magnetic trap setup (the mechanical reversal of the topoisomerase action), one can relate the size of the relaxation events to the change in the DNA's linking number: ΔLk. In contrast with topoIA, the change in linking number is exponentially distributed, see Fig. 9.5c. That is precisely the distribution expected from the proposed mechanism of topoIB: the enzyme creates a nick in the DNA, the molecule then swivels about that nick, releasing its torsional stress. At the completion of each turn, there is a probability p that the enzyme will religate the nick and a probability $1-p$ that the molecule will continue to swivel. The probability of relaxing ΔLk turns is thus:

$$P(\Delta Lk) = p(1-p)^{\Delta Lk - 1} \underset{p \to 0}{\longrightarrow} \exp(-\Delta Lk/\langle \Delta Lk \rangle)/\langle \Delta Lk \rangle$$

where $\langle \Delta Lk \rangle = 1/p$ is the average change in linking number. Since the torsional stress in the molecule is the driving force; this mechanism also predicts that the smaller the torque Γ in the molecule, the greater the probability of ligation. For a thermally activated (Arrhenius) process, one expects that $p = p_0 \exp(-\Gamma\delta\theta/k_B T)$, where $\delta\theta$ is the distance to the transition state, see section 7.6.2. As we have seen in section 4.3, the torque of a stretched supercoiled DNA molecule is related to the tension: $\Gamma = \sqrt{2\xi F k_B T}$, see Eq. 4.25. The model of action for topoIB, thus makes a definite prediction about the average change in linking number $\langle \Delta Lk \rangle$ and its dependence on the force, i.e.:

$$\langle \Delta Lk \rangle = e^{\delta\theta\sqrt{2\xi F/k_B T}} \times \langle \Delta Lk \rangle_{F=0}$$

This prediction was indeed verified in these single-molecule studies of supercoil relaxation by topoIB, see Fig. 9.5d, thus providing strong evidence for the proposed mechanism of action of topoIB.

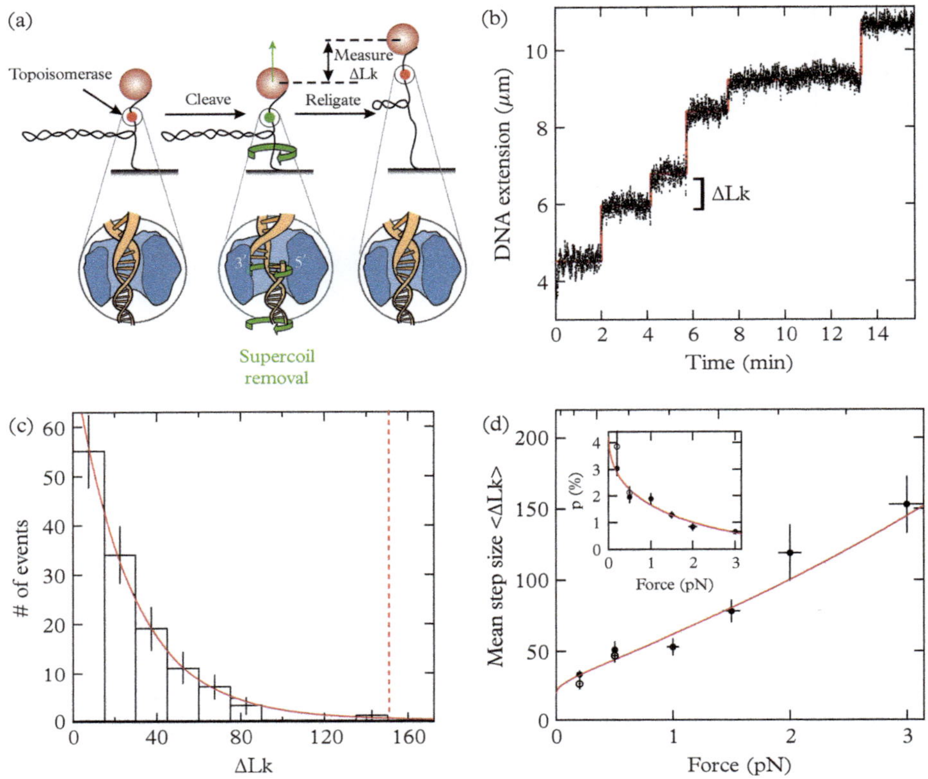

Fig. 9.5 Unlinking of (+)scDNA by a topoIB from vaccinia. *(a) Schematics of the experiment. (b) A variable increase in the DNA's extension is observed each time the enzyme removes a variable number of supercoils. (c) Distribution of the number of relaxed supercoils (ΔL_k) at each enzymatic cycle and exponential fit (the dashed line is the initial number of supercoils in the molecule). (d) Variation with the force of the average change in linking number ($\langle \Delta Lk \rangle$) and (inset) probability of religation p. The continuous lines are one-parameter fits to the prediction of the topoIB mechanical model, see section 9.3.2. Adapted from [415].*

Action of drugs on topoIB. Many drugs against cancer target topoisomerases, as these are often more active in rapidly dividing cells. Interfering with their function can result in cell death, but often the mechanism of interference of the drug is not well known. This is the case for topotecan, a drug targeting topoIB, which is supposed to stabilize the covalent DNA–topoisomerase complex, transforming it into an obstacle to DNA replication. In a seminal work, the action of this drug was studied using the single-molecule approach described above [416]. It was observed that the relaxation of supercoils by topoIB in the presence of topotecan was slowed down by a factor of

20! Hence not only does the drug stabilize the nicked state of DNA, its interactions with the enzyme and the DNA considerably hinder the rotation of the molecule about the un-nicked strand. The lifetime of the bound complex, estimated from the time during which DNA could be repeatedly coiled and slowly relaxed by a single topoIB was estimated to be about two minutes. This is a very long time on the scale of replication (with typical cycles of a few ms, see Chapter 7), which is consistent with the presumed role of topotecan as a roadblock in replication. Quite unexpected was the observation that the relaxation of negative supercoils occurred about three to five times faster than the relaxation of positive supercoils. A consequence of that different behaviour is that positive supercoils can accumulate on replicated DNA, an accumulation which generates a positive torque on DNA that might hinder the progression of replication. This consequence was indeed observed *in vivo* in experiments conducted in cell cultures treated (or not) with topotecan. This work was the first example of the power of single-molecule studies in analysing the mechanism of a cancer drug, revealing unexpected consequences of its interactions with the enzyme–DNA complex.

9.4 Single-molecule *in vitro* assays: type II topoisomerases

Type II topoisomerases designate a wide family of enzymes which alter the topology of DNA by consuming ATP. This family comprises enzymes such as the eukaryotic topoII or the prokaryotic topoIV that unlink and decatenate DNA, but also the bacterial gyrase. The role of gyrases in bacteria is to negatively supercoil DNA in order to compactify it and, by facilitating its local melting, to permit access to a variety of DNA-binding enzymes. Prokaryotic topoisomerases and gyrases are hetero-tetramers of the form A_2B_2, in which one unit (ParC for topoIV, GyrA for gyrase) is involved in DNA binding and cleavage/religation, while the other (ParE for topoIV and GyrB for gyrase) is involved in ATP binding and hydrolysis. Eukaryotic topoisomerases are homodimers in which these two units are domains of the same protein. Single-molecule experiments, while providing direct proof of the relaxation of supercoils in quanta of $|\Delta Lk| = 2$, have revealed significant differences between the eukaryotic topoII and prokaryotic topoIV, which we will discuss in this chapter.

9.4.1 DNA-binding properties of the eukaryotic type II topoisomerases

Binding of the eukaryotic type II topoisomerase to DNA, in the absence of ATP, results in hysteretic behaviour of the extension, as the degree of supercoiling is increased and decreased. The formation of DNA supercoils favours strong and largely irreversible binding of the protein onto DNA crossings: the enzyme–DNA complex is broken only by increasing the force pulling on the molecule, see Fig. 9.6. This data is compatible with the electron microscopy (EM) observations of topoII enzymes located at the crossings

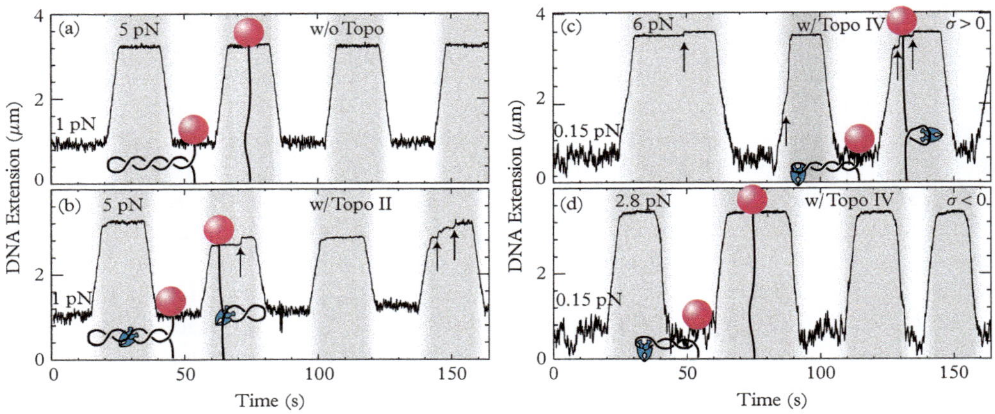

Fig. 9.6 Stretching supercoiled DNA in the presence of topoisomerases. *(a) When a supercoiled DNA molecule is stretched with a large force (5 pN), it lengthens continuously as the plectonemes are pulled out and the writhe is transferred into twist. (b) In the presence of the eukaryotic topoisomerase II, the supercoiled DNA molecule, when suddenly stretched with a force $F = 5$ pN, extends but displays pauses in extension (blockages) when pulled. This suggests that the enzyme bridges between opposite DNA segments on the plectoneme, transiently preventing its full unravelling. (c) In the presence of the prokaryotic enzyme, topoIV, small blockages are sometimes observed towards the end of plectonemic unravelling and on positively supercoiled DNA only. This suggests that topoIV may bind to and bend the apex of positively supercoiled DNA, from which it will be dislodged by a high tension on the molecule. Adapted from [417].*

between DNA segments. This stabilization of DNA crossings can be quantified by studying the interaction of topoII with DNA at the threshold for buckling (i.e., at the onset of a single supercoil). At that point, the molecule fluctuates rapidly between a state exhibiting a supercoil loop and one without. The presence of topoII increases the lifetime, τ_{on}, of the supercoiled state. The probability distribution of these lifetimes, $P(\tau_{on})$, displays a behaviour best fitted by a double exponential decay, suggesting that the enzyme–DNA complex exists in (at least) two different binding states. One can speculate about what these two states might be (possibly a covalent and a non-covalent binding of the enzyme to DNA), but the point is that these states can only be identified and studied with a single-molecule approach.

9.4.2 DNA binding properties of the prokaryotic topoIV

The bacterial enzyme, topoIV, generates a complex with DNA that (in the absence of ATP) has a much shorter lifetime than the topoII–DNA complex. In particular, the extension vs supercoiling curves, though different from the ones measured in the absence of an enzyme, are perfectly reversible. What these curves show, however, is a marked asymmetry between positive and negative supercoiling: the decrease in extension for overwound DNA, $\delta l_p^+ = 32$ nm/turn, is smaller than for underwound DNA, $\delta l_p^- = 48$ nm/turn (at $F = 0.12$ pN). These values have to be compared with the decrease in extension upon twisting DNA in the absence of proteins, which is symmetric for (+) and (−)scDNA and equal to $\delta l_0 = 68$ nm/turn. Remarkably, the decrease in DNA extension in the presence of enzymes, δl_p^+ and δl_p^-, varies little with force, whereas in their absence, δl_0 decreases as roughly the inverse square root of the force [417].

A simple mechanical model of the interaction between topoIV and DNA can account for the extension vs supercoiling data in presence of the enzyme. Let the interaction favour a bending of the molecule by a radius R_0 and a weak overtwisting of DNA, with local change in linking number $0 < \Delta \ll 1$, see also section 7.6.2. Conservation of linking number implies that this local overtwisting is compensated by a change in the writhe of the molecule by $-\Delta$ per bound enzyme. Thus, on (+)scDNA, enzyme binding reduces the amount of writhe, whereas it increases it on (−)scDNA. This qualitatively accounts for the largest decrease in extension in underwound as compared to overwound molecules: $\delta l_p^- > \delta l_p^+$. If binding of topoIV onto DNA favours a certain radius of bending R_0, the energy required to form a loop in the molecule at the onset of buckling [417, 219, 418] is:

$$\mathcal{E}^{prot} = \frac{1}{2}B\left(\frac{1}{R} - \frac{1}{R_0}\right)^2 2\pi R + 2\pi RF + \Delta G^0_{binding} \tag{9.1}$$

The first term in the RHS corresponds to the bending energy of a loop of radius R ($B \equiv k_B T\xi$ is the bending modulus, $\xi = 50$ nm is the DNA persistence length, and $k_B T = 4$ pN nm is the thermal energy). The second term is the work done against the force F while forming the loop. The third term is the free energy change upon binding of topoIV on relaxed DNA ($\sigma = 0$) at $F = 0$. The minimization of $\mathcal{E}^{prot}$ yields the radius R of the loop at the onset of buckling:

$$R = \frac{R_0}{\sqrt{1 + \frac{2FR_0^2}{k_B T\xi}}} \tag{9.2}$$

Past the buckling threshold, every extra turn of DNA changes its extension by typically $\delta \sim 2\pi R$. However, since the interaction of topoIV with DNA absorbs an amount $\Delta > 0$ of linking number, one needs to rotate the bead clockwise by $1 + \Delta$ turns to actually add one turn to DNA (i.e., to observe a decrease in extension by δ). Similarly, one needs to rotate the bead counterclockwise by $1 - \Delta$ turns to unwind DNA by one turn. As a result, the change in DNA extension per added turn of the magnets is:

$$\delta l_p^+ = \frac{\delta}{1 + \Delta} \tag{9.3}$$

$$\delta l_p^- = \frac{\delta}{1 - \Delta} \tag{9.4}$$

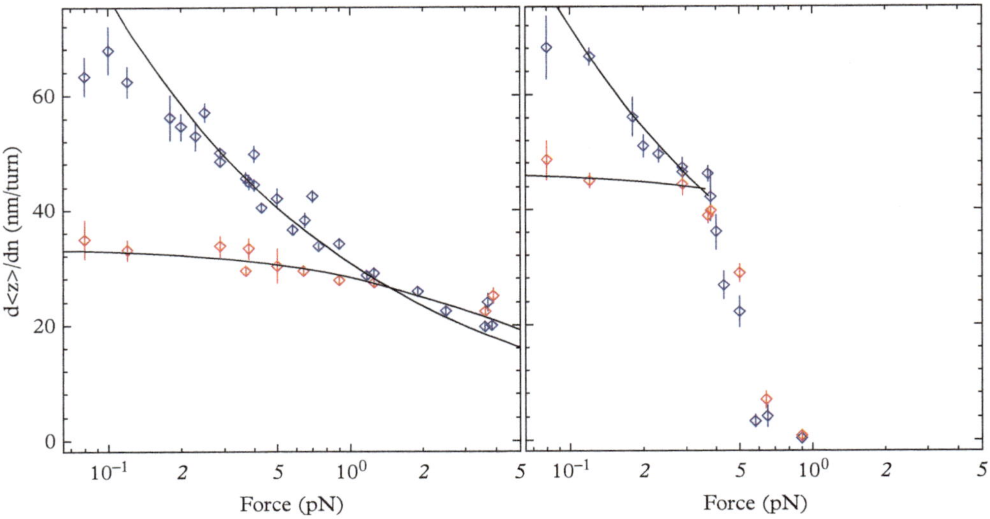

Fig. 9.7 Decrease in the DNA's extension per added twist *(turn of the magnets) as a function of the tension on (+) and (−) scDNA with and without enzyme ([topoIV] = 3 nM; in absence of ATP). (a)* δl_0 *(blue points) and* δl_p^+ *(red points) for (+)scDNA. (b)* δl_0 *and* δl_p^- *for (−)scDNA; above 0.4 pN the (−)scDNA unwinds. Error bars indicate the statistical error. The blue points* δl_0 *(data without enzymes) were fitted using a powerlaw function [219]:* $\delta l_0 = AF^{-0.4}$ *(black line) with* $A \approx 30$. *The red points (data with enzymes) were fitted (black line) using the model described in the text with* $2\pi R_0 = 40$ *nm and* $\Delta = 0.16$. *Adapted from [417].*

These predictions are in good agreement with the experimental data (for both (+) and (−)scDNA) and yield best-fitted values: $R_0 = 6.4$ nm and $\Delta = 0.16$, see Fig. 9.7. The value of R_0 is consistent with estimates of the Stokes radius of the enzyme [419] and with FRET data on the bending of DNA by the ParC subunits of topoIV [420]. The amount of supercoil induced in DNA upon binding to topoIV is small and has escaped bulk studies of the enzyme–DNA complex.

9.4.3 The prokaryotic topoisomerase IV binds cooperatively on supercoiled DNA

We have seen that the binding of topoIV onto DNA altered the reduction in DNA extension per added turn, δl_p^+ and δl_p^-, when comparing it to the decrease in absence of protein, δl_0. The data presented in section 9.4.2 were obtained at saturating concentrations of protein. One can use the decrease in DNA extension per added turn, δl, at various concentrations of topoIV to infer the probability p of binding of topoisomerase onto DNA [417]. This decrease in extension interpolates between the decrease in absence of enzyme δl_0 and the decrease in saturating conditions, δl_p^+ on (+)scDNA (or δl_p^- on (−)scDNA):

$$\delta l = p^+ \delta l_p^+ + (1 - p^+)\delta l_0 \tag{9.5}$$

The probability of binding on (+)scDNA p^+ can thus be inferred from the data:

$$p^+ = \frac{\delta l_0 - \delta l}{\delta l_0 - \delta l_p^+} \tag{9.6}$$

The probability p^+ and p^- of topoIV binding onto both (+) and (−)scDNA has been measured (at $F = 0.2$ pN) for increasing concentrations of topoIV, starting from 0.02 nM, see Fig. 9.8. Clearly, the binding of the enzyme is cooperative: at a concentration of topoIV ~ 0.2 nM it rapidly rises from a very small value to saturation ($p = 1$). The binding curves on (+) and (−)scDNA were fitted using a Michaelis–Hill function:

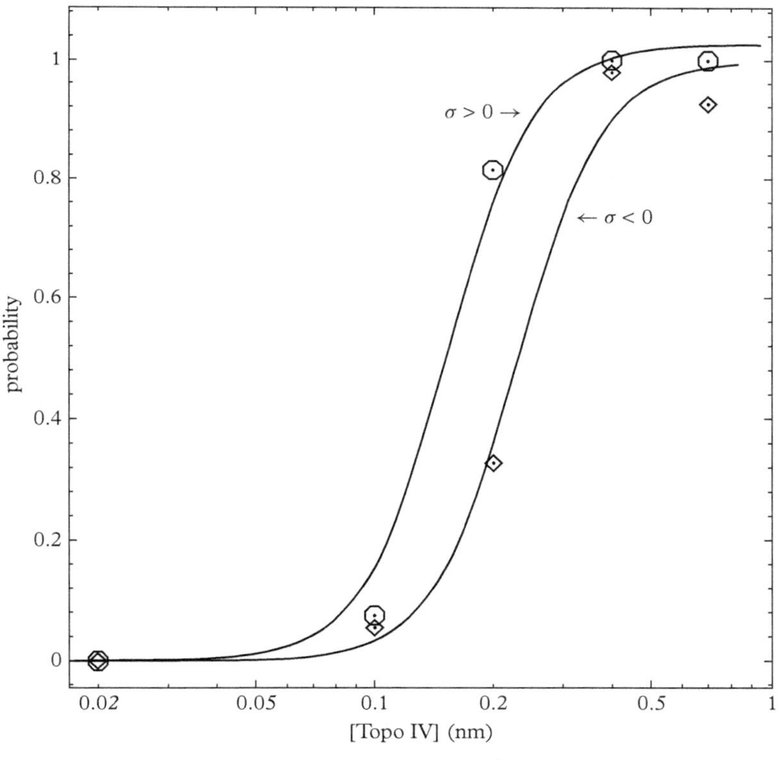

Fig. 9.8 Probability p of binding of topoIV onto (+) and (−)scDNA *versus the concentration of enzyme ($F = 0.2$ pN). The experimental points (∘ for $\sigma > 0$ and ⋄ for $\sigma < 0$) were fitted using a Michaelis–Hill function: $p = [\text{topoIV}]^\alpha / (K_d^\alpha + [\text{topoIV}]^\alpha)$; $\alpha = 4$ was found to yield the best agreement, with $K_d = 0.15$ nM for $\sigma > 0$ and $K_d = 0.23$ nM for $\sigma < 0$. Adapted from [417].*

$$p = \frac{[\text{topoIV}]^{\alpha}}{K_d^{\alpha} + [\text{topoIV}]^{\alpha}} \tag{9.7}$$

where $\alpha = 4$ and $K_d^+ = 0.15$ nM for (+)scDNA and $K_d^- = 0.23$ nM for (−)scDNA. The value of α (which was not known from bulk assays) suggests that the four subunits of topoIV ($ParC_2ParE_2$) assemble cooperatively onto DNA to generate an active complex. The values of the dissociation constants are very small and suggest that in bacteria all the topoisomerases are associated with DNA (one molecule in a volume of 1 μm^3 corresponds to a concentration of 1.6 nM), see, however, section 9.6 for a comparison with *in vivo* data. The model described in Eq. 9.1 can also be used to estimate the free energy of binding of topoIV onto supercoiled and relaxed DNA. From these estimates, one can deduce the dissociation constant of the topoIV–DNA complex, which is consistent with the values just quoted (that do not depend on a model of topoIV–DNA binding). It turns out that binding on relaxed DNA is 100 times weaker (i.e., requires a concentration of enzyme one hundred times larger) than on supercoiled DNA.

9.4.4 Uncoiling activity of eukaryotic topoisomerase II

To monitor the activity of a type II topoisomerase, one has to follow the same protocol used to study the type I enzymes, namely monitor the relaxation of supercoils following the mechanical coiling of the DNA within a rotating magnetic trap.

To ensure that the activity of a single enzyme is observed, the enzyme is diluted (typically to tens of pM) so that the interval between mechanical coiling and enzymatic uncoiling events (τ_w) is much larger than the time (τ_{on}) it takes the enzyme to undo the supercoils. From traces such as shown on Fig. 9.9, one can deduce the rate of uncoiling of a single topoisomerase as a function of [ATP] [421]. The variation of the topoII rate with [ATP] can be fitted, see Fig. 9.10c, by a model suggested by bulk quench flow measurements of ATP hydrolysis, which requires the asymmetric hydrolysis of two ATP molecules per enzymatic cycle [405, 406, 407]. The good agreement between the fits of this model to both single-molecule and bulk data is a testimony to the tight coupling between ATP hydrolysis and the mechanical cycle of the enzyme. Interestingly, the rate determined in single-molecule assays (3 cycles/s at saturating ATP concentrations) is more than ten times larger than the one measured in ensemble measurements [422]. While bulk experiments yield a value for the enzymatic activity that is averaged over active as well as inactive (or unbound) proteins, a single-molecule assay measures only the rate of a bound and active enzyme.

Individual enzymatic cycles of topoisomerase activity can be observed when the enzyme is slowed down by reducing the ATP concentration (typically [ATP]$< 100\ \mu$M). In that case, the time between relaxation events is compatible with the temporal resolution of the experimental setup (typically of $\sim$1 s). In the presence of topoII, the extension of an initially supercoiled molecule increases as it is uncoiled by well-defined steps of 90 nm (at $F = 0.7$ *pN*), see Fig. 9.10a. From the response of DNA to rotation of the magnets, see section 4.3, one knows that this step size corresponds to the simultaneous relaxation

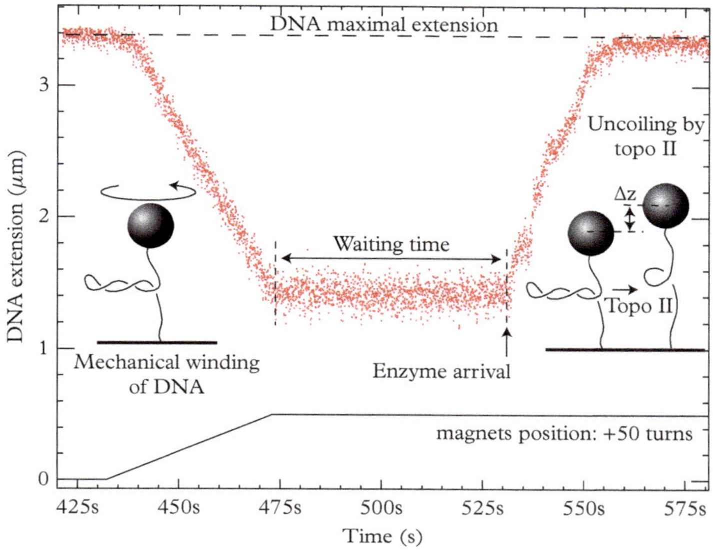

Fig. 9.9 Uncoiling activity of a eukaryotic (Yeast) topoII *in buffer containing 1 mM ATP. A bead bound by a single un-nicked DNA (11 kb) is rotated by fifty turns (the straight black line indicates the rotation position of the magnets), resulting in a decrease of DNA extension. After a variable waiting time, an enzyme binds to the molecule and uncoils it processively until the DNA reaches its maximal extension at a given force (here F = 1 pN). Adapted from [421].*

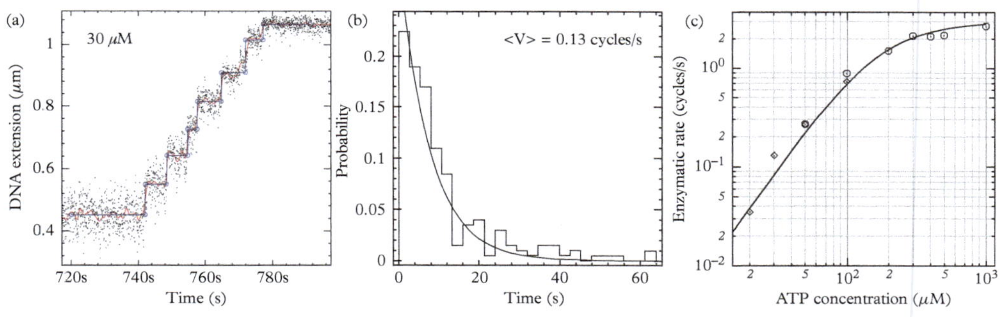

Fig. 9.10 Identification of single cycles and uncoiling rates of topoisomerase II. *(a) Relaxation of supercoils by topoII at 30 μM ATP. Dots: raw data; red curve: 2 Hz low pass filtering; blue line: best fit to a stepping signal. The size of the step corresponds to relaxation of two supercoils per enzymatic cycle:* $\Delta Lk = -2$. *(b) The distribution of the time between steps can be fitted to a single exponential decay that yields the mean uncoiling rate* $\langle v \rangle$ *of the enzyme (here* $\langle v \rangle = 0.13$ *cycles/s). (c) Uncoiling rate of a eukaryotic (fly (○) or Yeast (◇)) topoII as a function of ATP concentration. The solid line is a fit using a model proposed by Harkins et al. [406, 407] which assumes an asymmetric hydrolysis of two ATP molecules per enzymatic cycle.*

of two supercoils per enzymatic cycle, as expected from the mechanism of topoII, where a dsDNA T-segment is passed through a dsDNA G-segment (Fig. 9.2b), altering the intertwining of the two strands by $\Delta Lk = -2$.

9.4.5 Uncoiling activity of prokaryotic topoisomerase IV

The bacterial topoisomerase, topoIV, in spite of structural similarities with the eukaryotic topoII enzyme, exhibits some striking differences in its activity. While on positively supercoiled DNA the single-molecule uncoiling rate of topoIV is comparable to that observed with topoII, i.e., fast and processive, the relaxation of negative supercoils by topoIV is extremely slow and distributive, see Fig. 9.11, and observed only at much higher enzyme concentrations. This chiral discrimination of topoIV makes sense, but poses a peculiar paradox. It makes sense because, in contrast with eukaryotes, bacteria possess a second type II topoisomerase, gyrase (see section 9.5), whose role is to actively underwind DNA, so as to facilitate the insertion in the double helix of a variety of regulation factors and other proteins (e.g., the replication and transcription complexes). The lack of topoIV activity on (−) supercoiled DNA prevents interference with the unwinding activity of gyrase and a spurious consumption of ATP. On the other hand, the topoIV high rate of (+) supercoil removal allows for the efficient relaxation of torsion downstream from the replication complex, an essential requirement for replication.

The paradox arises when we consider the second role presumably played by topoIV during replication: sister chromosome decatenation [424, 425]. Since DNA is a right-

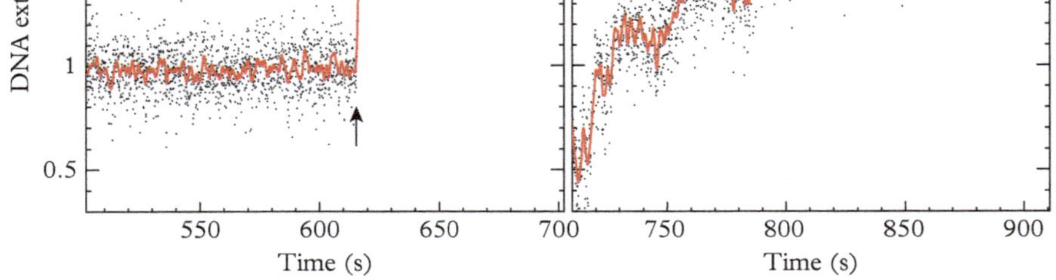

Fig. 9.11 Relaxation of positive and negative supercoils by topoIV. *(a) On positively supercoiled DNA, fast (~1 s) and processive relaxation events are observed (arrows) following mechanical overwinding of the DNA (which results in a decrease of its extension). (b) On negatively supercoiled DNA, at twenty times larger protein concentrations, a much slower and more distributive relaxation is observed. The continuous line is a low-pass filter at 1 Hz. Adapted from [423].*

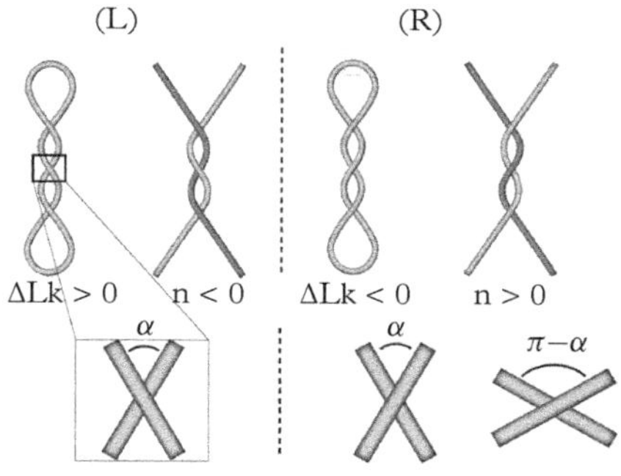

Fig. 9.12 Structure of braids and supercoils. *Left-handed braids are similar to (+) supercoiled DNA ($\Delta Lk > 0$), whereas right-handed braids are similar to (−) supercoiled DNA ($\Delta Lk < 0$). Notice that a right-handed crossing with an obtuse angle $\pi - \alpha$ is identical to a left-handed crossing with an acute angle α. Therefore, the local geometry of a given crossing may change due to thermal fluctuations.*

handed helix, the two daughter DNAs formed during replication may wind around each other as a right-handed braid very similar to the structure of a negatively supercoiled plectoneme, see Fig. 9.12.

If topoIV is incapable of relaxing negatively supercoiled DNA, how can it efficiently unlink the two sister chromosomes [424, 426, 427] arising during replication? To address that issue, one has to study the activity of topoIV on braided molecules [423, 428, 429].

9.4.6 Decatenation activity of type II topoisomerases

A bead tethered to a surface by two DNA molecules can be rotated, and the molecules braided, as described in section 4.6. Addition of a eukaryotic topoII in the presence of ATP results in the unlinking of the two molecules with a subsequent increase in their effective extension (i.e., in the bead-to-surface distance). As expected for this enzyme, the activity on left- or right-handed braid is identical (as it was on (+) and (−) supercoiled DNA), see Fig. 9.13a. The rate of DNA unlinking is very similar to the rate of uncoiling and varies similarly with force, see Fig. 9.14a.

Repeating this experiment with the bacterial topoIV led to different and surprising results. While, as expected, topoIV completely unlinks left-handed braids (which are similar to (+) supercoils) it does not unlink right-handed braids below a final catenation number n_S (which was different from bead to bead and varied with the force), see Fig. 9.13b. For $n > n_S$, the rate of unlinking of right-handed braids was about half the rate for left-handed braids, see Fig. 9.14b. Since at the concentrations used in this experiment, topoIV has no observable uncoiling activity on (−) supercoiled DNA, it is surprising that it so efficiently unlinks right-handed braids, if only partially. Comparison of n_S with the data on DNA braiding, see Fig. 4.15, section 4.6, revealed that it was actually equal to n_c, the number of turns above which the two braided molecules started to form plectonemes of braids [423]. This implies that topoIV could unlink right-handed braids when they

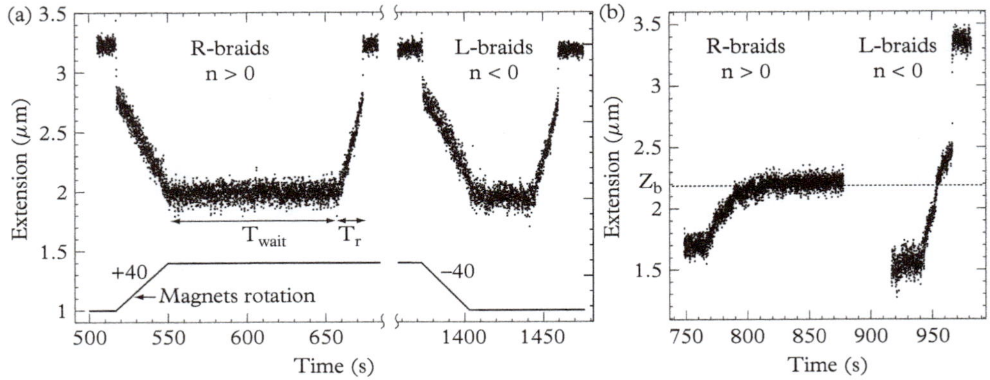

Fig. 9.13 DNA unlinking by type II topoisomerases. *(a) Record of unlinking by the Fly topoII. Upon rotation of the magnetic bead, the two molecules are braided by* $n = 40$ *turns. After a waiting time* T_{wait}*, an enzyme binds to DNA and completely unlinks the two molecules in a time* T_r *(notice the sharp increase in extension resulting from removal of the last link). A similar behaviour is observed on right- or left-handed braids. To ensure observation of single enzymatic activity, the concentration of enzyme is such that on average* $T_{wait} \gg T_r$*. (b) Record of unlinking by topoIV. While right-handed braids are only partially unlinked, left-handed one are completely relaxed. Adapted from [423].*

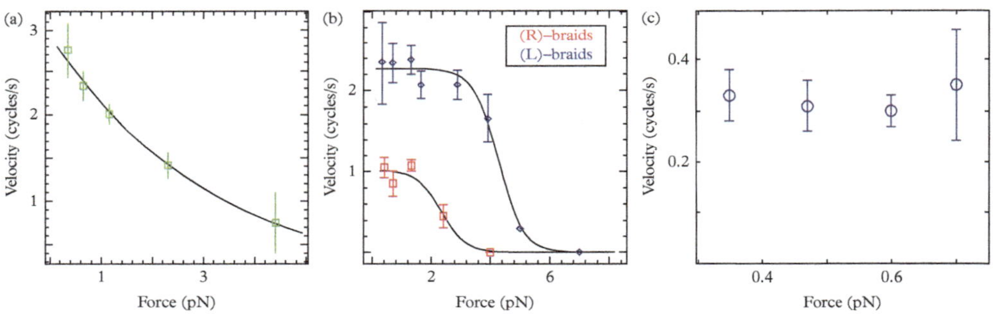

Fig. 9.14 Force dependence of the rates of unlinking of two braided DNAs *by a eukaryotic (Yeast) topoII (a), a prokaryotic (E.coli) topoIV (b), and of coiling of a single DNA by a prokaryotic (E.coli) gyrase (c) in buffer containing a saturating concentration of ATP (1 mM). Notice the strong force dependence of the eukaryotic enzyme, suggesting that its rate-limiting step is force-dependent, and the independence (at low force) of the rates of topoIV and gyrase, implying that their rate-limiting step is not force-dependent. Adapted from [423] and [238].*

buckled to form *left-handed* plectonemes of braids. These left-handed supercoils are similar to the (+) supercoils of DNA, which are the correct substrate for topoIV activity.

These results suggest that recognition of the chirality of braids by topoIV may depend on the crossing angle (α, see Fig. 9.12) made by the gate and transport segment, which (if $\alpha \neq 90°$) will be different for left-handed and right-handed crossings. However, precise single-molecule measurements of the preferred crossing angle of topoIV indicate that

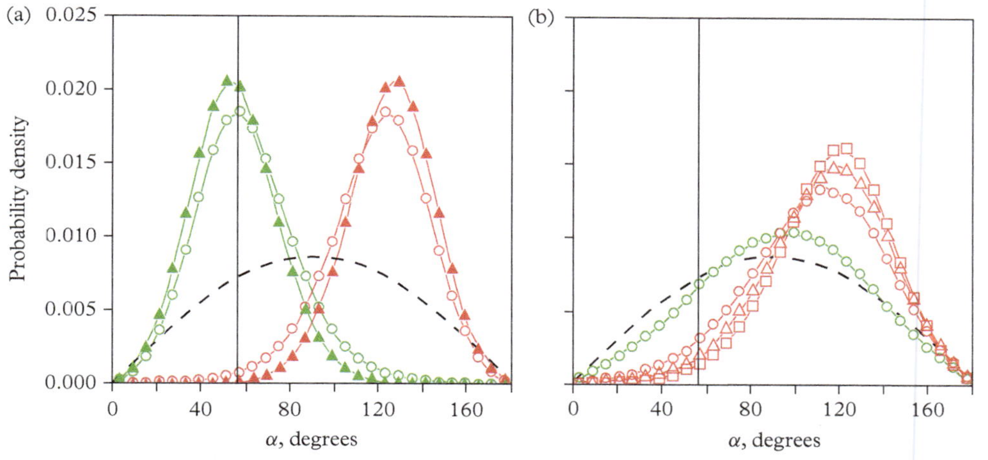

Fig. 9.15 Distribution of crossing angles in numerical simulations of supercoils and braids *for a 3.5 kbp DNA. (a) Distribution of the left-handed (green) and right-handed (red) crossing angle α in supercoiled DNA (circles) and braids (triangles) for $n = \pm 17$ turns. Notice the peak of the distribution at $\alpha = 57°$, which was suggested to be the preferred angle for unlinking by topoIV. (b) Distribution of the left-handed crossing angle for $n = 15, 12, 8$ turns (respectively red squares, triangles, and circles) and $n = 4$ turns (green circles) for 3.5 kbp braided circular molecules. Notice that for $n = 4$, the distribution is very close to the symmetric one expected for a random superposition of DNA segments (dashed line). Adapted from [428].*

this angle at 85° is too close to 90° to account for the difference in uncoiling activity of the enzyme on (+) and (−) supercoiled DNA [429].

The paradox of sister chromosome unlinking by topoIV may then have two solutions. Perhaps the sister chromosomes do not wind around each other, which may be the case if the two replication forks are anchored to each other or to the bacterial membrane. This anchoring creates a topological barrier that prevents the positive torsion ahead of the replication fork from inducing the two DNA behind to wind about each other. Alternatively, the two strands may wind around each other to such an extent as to form supercoils of right-handed braids that can be efficiently (if only partially) removed by topoIV. For removal of the last few braids, the enzyme may have to rely on angular fluctuations of right-handed DNA crossings to present the correct geometry for their unlinking, see Fig. 9.12 and Fig. 9.15. Numerical simulations of braided molecules do indeed suggest that the probability distribution of angular crossings becomes evenly distributed about $\alpha = 90°$ at low degrees of braiding [428], see Fig. 9.15.

9.5 Single-molecule study of the coiling activity of gyrases

DNA gyrases are type II topoisomerases whose main role in bacteria is not to relax torsional stresses in DNA but to negatively coil the molecule by wrapping it around

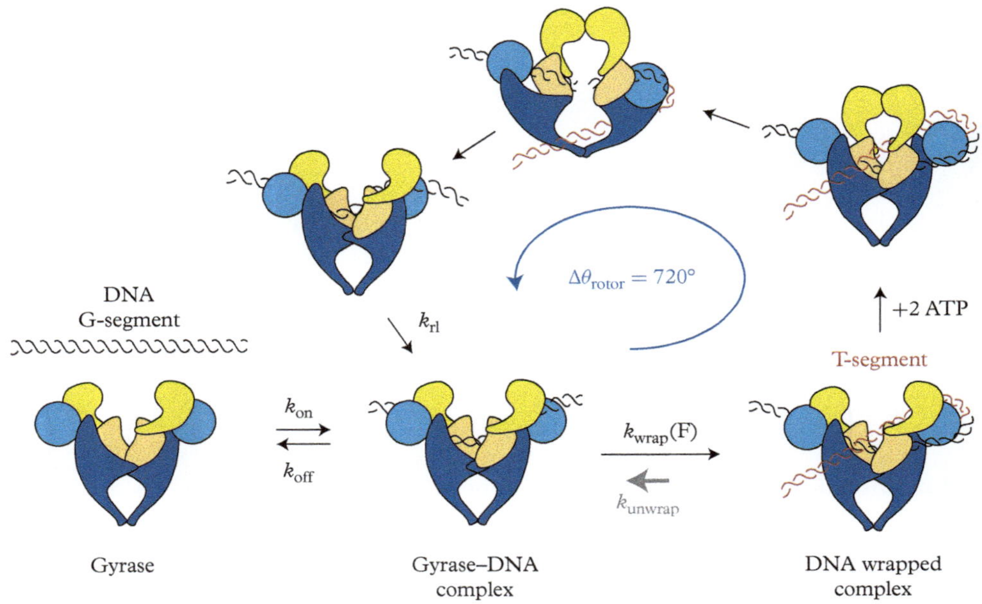

Fig. 9.16 Mechanistic model of the action of a DNA gyrase. *DNA wraps around a specific domain of the enzyme and is transferred through the gyrase gate segment as would a T-segment in a type II topoisomerase, see Fig. 9.2. Since wrapping is inhibited by DNA tension, the enzyme may dissociate from the state labelled gyrase–DNA. However, upon wrapping and ATP binding, the enzyme is committed to a full catalytic cycle in which two negative supercoils are introduced to the DNA, causing the rotor bead to spin by two turns in magnetic trap experiments. Adapted from [238].*

a specific domain of the enzyme before strand passage, see Fig. 9.16. This activity enhances the interaction of various enzymes (e.g., RNA polymerases) whose interactions with DNA require local melting of the double helix, see section 7.6.2. In a ground-breaking study [238] the mechanism of a single gyrase was studied using a magnetic trap to pull on a DNA molecule. A small "rotor" bead was attached to the middle of the DNA just below an engineered single-strand nick, which acts as a free swivel, and above a strong gyrase binding site. Every coil introduced in the DNA by the gyrase is thus released through the rotation of this bead, see Fig. 9.17a.

By monitoring the rotation of the bead over time, one can measure the rate of coiling by gyrase, see Fig. 9.14c, which is independent of force (as is the rate of uncoiling/unlinking by topoIV). Hence the rate-limiting step of the reaction cycle is force-independent. However, the processivity of the enzyme (the total number of catalytic cycles) is strongly dependent on the force, which implies that tension is favouring enzyme dissociation over catalysis. A model assuming a simple Arrhenius law for the force dependence of DNA wrapping around the enzyme (which competes with dissociation), see Fig. 9.16, yields a very good fit to the observed force dependence of the gyrase's processivity [238], see Fig. 9.17c.

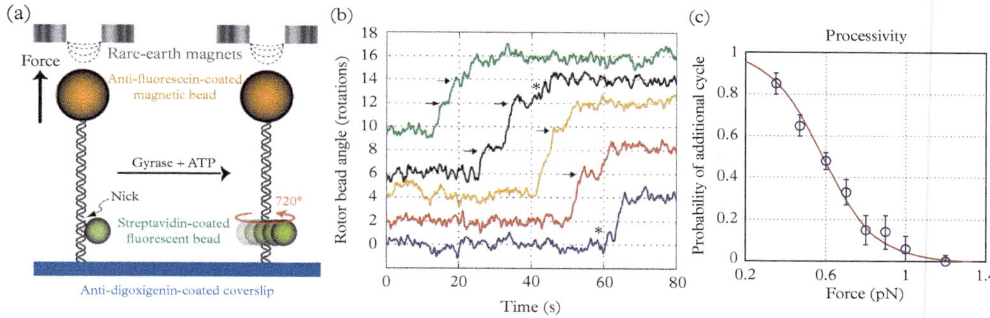

Fig. 9.17 The rotor-bead assay and monitoring of gyrase activity. *(a) A DNA molecule end-labelled with fluorescein (fluo) and digoxigenin (dig) is tethering an anti-fluo magnetic bead to an anti-dig coated surface. The magnetic bead is pulled by the field of small magnets positioned above it. A biotin inserted below a nick in the molecule binds to a small non-magnetic (rotor) bead, below which is a strong gyrase binding site. Supercoils introduced by the gyrase are relaxed by rotation of the rotor bead. (b) Monitoring the angle of the rotor bead with time allows one to monitor the activity of the gyrase. One notices pauses at multiples of two turns, corresponding to pauses just after (or before) the passage of one DNA strand through the other. (c) The number of observed successive enzymatic cycles (the processivity) depends strongly on the force. It can be described (continuous line) by assuming that the wrapping rate is decreased exponentially by the applied force (Arrhenius law), thus favouring the dissociation channel. Adapted from [238].*

9.6 *In vivo* studies of single topoIV localization and action in *E.coli*

As previously mentioned, the development of novel single-molecule visualization methods (see Chapter 3) have opened a new vista in the study of processes at the single-molecule level in live cells. While we previously exemplified the application of these methods to the study of replication, see section 7.4, we shall here present the recent use of these methods to the study of decatenation by topoIV in single *E.coli*. In these experiments [79], the native ParC and ParE subunits of topoIV were labelled with a photo-activable fluorescent protein: PAmCherry. By photo-activating only a small fraction of the labelled molecules, the authors could track them individually for about one second before they photobleached. From the mean square displacement of the individual traces (over only four successive steps) a diffusion constant for each molecule could be estimated (albeit with a large error, see section 3.5) and various subpopulations identified, see Fig. 9.18b. Of the observed ~260 molecules of ParC (or ParE) per cell, one third seemed to be localized at the replication forks (immobilized at one or two spots, see Fig. 9.18a). The remaining two thirds were diffusing by a factor of 20 slower than freely diffusing Green Fluorescent Protein in the *E.coli* cytoplasm [430], probably because of their transient association with DNA. The authors estimate that at any time there are about forty-five topoIV complexes actively decatenating the newly replicated molecules

(a)

(b)

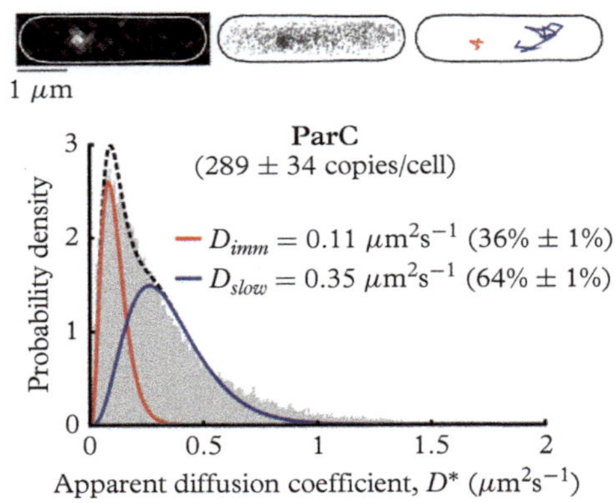

Fig. 9.18 *In vivo* study of the localization of the ParC units of topoIV. *(a) Left: image of a single ParC labelled with PAmCherry at 15 ms exposure time; Middle: super-resolved localization from all frames and all molecules in the imaged cell; Right: example of individual tracks of immobile (red) and slowly diffusing (blue) ParC units in the cell. (b) Apparent diffusion coefficient of ParC units deduced from the mean square displacement of only four successive steps in tracks, as shown in (a), and fitted to a two-population model. Adapted from [79].*

at a rate of about 0.55 cycle/s, not much different from the unlinking rate of R-braids at low forces measured in single-molecule experiments, see Fig. 9.14.

However, the estimated dissociation constant for the topoIV subunits *in vivo* [79] was more than two orders of magnitude larger than that measured *in vitro* on supercoiled DNA. This can be explained either by cellular factors that alter the affinity of these units to supercoiled DNA, or by a relaxation of DNA supercoiling during replication that results in a one hundred times lower affinity of topoIV for DNA, see section 9.4.3.

10 Conclusions

Ever since the 1990s, single-molecule experiments have revolutionized the field of biophysics. The stretching of biopolymers has ushered in a new age of force spectroscopy, which has provided new insight into complex interactions at the molecular level. The weak forces that determine the shape of biological molecules can be measured directly and accurately. Simple proteins can be unfolded, and their fine structure revealed. Single DNA and RNA molecules are now routinely manipulated, stretched, and twisted. These new techniques have brought valuable and unique information on the folding dynamics of biomolecules. Phase transitions between different molecular conformations have been induced and controlled by mechanical stress and observed by fluorescent techniques. Thus, simple RNA loops were observed to fold and unfold in real time, allowing one to probe theories of non-equilibrium statistical mechanics.

The single-molecule manipulation techniques have also allowed for a study of the function of many associated structural proteins and molecular motors under tensional or torsional load. These studies have shed a new light on the mechanism of these proteins/enzymes which is often relevant to their *in vivo* function.

In parallel, the development of single-molecule fluorescence and FRET measurements has allowed for a much more detailed understanding of the dynamics and mechanisms of enzymes at the nanometre scale. These methods are now being applied *in vivo*, to study the dynamics of biological networks within the cell. It is our hope and belief that the knowledge gained from the single-molecule *in vitro* studies described in this book will benefit these further investigations in a cellular context.

Single-Molecule Studies of Nucleic Acids and Their Proteins. David Bensimon, Vincent Croquette, Jean-François Allemand, Xavier Michalet, Terence Strick 2019. © Oxford University Press 2019. Published in 2019 by Oxford University Press.
DOI: 10.1093/oso/9780198530923.001.0001

References

1. Watson JD, Crick FHC. Molecular structure of nucleic acids: a structure for deoxyribose nucleic acid. *Nature*, 171:737–38, 1953.
2. Avery OT, MacLeod CM, McCarty M. Studies on the chemical nature of the substance inducing transformation of pneumococcal types: Induction of transformation by a deoxyribonucleic acid fraction isolated from pneumococcus type III. *The Journal of Experimental Medicine*, 79(2):137–58, 1944.
3. Hershey AD, Chase M. Independent functions of viral protein and nucleic acid in growth of bacteriophage. *The Journal of General Physiology*, 36(1):39–56, 1952.
4. Watson JC, Crick FHC. Genetic implications of the structure of deoxyribonucleic acid. *Nature*, 171:964–67, 1953.
5. de Gennes PG. *Scaling concepts in polymer physics*. Cornell University Press, 1979.
6. Saenger W. *Principles of nucleic acid structure*. Springer-Verlag, 1988.
7. Meselson M, Stahl FW. The replication of DNA in Escherichia coli. *Proc. Nat. Acad. Sci. USA*, 44(7):671–82, 1958.
8. Paulson JR, Laemmli UK. The structure of histone-depleted metaphase chromosomes. *Cell*, 12(3):817–28, 1977.
9. Felsenfeld G, Groudine M. Controlling the double helix. *Nature*, 421(6921):448–53, 2003.
10. Yusupov MM, Yusupova GZ, Baucom A, Lieberman K, Earnest TN, Cate JHD, Noller HF. Crystal structure of the ribosome at 5.5 Angstrom resolution. *Science*, 292:883–96, 2001.
11. Ogle JM, Brodersen DE, Clemons WM, Tarry MJ, Carter AP, Ramakrishnan V. Recognition of cognate transfer RNA by the 30S ribosomal subunit. *Science*, 292:897–902, 2001.
12. Xayaphoummine A, Bucher T, Thalmann F, Isambert H. Prediction and statistics of pseudoknots in RNA structures using exactly clustered stochastic simulations. *Proc. Nat. Acad. Sci. USA*, 100:15310–15, 2003.
13. Chee M, Yang R, Hubbell E, Berno A, Huang XC, Stern D, Winkler J, Lockhart DJ, Morris MS, Fodor SPA. Accessing genetic information with high-density DNA arrays. *Science*, 274:610–14, 1996.
14. Sano T, Smith CL, Cantor CR. Immuno-PCR: very sensitive antigen detection by means of specific antibody-DNA conjugates. *Science*, 258:120–22, 1992.
15. Smith SB, Finzi L, Bustamante C. Direct mechanical measurements of the elasticity of single DNA molecules by using magnetic beads. *Science*, 258:1122–26, 1992.
16. Anderson S, Bankier AT, Barrell BG, de Bruijn MHL, Coulson AR, Drouin J, Eperon IC, Nierlich DP, Roe BA, Sanger F, Schreier PH, Smith AJH, Staden R, Young IG. Sequence and organization of the human mitochondrial genome. *Nature*, 290:457–65, 1981.
17. Strick T, Allemand JF, Bensimon D, Bensimon A, Croquette V. The elasticity of a single supercoiled DNA molecule. *Science*, 271:1835–37, 1996.
18. Strick T, Allemand JF, Bensimon D, Croquette V. The behavior of supercoiled DNA. *Biophys. J.*, 74:2016–28, 1998.
19. Allemand JF, Bensimon D, Jullien L, Bensimon A, Croquette V. pH-dependent specific binding and combing of DNA. *Biophys. J.*, 73:2064–70, 1997.

20. Bensimon A, Simon AJ, Chiffaudel A, Croquette V, Heslot F, Bensimon D. Alignment and sensitive detection of DNA by a moving interface. *Science*, 265:2096–98, 1994.
21. Bensimon D, Simon AJ, Croquette V, Bensimon A. Stretching DNA with a receding meniscus: Experiments and models. *Phys. Rev. Lett.*, 74:4754–57, 1995.
22. Weier HUG, Wang M, Mullikin JC, Zhu Y, Cheng JF, Greulich KM, Bensimon A, Gray JM. Quantitative DNA fiber mapping. *Human Mol. Gen.*, 4:1903–10, 1995.
23. Michalet X, Ekong R, Fougerousse F, Rousseaux S, Schurra C, Hornigold N, van Slegtenhorst M, Wolfe J, Povey S, Beckmann JS, Bensimon A. Dynamic molecular combing: stretching the whole human genome for high-resolution studies. *Science*, 277:1518–23, 1997.
24. Finer JT, Simmons RM, Spudich JA. Single myosin molecule mechanics: piconewton forces and nanometre steps. *Nature*, 368:113–19, 1994.
25. Essevaz-Roulet B, Bockelmann U, Heslot F. Mechanical separation of the complementary strands of DNA. *Proc. Nat. Acad. Sci. USA*, 94:11935–40, 1997.
26. Bockelmann U, Essevaz-Roulet B, Heslot F. Molecular stick-slip revealed by opening DNA with piconewton forces. *Phys. Rev. Lett.*, 79(22):4489–92, 1997.
27. Florin EL, Moy VT, Gaub HE. Adhesion forces between individual ligand-receptor pairs. *Science*, 264:415–17, 1994.
28. Moy VT, Florin EL, Gaub HE. Intermolecular forces and energies between ligands and receptors. *Science*, 266:257–59, 1994.
29. Hon T, Williams JM, Beebe Jr. TP. Determination of single-bond forces from contact force variances in atomic force microscopy. *Langmuir*, 12:1291–95, 1996.
30. Dammer U, Popescu O, Wagner P, Anselmetti D, Günherodt H-J, Misevic GN. Binding strength between cell adhesion proteoglycans measured by atomic force microscopy. *Science*, 267:1173–75, 1995.
31. Lee GU, Chrisey LA, Colton RJ. Direct measurement of the forces between complementary strands of DNA. *Science*, 266:771–73, 1994.
32. Cluzel P, Lebrun A, Heller C, Lavery R, Viovy JL, Chatenay D, Caron F. DNA: an extensible molecule. *Science*, 271:792–94, 1996.
33. Smith SB, Cui Y, Bustamante C. Overstretching B-DNA: the elastic response of individual double-stranded and single-stranded DNA molecules. *Science*, 271:795–99, 1996.
34. Lebrun A, Lavery R. Modelling extreme stretching of DNA. *Nucl. Acid, Res.*, 24:2260, 1996.
35. Ishijima A, Doi T, Sakurada K, Yanagida T. Sub-piconewtown force fluctuations of actomyosin in vitro. *Nature*, 352:301–06, 1991.
36. Ashkin A, Dziedzic JM, Bjorkholm JE, Chu S. Observation of a single-beam gradient force optical trap for dielectric particles. *Opt. Lett.* 11: 288–290, 1986.
37. Gosse C, Croquette V. Magnetic tweezers. *Biophys. J.*, 82, 3314–29, 2002.
38. Crick F.H.C., A.F.W.Hughes, The physical properties of cytoplasm: A study by means of the magnetic particle method. Part I. Experimental, *Exp.Cell Res.* 1:37–80, 1950.
39. Evans E, Ritchie K, Merkel R. Sensitive force technique to probe molecular adhesion and structural linkages at biological interfaces. *Biophys. J.*, 68:2580–87, 1995.
40. Rief M, Gautel M, Oesterhelt F, Fernandez JM, Gaub HE. Reversible unfolding of individual titin immunoglobulin domains by AFM. *Science*, 276:1109–12, 1997.
41. Léger JF, Robert J, Bourdieu L, Chatenay D, Marko JF. RecA binding to a single double-stranded DNA molecule: a possible role of DNA conformational fluctuations. *Proc. Nat. Acad. Sci. USA*, 95:12295–96, 1998.
42. Lee J-B, Hite RK, Hamdan SM, Xie XS, Richardson CC, van Oijen AM. DNA primase acts as a molecular brake in DNA replication. *Nature*, 439:621–24, 2006.

43. Hill FR, Monachino E, van Oijen AM. The more the merrier: high-throughput single molecule techniques. *Biochem. Soc. Trans.*, 45:739–65, 2017.
44. Lewis JS, Spenkelink LM, Schauer GD, Hill FR, Georgescu RE, O'Donnell ME, van Oijen AM. Single-molecule visualization of *Saccharomyces cerevisiae* leading-strand synthesis reveals dynamic interaction between MTC and the replisome. *Proc. Nat. Acad. Sci. USA*, 2017.
45. Einstein A. *Investigation of the Brownian theory of movement.* Dover Publication, 1956.
46. Reif F. *Fundamentals of statistical and thermal physics.* McGraw-Hill, New York, 1965.
47. Smith SB, Cui YJ, Bustamante C. Optical-trap force transducer that operates by direct measurement of light momentum. *Meth. in Enzym.*, 361:134–62, 2003.
48. Smith SB. *Stretch transitions observed in single biopolymer molecules (DNA or protein) using laser tweezers.* Thesis, U. Twente, Netherlands, 1998.
49. Deufel C, Forth S, Simmons CR, Dejgosha S, Wang MD. Nanofabricated quartz cylinders for angular trapping: DNA supercoiling torque detection. *Nat. Meth.*, 4:223–25, 2007.
50. La Porta A, Wang MD. Optical torque wrench: angular trapping, rotation, and torque detection of quartz microparticles. *Phys. Rev. Lett.*, 92(19):190801, 2004.
51. Wu JR. Acoustical tweezers. *J. Acous. Soc. Am.*, 89(5):2140–43, 1991.
52. Ding X, Lin S-CS, Kiraly B, Yue H, Li S, Chiang I-K, Shi J, Benkovic SJ, Huang TJ. On-chip manipulation of single microparticles, cells, and organisms using surface acoustic waves. *Proc. Nat. Acad. Sci. USA*, 109(28):11105–09, 2012.
53. Sitters G, Kamsma D, Thalhammer G, Ritsch-Marte M, Peterman EJG, Wuite GJL. Acoustic force spectroscopy. *Nat. Meth.*, 12(1):47–50, 2015.
54. Ahmed D, Ozcelik A, Bojanala N, Nama N, Upadhyay A, Chen Y, Hanna-Rose W, Huang TJ. Rotational manipulation of single cells and organisms using acoustic waves. *Nat. Comm.*, 7:11085, 2016.
55. Léger J-F. L'ADN: une flexibilité structurale adaptée aux interactions avec les autres macromolécules de son environnement. PhD thesis, Université Louis Pasteur Strasbourg I, 1999.
56. Dammer U, Hegner M, Anselmetti D, Wagner P, Dreier M, Huber W, Guntherodt H-J. Specific antigen/antibody interactions measured by force microscopy. *Biophys. J.*, 70: 2437–41, 1996.
57. Basché T, Moerner WE, Orrit M, Wild UP. *Single-molecule optical detection, imaging and spectroscopy.* VCH, Weinheim, 1997.
58. Rigler R, Elson ES. *Fluorescence correlation spectroscopy: theory and applications.* Springer-Verlag, 2001.
59. Rigler R, Orrit M, Basché T. *Single molecule spectroscopy.* Springer-Verlag, Stockholm, 2001.
60. Zander CH, Enderlein J, Keller RA. *Single molecule detection in solution: methods and applications.* Wiley-VCH, 2002.
61. Gell C, Brockwell D, Smith A. *Handbook of single molecule fluorescence spectroscopy.* OUP, Oxford, 2006.
62. Selvin PR, Ha T. *Single molecule techniques: a laboratory manual.* CSHL Press, 2008.
63. Moerner WE, Kador L. Optical detection and spectroscopy of single molecules in a solid. *Phys. Rev. Lett.*, 62(21):2535–38, 1989.
64. Orrit M, Bernard J. Single pentacene molecules detected by fluorescence excitation in a p-terphenyl crystal. *Phys. Rev. Lett.*, 65(21):2716–19, 1990.
65. Betzig E, Chichester RJ. Single molecules observed by near-field scanning optical microscopy. *Science*, 262(5138):1422–25, 1993.

66. Basché T, Moerner WE, Orrit M, Talon H. Photon antibunching in the fluorescence of a single dye molecule trapped in a solid. *Phys. Rev. Let.*, 69:1516–19, 1992.
67. Trautman JK, Macklin JJ, Brus LE, Betzig E. Near-field spectroscopy of single molecules at room temperature. *Science*, 369:40–42, 1994.
68. Ambrose WP, Moerner WE. Fluorescence spectroscopy and spectral diffusion of single impurity molecules in a crystal. *Nature*, 349:225–27, 1991.
69. Shera EB, Seitzinger NK, Davis LM, Keller RA, Soper SA. Detection of single fluorescent molecules. *Chem. Phys. Lett.*, 174(6):553–57, 1990.
70. Schmidt T, Schütz GJ, Baumgartner W, Gruber HJ, Schindler H. Imaging of single molecule diffusion. *Proc. Nat. Acad. Sci. USA*, 93:2926–29, 1996.
71. Schütz GJ, Kada G, Pastushenko VP, Schindler H. Properties of lipid microdomains in a muscle cell membrane visualized by single molecule microscopy. *Embo. J.*, 19(5):892–901, 2000.
72. Sako Y, Hibino K, Miyauchi T, Miyamoto Y, Ueda M, Yanagida T. Single-molecule imaging of signaling molecules in living cells. *Single Mol.*, 1(2):159–63, 2000.
73. Moerner WE. Examining nanoenvironments in solids on the scale of a single, isolated impurity molecule. *Science*, 265:46–53, 1994.
74. Moerner WE, Orrit M. Illuminating single molecules in condensed matter. *Science*, 283: 1670–76, 1999.
75. English DS, Harbron EJ, Barbara PF. Role of rare sites in single molecule spectroscopy measurements of spectral diffusion. *J. Chem. Phys.*, 114(23):10479–85, 2001.
76. A.Gaiduk, M.Yorulmaz, P.V.Ruijgrok and M.Orrit, Room temperature detection of a single molecule's absorption by photothermal contrast, *Science*, 330:353–356, 2010.
77. Kapanidis AN, Weiss S. Fluorescent probes and bioconjugation chemistries for single-molecule fluorescence analysis of biomolecules. *J. Chem. Phys.*, 117(24):10953–64, 2002.
78. Reyes-Lamothe R, Sherratt DJ, Leake MC. Stoichiometry and architecture of active DNA replication machinery in Escherichia coli. *Science*, 328(5977):498–501, 2010.
79. Zawadzki P, Stracy M, Ginda K, Zawadzka K, Lesterlin C, Kapanidis AN, Sherratt DJ. The localization and action of topoisomerase IV in Escherichia coli chromosome segregation is coordinated by the SMC complex, MukBEF. *Cell*, 13:2587–96, 2015.
80. Weiss S. Fluorescence spectroscopy of single biomolecules. *Science*, 283:1676–83, 1999.
81. Weiss S. Shattering the diffraction limit of light: a revolution in fluorescence microscopy? *Proc. Nat. Acad. Sci. USA*, 97:8747–49, 2000.
82. Ishijima A, Yanagida T. Single molecule nanobioscience. *Trends Biochem. Sci.*, 26(7):438–44, 2001.
83. Nie S, Zare RN. Optical detection of single molecules. *Ann. Rev. Biophys. Biomolec. Struct.*, 26:567–96, 1997.
84. Plakhotnik T, Donley EA, Wild UP. Single-molecule spectroscopy. *Ann. Rev. Phys. Chem.*, 48:181–212, 1997.
85. Xie XS, Trautman JK. Optical studies of single molecules at room temperature. *Ann. Rev. of Phys. Chem.*, 49:441–80, 1998.
86. Tamarat P, Maali A, Lounis B, Orrit M. Ten years of single-molecule spectroscopy. *J. Phys. Chem. A*, 104(1):1–16, 2000.
87. Michalet X, Weiss S. Single-molecule spectroscopy and microscopy. *C. R. Phys.*, 3(5):619–44, 2002.
88. Funatsu T, Harada Y, Tokunaga M, Saito K, Yanagida T. Imaging of single fluorescent molecules and individual ATP turnovers by single myosin molecules in aqueous solution. *Nature*, 374:555–59, 1995.

89. Bouchaud JP. Weak ergodicity breaking and aging in disordered systems. *Journal De Physique I*, 2(9):1705–13, 1992.
90. Beck VA, Shaqfeh ESG. Ergodicity-breaking and the unraveling dynamics of a polymer in linear and nonlinear extensional flows. *Journal of Rheology*, 51(3):561–74, 2007.
91. Lubelski A, Sokolov IM, Klafter J. Nonergodicity mimics inhomogeneity in single particle tracking. *Phys. Rev. Lett.*, 100(25):250602, 2008.
92. Riggsbee CW, Deiters A. Recent advances in the photochemical control of protein function. *Trends in Biotechnology*, 28(9):468–75, 2010.
93. Zhou XX, Chung HK, Lam AJ, Lin MZ. Optical control of protein activity by fluorescent protein domains. *Science*, 338(6108):810–14, 2012.
94. Perrin J. La fluorescence. *Ann. Phys. Paris*, 10(9):133–59, 1918.
95. Hirschfeld T. Optical microscopic observation of single small molecules. *Appl. Optics*, 15(12):2965–66, 1976.
96. Harada Y, Yanagida T. Direct observation of molecular motility by light microscopy. *Cell Motility and the Cytoskeleton*, 10(1–2):71–76, 1988.
97. Li LQ, Davis LM. Single photon avalanche diode for single molecule detection. *Review of Scientific Instruments*, 64(6):1524–29, 1993.
98. Nie S, Chiu DT, Zare RN. Probing individual molecules with confocal fluorescence microscopy. *Science*, 266:1018–21, 1994.
99. Zander C, Sauer M, Drexhage KH, Ko DS, Schulz A, Wolfrum J, Brand L, Eggeling C, Seidel CAM. Detection and characterization of single molecules in aqueous solution. *Applied Physics B-Lasers and Optics*, 63(5):517–23, 1996.
100. Rigler R, Mets U. Diffusion of single molecules through a Gaussian laser beam. *SPIE Proceedings*, 1921:239–48, 1992.
101. Widengren J, Metz U., Rigler R. Fluorescence correlation spectroscopy of triplet states in solution: a theoretical and experimental study. *J.Phys.Chem*, 99:13368–13379, 1995.
102. Krichevsky O, Bonnet G. Fluorescence correlation spectroscopy: the technique and its applications. *Rep. Prog. Phys.*, 65:251–97, 2002.
103. Macklin JJ, Trautman JK, Harris TD, Brus LE. Imaging and time-resolved spectroscopy of single molecules at an interface. *Science*, 272:255–58, 1996.
104. Bartko AP, Dickson RM. Imaging three-dimensional single molecule orientations. *J. Phys. Chem. B*, 103:11237–41, 1999.
105. Nie S, Emory SR. Probing single molecules and single nanoparticles with surface-enhanced raman scattering. *Science*, 275:1102–06, 1997.
106. Lounis B, Jelezko F, Orrit M. Single molecules driven by strong resonant fields: Hyper-raman and subharmonic resonances. *Physical Review Letters*, 78(19):3673–76, 1997.
107. Empedocles SA, Bawendi MG. Quantum-confined stark effect in single CdSe nanocrystallite quantum dots. *Science*, 278:2114–17, 1997.
108. Celebrano M, Kukura P, Renn A, Sandoghdar V. Single-molecule imaging by optical absorption. *Nature Photonics*, 5(2):95–98, 2011.
109. Hildner R, Brinks D, van Hulst NF. Femtosecond coherence and quantum control of single molecules at room temperature. *Nature Physics*, 7(2):172–77, 2011.
110. Weiss S. Measuring conformational dynamics of biomolecules by single molecule fluorescence spectroscopy. *Nat. Struct. Biol.*, 7(9):724–29, 2000.
111. Atkins P. *Physical Chemistry, Vol. II.* Oxford University Press, New York, 10th edition, 2015.
112. Lakowicz JR. *Principles of Fluorescence Spectroscopy*. Plenum, New York, 3rd Edition, 2006.
113. Einstein A. The quantum theory of radiation. *Phys. Zeitschr.*, 18:121–28, 1917.

114. Hilborn RC. Einstein coefficients, cross sections, f values, dipole moments, and all that. *Am. J. Phys.*, 50:982–86, 1982.
115. Yariv A. *Quantum electronics*. Wiley, New York, 2nd edition 1975.
116. Cohen-Tannoudji C, Dupont-Roc J, Grynberg G. *Atom-photon interactions. Basic processes and applications*. Wiley, NY, 1992.
117. Karu T. Primary and secondary mechanisms of action of visible to near-IR radiation on cells. *Journal of Photochemistry and Photobiology B: Biology*, 49(1):1–17, 1999.
118. Tinevez JY, Dragavon J, Baba-Aissa L, Roux P, Perret E, Canivet A, Galy V, Shorte S. A quantitative method for measuring phototoxicity of a live cell imaging microscope. *Meth. Enzym.*, 506:291–309, 2012.
119. Faris GW, Copeland RA. Wavelength dependence of the Raman cross section for liquid water. *App. Optics*, 36(12):2686–88, 1997.
120. Strickler SJ, Berg RA. Relationship between absorption intensity and fluorescence lifetime of molecules. *J. Chem. Phys.*, 37(4):814–22, 1962.
121. Valeur B, Berberan-Santos MN. *Molecular fluorescence: Principles and applications*. Wiley-VCH, 2nd edition, 2012.
122. Dirac PAM. The quantum theory of emission and absorption of radiation. *Proc. Roy. Soc. London A*, 114:243–65, 1927.
123. Byron CM, Werner TC. Experiments in synchronous fluorescence spectroscopy for the undergraduate instrumental chemistry course. *Journal of Chemical Education*, 68(5):433–36, 1991.
124. Xie SX, Dunn RC. Probing single molecule dynamics. *Science*, 265(5170):361–64, 1994.
125. Ambrose WP, Goodwin PM, Martin JC, Keller RA. Alterations of single molecule fluorescence lifetimes in near-field optical microscopy. *Science*, 265(5170):364–67, 1994.
126. Gersen H, Garcia-Parajo MF, Novotny L, Veerman JA, Kuipers L, van Hulst NF. Influencing the angular emission of a single molecule. *Phys. Rev. Lett.*, 85(25):5312–15, 2000.
127. Herten DP, Tinnefeld P, Sauer M. Identification of single fluorescently labelled mononucleotide molecules in solution by spectrally resolved time-correlated single-photon counting. *Applied Physics B-lasers and Optics*, 71(5):765–71, 2000.
128. Eggeling C, Berger S, Brand L, Fries JR, Schaffer J, Volkmer A, Seidel CAM. Data registration and selective single-molecule analysis using multi-parameter fluorescence detection. *Journal of Biotechnology*, 86:163–80, 2001.
129. Ha T, Enderle Th, Chemla DS, Selvin PR, Weiss S. Quantum jumps of single molecules at room temperature. *Chem. Phys. Lett.*, 271:1–5, 1997.
130. Dickson RM, Cubitt AB, Tsien RY, Moerner WE. On/off blinking and switching behaviour of single molecules of green fluorescent protein. *Nature*, 388(6640):355–58, 1997.
131. Empedocles SA, Norris DJ, Bawendi MG. Photoluminescence spectroscopy of single CdSe nanocrystallite quantum dots. *Phys. Rev. Lett.*, 77(18):3873–76, 1996.
132. Efros AL, Rosen M. Random telegraph signal in the photoluminescence intensity of a single quantum dot. *Phys. Rev. Lett.*, 78(6):1110–13, 1997.
133. Banin U, Bruchez M, Alivisatos AP, Ha T, Weiss S, Chemla DS. Evidence for a thermal contribution to emission intermittency in single CdSe/CdS core/shell nanocrystals. *J. Chem. Phys.*, 110(2):1195–201, 1999.
134. Roy R, Hohng S, Ha T. A practical guide to single-molecule FRET. *Nat. Meth.*, 5(6):507–16, 2008.
135. Rasnik I, Myong S, Ha T. Unraveling helicase mechanisms one molecule at a time. *Nucl. Acid. Res.*, 34:4225–31, 2006.
136. Ha T, Laurence TA, Chemla DS, Weiss S. Polarization spectroscopy of single fluorescent molecules. *J. Phys. Chem. B*, 103(33):6839–50, 1999.

137. Forkey JN, Quinlan ME, Goldman YE. Protein structural dynamics by single-molecule fluorescence polarization. *Progr. Biophy. Mol. Biol.*, 74:1–35, 2000.
138. Ha T, Glass J, Enderle T, Chemla DS, Weiss S. Hindered rotational diffusion and rotational jumps of single molecules. *Phys. Rev. Lett.*, 80(10):2093–96, 1998.
139. Förster T. Zwischenmolekulare Energiewanderung und Fluoreszenz. *Annalen der Physik*, 2: 55–75, 1948.
140. Wieb Van Der Meer B, Coker G, Chen III SYS. *Resonance energy transfer. Theory and data.* Wiley-VCH, 1991.
141. Selvin PR. The renaissance of fluorescence resonance energy transfer. *Nat. Struct. Biol.*, 7(9):730–34, 2000.
142. Ha T. Single-molecule fluorescence resonance energy transfer. *Methods*, 25(1):78–86, 2001.
143. Andrews DL. *Tutorials in complex photonic media.* SPIE Press, Bellingham, 2009.
144. Andrews DL, Rodriguez J. Resonance energy transfer: Spectral overlap, efficiency, and direction. *J. Chem. Phys.*, 127(8):084509, 2007.
145. Stryer L, Haugland RP. Energy transfer: a spectroscopic ruler. *Proc. Nat. Acad. Sci. USA*, 58:719–26, 1967.
146. Kalinin S, Peulen T, Sindbert S, Rothwell PJ, Berger S, Restle T, Goody RS, Gohlke H, Seidel CAM. A toolkit and benchmark study for FRET-restrained high-precision structural modeling. *Nature Methods*, 9(12):1218–29, 2012.
147. Kalinin S, Felekyan S, Antonik M, Seidel CAM. Probability distribution analysis of single-molecule fluorescence anisotropy and resonance energy transfer. *J. Phys. Chem. B*, 111(34):10253–62, 2007.
148. Iqbal A, Arslan S, Okumus B, Wilson TJ, Giraud G, Norman DG, Ha T, Lilley DMJ. Orientation dependence in fluorescent energy transfer between Cy3 and Cy5 terminally attached to double-stranded nucleic acids. *Proc. Nat. Acad. Sci. USA*, 105(32):11176–81, 2008.
149. Kalinin S, Sisamakis E, Magennis SW, Felekyan S, Seidel CAM. On the origin of broadening of single-molecule FRET efficiency distributions beyond shot noise limits. *J. Phys. Chem. B*, 114(18):6197–206, 2010.
150. Lipman EA, Schuler B, Bakajin O, Eaton WA. Single-molecule measurement of protein folding kinetics. *Science*, 301(5637):1233–35, 2003.
151. Kapanidis AN, Margeat E, Laurence TA, Doose S, Ho SO, Mukhopadhyay J, Mekler V, Ebright RH, Retention WS. Retention of transcription initiation factor sigma70 in transcription elongation: single-molecule analysis. *Mol. Cell*, 20:347–59, 2005.
152. Ha T, Tinnefeld P. Photophysics of fluorescent probes for single-molecule biophysics and super-reslution imaging. *Ann. Rev. Phys. Chem.*, 63:595–617, 2012.
153. Duveneck GL, Abel AP, Bopp MA, Kresbach GM, Ehrat M. Planar waveguides for ultra-high sensitivity of the analysis of nucleic acids. *Anal. Chim. Acta*, 469(1):49–61, 2002.
154. Mukundan H, Anderson AS, Grace WK, Grace KM, Hartman N, Martinez JS, Swanson BI. Waveguide-based biosensors for pathogen detection. *Sensors*, 9(7): 5783–809, 2009.
155. Levene MJ, Korlach J, Turner SW, Foquet M, Craighead HG, Webb WW. Zero-mode waveguides for single-molecule analysis at high concentrations. *Science*, 299(5607): 682–86, 2003.
156. Abbe E. Beitröge zur Theorie der Mikroskops und der Mikroskopischen Wahrnehmung. *Schultze's Archiv für mikroskopische Anatomie*, 9:413–68, 1873.
157. Hell SW, Wichmann J. Breaking the diffraction resolution limit by stimulated emission: stimulated-emission-depletion fluorescence microscopy. *Opt. Lett.*, 19(11):780–82, 1994.

158. Klar TA, Hell SW. Subdiffraction resolution in far-field fluorescence microscopy. *Opt. Lett.*, 24(14):954–56, 1999.
159. Klar TA, Jakobs S, Dyba M, Egner A, Hell SW. Fluorescence microscopy with diffraction resolution barrier broken by stimulated emission. *Proc. Nat. Acad. Sci. USA*, 97: 8206–10, 2000.
160. Honigmann A, Eggeling C, Schulze M, Lepert A. Super-resolution STED microscopy advances with yellow CW OPSL. *Laser Focus World*, 48(1):75–79, 2012.
161. Shusterman R, Alon S, Gavrinyov T, Krichevsky O. Monomer dynamics in double- and single-stranded DNA polymers. *Phys. Rev. Lett.*, 92:048303, 2004.
162. Deniz AA, Dahan M, Grunwell JR, Ha T, Faulhaber AE, Chemla DS, Weiss S, Schultz PG. Single-pair fluorescence resonance energy transfer on freely diffusing molecules: observation of Förster distance dependence and subpopulations. *Proc. Nat. Acad. Sci. USA*, 96(7): 3670–75, 1999.
163. Tinnefeld P, Buschmann V, Herten D-P, Han K-T, Sauer M. Confocal fluorescence lifetime imaging microscopy (FLIM) at the single molecule level. *Single Molecules*, 1(3):215–223, 2000.
164. van Oijen AM, Köhler J, Schmidt J, Müller M, Brakenhoff GJ. 3-dimensional super-resolution by spectrally selective imaging. *Chem. Phys. Lett.*, 292:183–87, 1998.
165. Lacoste TD, Michalet X, Pinaud F, Chemla DS, Alivisatos AP, Weiss S. Ultrahigh-resolution multicolor colocalization of single fluorescent probes. *Proc. Nat. Acad. Sci. USA*, 97(17): 9461–66, 2000.
166. Shroff H, Galbraith CG, Galbraith JA, Betzig E. Live-cell photoactivated localization microscopy of nanoscale adhesion dynamics. *Nat. Meth.*, 5(5):417–23, 2008.
167. Huang B, Wang W, Bates M, Zhuang X. Three-dimensional super-resolution imaging by stochastic optical reconstruction microscopy. *Science*, 319(5864):810–13, 2008.
168. Neuhauser RG, Shimizu KT, Woo WK, Empedocles SA, Bawendi MG. Correlation between fluorescence intermittency and spectral diffusion in single semiconductor quantum dots. *Phys. Rev. Lett.*, 85(15):3301–04, 2000.
169. Perrin J. Mouvement brownien et réalité moléculaire. *Ann. Chim. et Phys.*, 8(9):5–114, 1909.
170. Brown R. On the particles contained in the pollen of plants and the general existence of active molecules. *Edinburgh New Philosophical Journal*, 358–71, 1828.
171. Gelles J, Schnapp BJ, Sheetz M. Tracking kinesin-driven movements with nanometre-scale precision. *Nature*, 331:450–53, 1988.
172. Enderlein J. Path integral approach to fluorescence correlation experiments. *Physics Letters A*, 221(6):427–33, 1996.
173. Heilemann M, Herten DP, Heintzmann R, Cremer C, Muller C, Tinnefeld P, Weston KD, Wolfrum J, Sauer M. High-resolution colocalization of single dye molecules by fluorescence lifetime imaging microscopy. *Anal. Chem.*, 74(14):3511–17, 2002.
174. Ulbrich MH, Isacoff EY. Subunit counting in membrane-bound proteins. *Nat. Meth.*, 4(4):319–21, 2007.
175. Peters R. Single-molecule fluorescence analysis of cellular nanomachinery components. *Ann. Rev. Biophys. Biomol. Struct.*, 36:371–94, 2007.
176. Huang F, Hartwich TMP, Rivera-Molina FE, Lin Y, Duim WC, Long JJ, Uchil PD, Myers JR, Baird MA, Mothes W, Davidson MW, Toomre D, Bewersdorf J. Video-rate nanoscopy using sCMOS camera-specific single-molecule localization algorithms. *Nat. Meth.*, 10(7):653, 2013.
177. Betzig E. Proposed method for molecular optical imaging. *Opt. Lett.*, 20(3):237–39, 1995.

178. Michalet X, Lacoste TD, Pinaud F, Chemla DS, Alivisatos AP, Weiss S. Ultrahigh-resolution multicolor colocalization of single fluorescent probes. *Proc. Nat. Acad. Sci. USA*, 97(17):9461–66, 2000.
179. Thompson RE, Larson DR, Webb WW. Precise nanometer localization analysis for individual fluorescent probes. *Biophys. J.*, 82(5):2775–83, 2002.
180. Ober RJ, Ram S, Ward ES. Localization accuracy in single-molecule microscopy. *Biophy. J.*, 86(2):1185–1200, 2004.
181. Mortensen KI, Churchman LS, Spudich JA, Flyvbjerg H. Optimized localization analysis for single-molecule tracking and super-resolution microscopy. *Nat. Meth.*, 7(5):377–81, 2010.
182. Chao J, Ram S, Ward ES, Ober RJ. A comparative study of high resolution microscopy imaging modalities using a three-dimensional resolution measure. *Optics Express*, 17(26): 24377–402, 2009.
183. Deschout H, Zanacchi FC, Mlodzianoski M, Diaspro A, Bewersdorf J, Hess ST, Braeckmans K. Precisely and accurately localizing single emitters in fluorescence microscopy. *Nat. Meth.*, 11(3):253–66, 2014.
184. Izeddin I, Recamier V, Bosanac L, Cisse II, Boudarene L, Dugast-Darzacq C, Proux F, Benichou O, Voituriez R, Bensaude O, Dahan M, Darzacq X. Single-molecule tracking in live cells reveals distinct target-search strategies of transcription factors in the nucleus. *eLife*, 3, 2014.
185. Shav-Tal Y, Darzacq X, Shenoy SM, Fusco D, Janicki SM, Spector DL, Singer RH. Dynamics of single mRNPs in nuclei of living cells. *Science*, 304(5678):1797–800, 2004.
186. Bronstein I, Israel Y, Kepten E, Mai S, Shav-Tal Y, Barkai E, Garini Y. Transient anomalous diffusion of telomeres in the nucleus of mammalian cells. *Phy. Rev. Lett.*, 103(1), 2009.
187. Qian H, Sheetz MP, Elson EL. Single particle tracking. Analysis of diffusion and flow in two-dimensional systems. *Biophys. J.*, 60(4):910–21, 1991.
188. Saxton MJ. Single-particle tracking: the distribution of diffusion coefficients. *Biophys. J.*, 72(4):1744–53, 1997.
189. Michalet X, Berglund AJ. Optimal diffusion coefficient estimation in single-particle tracking. *Phys. Rev. E*, 85:061916, 2012.
190. Michalet X. Mean square displacement analysis of single-particle trajectories with localization error: Brownian motion in an isotropic medium. *Phys. Rev. E*, 83(5):059904, 2011.
191. Berglund AJ. Statistics of camera-based single-particle tracking. *Phys. Rev. E*, 82(1):011917, 2010.
192. Vestergaard CL, Blainey PC, Flyvbjerg H. Optimal estimation of diffusion coefficients from single-particle trajectories. *Phys. Rev. E*, 89(2):022726, 2014.
193. Backlund MP, Joyner R, Moerner WE. Chromosomal locus tracking with proper accounting of static and dynamic errors. *Phys. Rev. E*, 91(6):062716, 2015.
194. Ashley TT, Andersson SB. Method for simultaneous localization and parameter estimation in particle tracking experiments. *Phys. Rev. E*, 92(5), 2015.
195. Saxton MJ, Jacobson K. Single-particle tracking: applications to membrane dynamics. *Ann. Rev. Biophys. Biomol. Struct.*, 26:373–99, 1997.
196. Miesenböck G. The optogenetic catechism. *Science*, 326:395–99, 2009.
197. Fenno L, Yizhar O, Deisseroth K. The development and application of optogenetics. *Ann. Rev. Neurosci.*, 34:389–412, 2011.
198. Gautier A, Gauron C, Volovitch M, Bensimon D, Jullien L, Vriz S. "How to control proteins with light in living systems". *Nat. Chem. Biol.*, 10:533–41, 2014.

199. Miesenböck G, De Angelis DA, Rothman JE. Visualizing secretion and synaptic transmission with pH-sensitive green fluorescent proteins. *Nature*, 394:192–95, 1998.
200. Miyawaki A. Fluorescence imaging of physiological activity in complex systems using GFP-based probes. *Curr. Opin. Neurobiol.*, 13:591–96, 2003.
201. Akerboom J, Chen TW, Wardill TJ, Tian L, Marvin JS, Mutlu S, Calderon NC, Esposti F, *et al.* Optimization of a GCaMP calcium indicator for neural activity imaging. *J. Neurosci.*, 32:13819–40, 2012.
202. Kralj J, Douglass AD, Hochbaum DR, Maclaurin D, Cohen AE. Optical recording of action potentials in mammalian neurons using a microbial rhodopsin. *Nat. Met.*, 9:90–95, 2012.
203. Cantor CR, Schimmel PR. *Biophysical chemistry, Part III: The behaviour of biological macromolecules.* W.H. Freeman, San Francisco, 1980.
204. Fisher ME. Magnetism in one-dimensional systems—the Heisenberg model for infinite spin. *Am. J. Phys.*, 32:343–46, 1964.
205. Hagerman PJ. Flexibility of DNA. *Ann. Rev. Biophys. Chem.*, 17:265–68, 1988.
206. Fixman M, Kovac J. Polymer conformational statistics III: modified Gaussian models of the stiff chains. *J. Chem. Phys.*, 58:1564–68, 1973.
207. Bustamante C, Marko JF, Siggia ED, Smith S. Entropic elasticity of λ-phage DNA. *Science*, 265:1599–600, 1994.
208. Vologodskii AV. DNA extension under the action of an external force. *Macromolecules*, 27:5623–25, 1994.
209. Marko JF, Siggia E. Statistical mechanics of supercoiled DNA. *Phys. Rev. E*, 52(3):2912–38, 1995.
210. Marko JF, Siggia E. Stretching DNA. *Macromolecules*, 28(26):8759–70, 1995.
211. Bouchiat C, Wang MD, Block SM, Allemand JF, Strick TR, Croquette V. Estimating the persistence length of a worm-like chain molecule from force-extension measurements. *Biophys. J.*, 76:409–13, 1999.
212. Flory P. *Principles of polymer chemistry*. Cornell University Press, Ithaca, 1975.
213. Zimm BH. Dynamics of polymer molecules in dilute solution: viscoelasticity, flow birefringence and dielectric loss. *J. Chem. Phys.*, 24:269–78, 1956.
214. Rouse PE. A theory of the linear viscoelastic properties of dilute solutions of coiling polymers. *J. Chem. Phys.*, 21:1272–79, 1953.
215. Perkins TT, Smith DE, Chu S. Direct observation of tube-like motion of a single polymer chain. *Science*, 264:819–22, 1994.
216. Calladine CR, Drew HR, Luisi BF, Travers AA. *Understanding DNA: the molecule and how it works*. Elsevier Academic Press, 3rd edition, 2004.
217. Kornberg A, Baker T. *DNA replication*. W. H. Freeman, San Francisco, 1992.
218. White JH. Self linking and the Gauss integral in higher dimensions. *Am. J. Math.*, 91:693–728, 1969.
219. Charvin G, Allemand JF, Strick TR, Bensimon D, Croquette V. Twisting DNA: single molecule studies. *Contemp. Phys.*, 45(5):383–403, 2004.
220. Moroz JD, Nelson P. Torsional directed walks, entropic elasticity, and DNA twist stiffness. *Proc. Nat. Acad. Sci. USA*, 94:14418–22, 1998.
221. Bouchiat C, Mézard M. Elasticity theory of a supercoiled DNA molecule. *Phys. Rev. Lett.*, 80:1556–59, 1998.
222. Moroz JD, Nelson P. Entropic elasticity of twist-storing polymers. *Macromolecules*, 31: 6333–47, 1998.
223. Rossetto V, Maggs AC. Writhing geometry of open DNA. *J. Chem. Phys.*, 118:9864–74, 2003.

224. Strick TR, Dessinges MN, Charvin G, Dekker NH, Allemand JF, Bensimon D, Croquette V. Stretching of macromolecules and proteins. *Rep. Prog. Phys.*, 66:1–45, 2003.
225. Van Loenhout MTJ, de Grunt MV, Dekker C. Dynamics of DNA supercoils. *Science*, 338(6103):94–97, 2012.
226. Ganji M, Kim SH, van der Torre J, Abbondanzieri E, Dekker C. Intercalation-based single-molecule fluorescence assay to study DNA supercoil dynamics. *Nano Lett.*, 16(7):4699–707, 2016.
227. Forth S, Deufel C, Sheinin MY, Daniels B, Sethna JP, Wang MD. Abrupt buckling transition observed during the plectoneme formation of individual DNA molecules. *Phys. Rev. Lett.*, 100:148301, 2008.
228. Mosconi F, Allemand JF, Bensimon D, Croquette V. Measurement of the torque on a single stretched and twisted DNA using magnetic tweezers. *Phys. Rev. Lett.*, 102: 078301, 2009.
229. Bryant Z, Stone MD, Gore J, Smith SB, Cozzarelli NR, Bustamante C, *et al.* Structural transitions and elasticity from torque measurements on DNA. *Nature*, 424(6946):338–41, 2003.
230. Selvin PR, Cook DN, Pon NG, Bauer WR, Klein MP, Hearst JE. Torsional rigidity of positively and negatively supercoiled DNA. *Science*, 255:82–85, 1992.
231. Heath PJ, Clendenning JB, Fujimoto BS, Schurr MJ. Effect of bending strain on the torsion elastic constant of DNA. *J. Mol. Biol.*, 260:718–30, 1996.
232. Horowitz DS, Wang JC. Torsional rigidity of DNA and length dependence of the free energy of DNA supercoiling. *J. Mol. Biol.*, 173:75–91, 1984.
233. Shore D, Baldwin RL. Energetics of DNA twisting: II. topoisomer analysis. *J. Mol. Biol.*, 170:983–1007, 1983.
234. Vologodskii AV, Cozzarelli NR. Conformational and thermodynamic properties of supercoiled DNA. *Ann. Rev. Biophys. Biomol. Struct.*, 23:609–43, 1994.
235. Strick TR, Bensimon D, Croquette V. Micro-mechanical measurement of the torsional modulus of DNA. *Genetica, proceedings of NATO ARW on "Structural Biology and Functional Genomics"*, 106:57–62, 1999.
236. Revyakin A, Ebright RH, Strick TR. Promoter unwinding and promoter clearance by RNA polymerase: detection by single-molecule DNA nanomanipulation. *Proc. Nat. Acad. Sci. USA*, 101:4776–80, 2004.
237. Lionnet T, Joubaud S, Lavery R, Bensimon D, Croquette V. Wringing out DNA. *Phys. Rev. Lett.*, 96:178102, 2006.
238. Gore J, Bryant Z, Nöllmann M, Le MU, Cozzarelli NR, Bustamante C. DNA overwinds when stretched. *Nature*, 442:836–39, 2006.
239. Charvin G, Vologodskii A, Bensimon D, Croquette V. Braiding DNA: experiments, simulations, and models. *Biophys. J.*, 88(6):4124–36, 2005.
240. Marko JF. Supercoiled and braided DNA under tension. *Phys. Rev. E*, 55:1758–72, 1997.
241. Vologodskii A, Marko JF. Extension of torsionally stressed DNA by external force. *Biophys. J.*, 73:123–32, 1997.
242. Vologodskii AV, Frank-Kamenetskii MD. Modeling supercoiled DNA. *Meth. Enzymol.*, 211:468–72, 1992.
243. Frank-Kamenetskii MD, Lukashin AV, Anshelevich VV, Vologodskii AV. Torsional and bending rigidity of the double helix from data on small DNA rings. *J. Biomol. Struct. and Dyn.*, 2(5):1005–12, 1985.
244. Metropolis N, Rosenbluth AW, Rosenbluth MN, Teller AH, Teller E. Equation of state calculations by fast computing machines. *J. Chem. Phys.*, 21(6):1087–92, 1953.

245. Frank-Kamenetskii- MD, Vologodskii- AV. Topological aspects of the physics of polymers: the theory and its biophysical applications. *Soviet Physics Uspekhi*, 24(8):679, 1981.
246. Stigter D. Interactions of highly charged colloidal cylinders with applications to double stranded DNA. *Biopolymers*, 16(7):1435–48, 1977.
247. Rybenkov VV, Cozzarelli NR, Vologodskii AV. Probability of DNA knotting and the effective diameter of the DNA double helix. *Proc. Nat. Acad. Sci. USA*, 90:5307–11, 1993.
248. Boles TC, White JH, Cozzarelli MR. Structure of plectonemically supercoiled DNA. *J. Mol. Biol.*, 213:931–51, 1990.
249. Maier B, Bensimon D, Croquette V. Replication by a single DNA-polymerase of a stretched single strand DNA. *Proc. Nat. Acad. Sci. USA*, 97:12002–07, 2000.
250. Wuite GJL, Smith SB, Young M, Keller D, Bustamante C. Single-molecule studies of the effect of template tension on T7 DNA polymerase activity. *Nature*, 404:103–06, 2000.
251. Dessinges MN, Maier B, Zhang Y, Peliti M, Bensimon D, Croquette V. Stretching single stranded DNA, a model polyelectrolyte. *Phys. Rev. Lett.*, 89:248102, 2002.
252. Saleh OA, McIntosh D, Pincus P, Ribeck N. The low-force elasticity of single-stranded DNA. *Phys. Rev. Lett.*, 102:068301, 2009.
253. Zhang Y, Zhou H, Ou-Yang ZC. Stretching single-stranded DNA: interplay of electrostatic, base-pairing and base-stacking interactions. *Biophys. J.*, 81:1133–43, 2001.
254. Montanari A, Mézard M. Hairpin formation and elongation of biomolecules. *Phys. Rev. Lett.*, 86:2178–81, 2001.
255. Bouchiat C. Hartree-Fock model of a self-avoiding flexible polymer. *ENS preprint*, cond-mat/0201445, 2002.
256. Liphardt J, Onoa B, Smith SB, Tinoco Jr I, Bustamante C. Reversible unfolding of single RNA molecules by mechanical force. *Science*, 292:733–77, 2001.
257. Collin D, Ritort F, Jarzynski C, Smith SB, Tinoco Jr I, Bustamante C. Verification of the Crooks fluctuation theorem and recovery of RNA folding free energies. *Nature*, 437(7056): 231–34, 2005.
258. Jarzynski C. Nonequilibrium equality for free energy differences. *Phys. Rev. Lett.*, 78:2690, 1997.
259. Ritort F, Bustamante C, Tinoco Jr I. A two-state kinetic model for the unfolding of single molecules by mechanical force. *Proc. Nat. Acad. Sci. USA*, 99(21):13544–48, 2002.
260. Liphardt J, Dumont S, Smith SB, Tinoco Jr I, Bustamante C. Equilibrium information from nonequilibrium measurements in an experimental test of Jarzynski's equality. *Science*, 296(5574):1832–35, 2002.
261. Allemand JF, Bensimon D, Lavery R, Croquette V. Stretched and overwound DNA form a Pauling-like structure with exposed bases. *Proc. Nat. Acad. Sci. USA*, 95:14152–7, 1998.
262. Pauling L, Corey RB. A proposed structure for the nucleic acids. *Proc. Nat. Acad. Sci. USA*, 39:84–97, 1953.
263. Lebrun A, Lavery R. Unusual DNA conformations. *Curr. Opin. Struct. Biol.*, 7:348–54, 1997.
264. Wilkins MHF, Gosling RG, Seeds WE. Nucleic acid: an extensible molecule? *Nature*, 167:759–60, 1951.
265. Wenner JR, Williams MC, Rouzina I, Bloomfield VA. Salt dependence of the elasticity and overstretching transition of single DNA molecules. *Biophys. J.*, 82(6):3160–69, 2002.
266. Cocco S, Yan J, Leger JF, Chatenay D, Marko JF. Overstretching and force-driven strand separation of double-helix DNA. *Phys. Rev. E*, 70(1, 1):011910, 2004.
267. Shokri L, McCauley MJ, Rouzina L, Williams MC. DNA overstretching in the presence of glyoxal: structural evidence of force-induced DNA melting. *Biophys. J.*, 95(3):1248–55, 2008.

268. Hern-Paik D, Perkins TT. Overstretching DNA at 65 pN does not require peeling from free ends or nicks. *J. Am. Chem. Soc.*, 133(10):3219–21, 2011.
269. Fu H, Chen H, Zhang X, Qu Y, Marko JF, Yan J. Transition dynamics and selection of the distinct S-DNA and strand unpeeling modes of double helix overstretching. *Nucl. Acid. Res.*, 39(8):3473–81, 2011.
270. Zhang X, Chen H, Fu H, Doyle PS, Yan J. Two distinct overstretched DNA structures revealed by single-molecule thermodynamics measurements. *Proc. Nat. Acad. Sci. USA*, 109(21):8103–08, 2012.
271. Zhang X, Chen H, Le S, Rouzina I, Doyle PS, Yan J. Revealing the competition between peeled ssDNA, melting bubbles, and S-DNA during DNA overstretching by single-molecule calorimetry. *Proc. Nat. Acad. Sci. USA*, 110(10):3865–70, 2013.
272. Rouzina I, Bloomfield VA. Force-induced melting of the DNA double helix 1. Thermodynamic analysis. *Biophys. J.*, 80(2):882–93, 2001.
273. Rouzina I, Bloomfield VA. Force-induced melting of the DNA double helix 2. Effect of solution conditions. *Biophys. J.*, 80(2):894–900, 2001.
274. Leger JF, Romano G, Sarkar A, Robert J, Bourdieu L, Chatenay D, Marko JF. Structural transitions of a twisted and stretched DNA molecule. *Phys. Rev. Lett.*, 83:1066–69, 1999.
275. Sarkar A, Leger JF, Chatenay D, Marko JF. Structural transitions in DNA driven by external force and torque. *Phys. Rev. E.*, 63:051903–1–051903–10, 2001.
276. Viovy JL, Heller C, Caron F, Cluzel P, Chatenay D. Ultrafast sequencing of DNA by mechanical opening of the double helix: a theoretical investigation. *C.R. Acad. Sci. Paris*, 317:795–800, 1994.
277. Thompson RE, Siggia ED. Physical limits on the mechanical measurement of the secondary structure of bio-molecules. *Europhys. Lett.*, 31:335–40, 1995.
278. Clausen-Schaumann H, Rief M, Gaub HE. Sequence-dependent mechanics of single DNA molecules. *Nat. Struct. Biol.*, 6:346–49, 1999.
279. Bell GI. Models for the specific adhesion of cells to cells. *Science*, 200:618–27, 1978.
280. Evans E, Ritchie K. Dynamic strength of molecular adhesion bonds. *Biophys. J.*, 72:1541–55, 1997.
281. Hegner M, Smith SB, Bustamante C. Polymerization and mechanical properties of single RecA–DNA filaments. *Proc. Nat. Acad. Sci. USA*, 96:10109–14, 1999.
282. Ding F, Manosas M, Spiering MM, Benkovic SJ, Bensimon D, Allemand JF, Croquette V. Single-molecule sequencing by magnetic trap manipulation of a DNA hairpin. *Nat. Meth.*, 9:367–72, 2012.
283. Galletto R, Amitani I, Baskin RJ, Kowalczykowski SC. Direct observation of individual RecA filaments assembling on single DNA molecules. *Nature*, 443:875–78, 2006.
284. Van Mameren J, Gross P, Farge G, Hooijman P, Modesti M, Falkenberg M, Wuite GJL, Peterman EJG. Unraveling the structure of dna during overstretching by using multicolor, single-molecule fluorescence imaging. *Proc. Nat. Acad. Sci. USA*, 106:18231–36, 2009.
285. Dufrêne YF, Evans E, Engel A, Helenius J, Gaub HE, Müller DJ. Five challenges to bringing single-molecule force spectroscopy into living cells. *Nat. Meth.*, 8:123–27, 2011.
286. Li GW, Xie XS. Central dogma at the single-molecule level in living cells. *Nature*, 475:308–15, 2011.
287. Furth JJ, Hurwitz J, Krug R, Alexander M. The incorporation of adenylic and cytidylic acids into ribonucleic acid. *J. Biol. Chem.*, 236:3317–22, 1961.
288. Weiss SB, Nakamoto T. Net synthesis of ribonucleic acid with a microbial enzyme requiring deoxyribonucleic acid and four ribonucleoside triphosphates. *J. Biol. Chem.*, 236:18–20, 1961.

289. Lehman IR, Bessman MJ, Simms ES, Kornberg A. Enzymatic synthesis of deoxyribonucleic acid. I. preparation of substrates and partial purification of an enzyme from Escherichia coli. *J. Biol. Chem.*, 233:163–70, 1958.
290. Steitz TA. DNA polymerases: Structural diversity and common mechanisms. *J. Biol. Chem*, 274:17395–98, 1999.
291. Joyce CM, Steitz TA. Polymerase structures and function—variations on a theme. *J. Bacteriol.*, 177:6321–29, 1995.
292. Baker TA, Bell SP. Polymerases and the replisome: machines within machines. *Cell*, 92: 295–305, 1998.
293. Manosas M, Spiering MM, Ding F, Bensimon D, Allemand JF, Benkovic SJ, Croquette V. Mechanism of strand displacement synthesis by DNA replicative polymerases. *Nucl. Acid. Res.*, 40(13):6174–86, 2012.
294. Manosas M, Spiering MM, Ding F, Croquette V, Benkovic SJ. Collaborative coupling between polymerase and helicase for leading-strand synthesis. *Nucl. Acid. Res.*, 40(13):6187–88, 2012.
295. Harris TD, Buzby PR, Babcock H, Beer E, Bowers J, Braslavsky I, Causey M, Colonell J, *et al.* Single-molecule DNA sequencing of a viral genome. *Science*, 320(5872):106–09, 2008.
296. Eid J, Fehr A, Gray J, Luong K, Lyle J, Otto G, Peluso P, Rank D, *et al.* Real-time DNA sequencing from single polymerase molecules. *Science*, 323(5910):133–38, 2009.
297. Laszlo AH, Derrington IM, Ross BC, Brinkerhoff H, Adey A, Nova IC, Craig JM, Langford KW, Samson JM, Daza R, Doering K, Shendure J, Gundlach JH. Decoding long nanopore sequencing reads of natural DNA. *Nat. Biotech.*, 32(8):829–33, 2014.
298. Branton D, Deamer DW, Marziali A, Bayley H, Benner SA, Butler T, Di Ventra M, Garaj S, *et al.* The potential and challenges of nanopore sequencing. *Nat. Biotech.*, 26:1146–53, 2008.
299. Kasianowicz JJ, Brandin E, Branton D, Deamer DW. Characterization of individual polynucleotide molecules using a membrane channel. *Proc. Nat. Acad. Sci. USA*, 93(24):13770–73, 1996.
300. Derrington IM, Butler TZ, Collins MD, Manrao E, Pavlenok M, Niederweis M, Gundlach JH. Nanopore DNA sequencing with MspA. *Proc. Nat. Acad. Sci. USA*, 107(37):16060–65, 2010.
301. Alberts B, Bray D, Lewis J, Raff M, Roberts K, Watson JD. *Molecular biology of the cell.* Garland Publishing Inc., New York, 3rd edition, 1994.
302. Galas DJ, Schmitz A. DNAse footprinting: a simple method for the detection of protein-DNA binding specificity. *Nucl. Acid. Res.*, 5(9):3157–70, 1978.
303. Schmitz A, Galas DJ. The interaction of RNA polymerase and lac repressor with the lac control region. *Nucl. Acid. Res*, 6(1):111–37, 1979.
304. Siebenlist U. RNA polymerase unwinds an 11-base pair segment of a phage T7 promoter. *Nature*, 279(5714):651–52, 1979.
305. Kirkegaard K, Buc H, Spassky A, Wang JC. Mapping of single-stranded regions in duplex DNA at the sequence level: single-strand-specific cytosine methylation in RNA polymerase-promoter complexes. *Proc. Nat. Acad. Sci. USA*, 80(9):2544–48, 1983.
306. Amouyal M, Buc H. Topological unwinding of strong and weak promoters by RNA polymerase. A comparison between the lac wild-type and the UV5 sites of Escherichia coli. *J. Mol. Biol.*, 195(4):795–808, 1987.
307. McClure WR, Cech CL, Johnston DE. A steady state assay for the RNA polymerase initiation reaction. *J. Biol. Chem.*, 253(24):8941–48, 1978.
308. Carpousis AJ, Gralla JD. Cycling of ribonucleic acid polymerase to produce oligonucleotides during initiation in vitro at the lac UV5 promoter. *Biochemistry*, 19(14):3245–53, 1980.

309. Gralla JD, Carpousis AJ, Stefano JE. Productive and abortive initiation of transcription in vitro at the lac UV5 promoter. *Biochemistry*, 19(25):5864–9, 1980.
310. Krummel B, Chamberlin MJ. RNA chain initiation by Escherichia coli RNA polymerase. Structural transitions of the enzyme in early ternary complexes. *Biochemistry*, 28(19): 7829–42, 1989.
311. Hsu LM. Promoter clearance and escape in prokaryotes. *Biochim. Biophys. Acta*, 1577(2):191–207, 2002.
312. Straney DC, Crothers DM. A stressed intermediate in the formation of stably initiated RNA chains at the Escherichia coli lac UV5 promoter. *J. Mol. Biol.*, 193(2):267–78, 1987.
313. Rhodes G, Chamberlin MJ. Ribonucleic acid chain elongation by Escherichia coli ribonucleic acid polymerase. I. Isolation of ternary complexes and the kinetics of elongation. *J. Biol. Chem.*, 249(20):6675–83, 1974.
314. Solage A, Cedar H. The kinetics of E. coli RNA polymerase. *Nucl. Acid. Res.*, 3(9):2207–22, 1976.
315. Wilson KS, von Hippel PH. Stability of Escherichia coli transcription complexes near an intrinsic terminator. *J. Mol. Biol.*, 244(1):36–51, 1994.
316. Wilson KS, von Hippel PH. Transcription termination at intrinsic terminators: the role of the RNA hairpin. *Proc. Nat. Acad. Sci. USA*, 92(19):8793–97, 1995.
317. Gusarov I, Nudler E. The mechanism of intrinsic transcription termination. *Mol. Cell*, 3:495–504, 1999.
318. Santangelo TJ, Roberts JW. Forward translocation is the natural pathway of RNA release at an intrinsic terminator. *Mol. Cell*, 14(1):117–26, 2004.
319. Ebright RH. RNA polymerase: Structural similarities between bacterial RNA polymerase and eukaryotic RNA polymerase II. *J. Mol. Biol.*, 304(5):687–98, 2000.
320. Minakhin L, Bhagat S, Brunning A, Campbell EA, Darst SA, Ebright RH, Severinov K. Bacterial RNA polymerase subunit omega and eukaryotic RNA polymerase subunit RPB6 are sequence, structural, and functional homologs and promote RNA polymerase assembly. *Proc. Nat. Acad. Sci. USA*, 98(3):892–97, 2001.
321. Severinova E, Severinov K, Feny D, Marr M, Brody EN, Roberts JW, Chait BT, Darst SA. Domain organization of the Escherichia coli RNA polymerase sigma 70 subunit. *J. Mol. Biol.*, 263(5):637–47, 1996.
322. Landick R, Vaughn V, Lau ET, VanBogelen RA, Erickson JW, Neidhardt FC. Nucleotide sequence of the heat shock regulatory gene of E. coli suggests its protein product may be a transcription factor. *Cell*, 38(1):175–82, 1984.
323. Grossman AD, Erickson JW, Gross CA. The htpR gene product of E. coli is a sigma factor for heat-shock promoters. *Cell*, 38(2):383–90, 1984.
324. Hirschman J, Wong PK, Sei K, Keener J, Kustu S Products of nitrogen regulatory genes ntrA and ntrC of enteric bacteria activate glnA transcription in vitro: evidence that the ntrA product is a sigma factor. *Proc. Nat. Acad. Sci. USA*, 82(22):7525–29, 1985.
325. Chen J, Darst SA, Thirumalai D. Promoter melting triggered by bacterial RNA polymerase occurs in three steps. *Proc. Nat. Acad. Sci. USA*, 107(28):12523–28, 2010.
326. Carpousis AJ, Gralla JD. Interaction of RNA polymerase with lacUV5 promoter DNA during mRNA initiation and elongation. Footprinting, methylation, and rifampicin-sensitivity changes accompanying transcription initiation. *J. Mol. Biol.*, 183(2):165–77, 1985.
327. Revyakin A, Liu CY, Ebright RH, Strick TR. Abortive initiation and productive initiation by RNA polymerase involve DNA scrunching. *Science*, 314:1139–43, 2006.
328. Kapanidis AN, Margeat E, Ho SO, Kortkhonjia E, Weiss S, Ebright RH. Initial transcription by RNA polymerase proceeds through a DNA-scrunching mechanism. *Science*, 314:1144–47, 2006.

329. Murakami KS, Darst SA. Bacterial RNA polymerases: the wholo story. *Curr. Opin. Struct. Biol.*, 13(1):31–39, 2003.
330. Kassavetis GA, Chamberlin MJ. Pausing and termination of transcription within the early region of bacteriophage T7 DNA in vitro. *J. Biol. Chem.*, 256(6):2777–86, 1981.
331. Kingston RE, Nierman WC, Chamberlin MJ. A direct effect of guanosine tetraphosphate on pausing of Escherichia coli RNA polymerase during RNA chain elongation. *J. Biol. Chem.*, 256(6):2787–97, 1981.
332. Kashlev M, Komissarova N. Transcription termination: primary intermediates and secondary adducts. *J. Biol. Chem.*, 277(17):14501–08, 2002.
333. Yin H, Artsimovitch I, Landick R, Gelles J. Nonequilibrium mechanism of transcription termination from observations of single RNA polymerase molecules. *Proc. Nat. Acad. Sci. USA*, 96(23):13124–29, 1999.
334. Roberts JW. Termination factor for RNA synthesis. *Nature*, 224(5225):1168–74, 1969.
335. Harada Y, Funatsu T, Murakami K, Nonoyama Y, Ishihama A, Yanagida T. Single-molecule imaging of RNA Polymerase-DNA interactions in real time. *Biophys. J.*, 76:709–15, 1999.
336. Friedman LJ, Gelles J. Mechanism of transcription initiation at an activator-dependent promoter defined by single-molecule observation. *Cell*, 148(4):679–89, 2012.
337. Wang F, Redding S, Finkelstein IJ, Gorman J, Reichman DR, Greene EC. The promoter-search mechanism of Escherichia coli RNA polymerase is dominated by three-dimensional diffusion. *Nat. Struct. Mol. Biol.*, 2012.
338. Sakata-Sogawa K, Shimamoto N. RNA polymerase can track a DNA groove during promoter search. *Proc. Nat. Acad. Sci. USA*, 101:14731–35, 2004.
339. Opalka N, Brown J, Lane WJ, Twist KA, Landick R, Asturias FJ, Darst SA. Complete structural model of Escherichia coli RNA polymerase from a hybrid approach. *PLoS Bio.*, 8(9): e1000483, 2010.
340. Rivetti C, Guthold M, Bustamante C. Wrapping of DNA around the E.coli RNA polymerase open promoter complex. *EMBO J.*, 18(16):4464–75, 1999.
341. Greenleaf WJ, Woodside MT, Abbondanzieri EA, Block SM. Passive all-optical force clamp for high-resolution laser trapping. *Phys. Rev. Lett.*, 95:208102, 2005.
342. Schafer DA, Gelles J, Sheetz MP, Landick R. Transcription by single molecules of RNA polymerase observed by light microscopy. *Nature*, 352:444–48, 1991.
343. Yin H, Landick R, Gelles J. Tethered particle motion method for studying transcript elongation by a single RNA polymerase molecule. *Biophys. J.*, 67:2468–78, 1994.
344. Tolic-Norrelykke SF, Engh AM, Landick R, Gelles J. Diversity in the rates of transcript elongation by single RNA polymerase molecules. *J. Biol. Chem.*, 279:3292–99, 2004.
345. Abbondanzieri EA, Greenleaf WJ, Shaevitz JW, Landick R, Block SM. Direct observation of base-pair stepping by RNA polymerase. *Nature*, 438:460–65, 2005.
346. Wang MD, Schnitzer MJ, Yin H, Landick R, Gelles J, Block S. Force and velocity measured for single molecules of RNA polymerase. *Science*, 282:902–07, 1998.
347. Neuman KC, Abbondanzieri EA, Landick R, Gelles J, Block SM. Ubiquitous transcriptional pausing is independent of RNA polymerase backtracking. *Cell*, 115:437–47, 2003.
348. Adelman K, Yuzenkova J, La Porta A, Zenkin N, Lee J, Lis JT, Borukhov S, Wang MD, Severinov K. Molecular mechanism of transcription inhibition by peptide antibiotic Microcin J25. *Mol. Cell*, 14:753–62, 2004.
349. Davenport RJ, Wuite GJL, Landick R, Bustamante C. Single-molecule study of transcriptional pausing and arrest by E. coli RNA polymerase. *Science*, 287(5462):2497–500, 2000.

350. Forde NR, Izhaky D, Woodcock GR, Wuite GJL, Bustamante C. Using mechanical force to probe the mechanism of pausing and arrest during continuous elongation by Escherichia coli RNA polymerase. *Proc. Nat. Acad. Sci. USA*, 99(18):11682–87, 2002.
351. Bai L, Shundrovsky A, Wang MD. Sequence-dependent kinetic model for transcription elongation by RNA polymerase. *J. Mol. Biol.*, 344:335–49, 2004.
352. Shundrovsky A, Santangelo TJ, Roberts JW, Wang MD. A single-molecule technique to study sequence-dependent transcription pausing. *Biophys. J.*, 87(6):3945–53, 2004.
353. Herbert KM, La Porta A, Wong BJ, Mooney RA, Neuman KC, Landick R, Block SM. Sequence-resolved detection of pausing by single RNA polymerase molecules. *Cell*, 125:1083–94, 2006.
354. Harada Y, Ohara O, Takatsuki A, Itoh H, Shimamoto N, Kinosita Jr K. Direct observation of DNA rotation during transcription by Escherichia coli RNA polymerase. *Nature*, 409:113–15, 2001.
355. Ma J, Bai L, Wang MD. Transcription under torsion. *Science*, 340:1580–83, 2013.
356. Larson DR, Singer RH, Zenklusen D. A single molecule view of gene expression. *Trends Cell Biol.*, 19(11):630–37, 2009.
357. Chubb JR, Trcek T, Shenoy SM, Singer RH. Transcriptional pulsing of a developmental gene. *Curr. Biol.*, 16(10):1018–25, 2006.
358. Darzacq X, Shav-Tal Y, de Turris V, Brody Y, Shenoy SM, Phair RD, Singer RH. In vivo dynamics of RNA polymerase II transcription. *Nat. Struct. & Mol. Biol.*, 14(9):796–806, 2007.
359. Van Oijen AM, Loparo JJ. Single molecule studies of the replisome. *Ann. Rev. Biophys.*, 39:429–48, 2010.
360. Howan K, Smith AJ, Westblade LF, Joly N, Grange W, Zorman S, Darst SA, Savery NJ, Strick TR. Initiation of transcription-coupled repair characterized at single-molecule resolution. *Nature*, 490(7420):431–34, 2012.
361. Delagoutte E, von Hippel PH. Helicase mechanisms and the coupling of helicases within macromolecular machines. Part 1: Structures and properties of isolated helicases. *Quart. Rev. Biophys.*, 35:431–78, 2002.
362. Delagoutte E, von Hippel PH. Helicase mechanisms and the coupling of helicases within macromolecular machines. Part II: Integration of helicases into cellular processes. *Quart. Rev. Biophys.*, 36:1–69, 2003.
363. Lohman TM, Bjornson KP. Mechanisms of helicase-catalysed unwinding. *Ann. Rev. Biochem.*, 65:169–214, 1996.
364. West SC. DNA helicases: new breeds of translocating motors and molecular pumps. *Cell*, 86:177–80, 1996.
365. Soultanas P, Wigley DB. Unwinding the 'Gordian knot' of helicase action. *Trends in Biochem. Sci.*, 26:47–54, 2001.
366. JD Watson, Baker TA, Bell SP, Gann A, Levine M, Losick R. *Molecular Biology of the Gene*. CSHL Press, 2004.
367. Patel SS, Picha KM. Structure and function of hexameric helicases. *Ann. Rev. Biochem.*, 69:651, 2000.
368. Henn A, Medalia O, Shi S-P, Steinberg M, Franceschi F, Sagi I. Visualization of unwinding activity of duplex RNA by DbpA, a DEAD box helicase, at single-molecule resolution by atomic force microscopy. *Proc. Nat. Acad. Sci. USA*, 98:5007, 2001.
369. Spies M. DNA helicases and DNA motor proteins. *Advances in experimental medicine and biology*, 973, 2012.

370. Von Hippel PH, Delagoutte E. A general model for nucleic acid helicases and their "coupling" within macromolecular machines. *Cell*, 104:177–90, 2001.
371. Marians KJ. Crawling and wiggling on DNA: structural insights to the mechanism of DNA unwinding by helicases. *Structure*, 8:R227–35, 2000.
372. Hodeib S, Raj S, Manosas M, Zhang W, Bagchi D, Ducos B, Fiorini F, Kanaan J, Le Hir H, Allemand JF, Bensimon D, Croquette V. A mechanistic study of helicases with magnetic traps. *Prot. Sci.*, 26:1314–36, 2017.
373. Yasuda R, Noji H, Kinosita K, Yoshida M. F-1-ATPase is a highly efficient molecular motor that rotates with discrete 120 degree steps. *Cell*, 93(7):1117–24, 1998.
374. Roman LJ, Kowalczykowski SC. Characterization of the helicase activity of the Escherichia coli recBCD enzyme using a novel helicase assay. *Biochemistry*, 28:2863–73, 1989.
375. Roman LJ, Eggleston AK, Kowalczykowski SC. Processivity of the DNA helicase activity of Escherichia coli RecBCD enzyme. *J. Biol. Chem.*, 267:4207–14, 1992.
376. Roman LJ, Kowalczykowski SC. Characterization of the adenosinetriphosphatase activity of the Escherichia coli RecBCD enzyme: relationship of ATP hydrolysis to the unwinding of duplex DNA. *Biochemistry*, 28:2873–81, 1989.
377. Bianco PR, Brewer LR, Corzett M, Balhorn R, Yeh Y, Kowalczykowski SC, Baskin RJ. Processive translocation and DNA unwinding by individual RecBCD enzyme molecules. *Nature*, 409:374–78, 2001.
378. Handa N, Bianco PR, Baskin RJ, Kowalczykowski SC. Direct visualization of RecBCD movement reveals cotranslocation of the RecD motor after chi recognition. *Mol. Cell*, 17:745–50, 2005.
379. Spies M, Bianco PR, Dillingham MS, Handa N, Baskin RJ, Kowalczykowski SC. A molecular throttle: the recombination hotspot chi controls DNA translocation by the RecBCD helicase. *Cell*, 114:647–54, 2003.
380. Lionnet T, Spiering MM, Benkovic SC, Bensimon D, Croquette V. Real-time observation of T4 gp41 helicase reveals a passive unwinding mechanism. *Proc. Nat. Acad. Sci. USA*, 104:19790–95, 2007.
381. Manosas M, Xi X, Bensimon D, Croquette V. Active and passive mechanisms of helicases. *Nucl. Acid. Res.*, 38(16):5518–26, 2010.
382. Matson SW. Escherichia coli helicase II (urvD gene product) translocates unidirectionally in a 3′ to 5′ direction. *J. Biol. Chem.*, 261:10169–75, 1986.
383. Runyon GT, Lohman TM. Escherichia coli helicase II (uvrD) protein can completely unwind fully duplex linear and nicked circular DNA. *J. Biol. Chem.*, 264:17502–12, 1989.
384. Ali JA, Lohman TM. Kinetic measurement of the step size of DNA unwinding by Escherichia coli UvrD helicase. *Science*, 275(5298):377–80, 1997.
385. Neuman KC, Saleh OA, Lionnet T, Lia G, Allemand JF, Bensimon D, Croquette V. Statistical determination of the step-size of molecular motors. *J.Phys.: Condens. Matter*, 17:3811–20, 2005.
386. Dessinges MN, Lionnet T, Xi X, Bensimon D, Croquette V. Single-molecule assay reveals strand switching and enhanced processivity of UvrD. *Proc. Nat. Acad. Sci. USA*, 101:6439–44, 2004.
387. Wong I, Chao KL, Bujalowski W, Lohman TM. DNA-induced dimerization of the Escherichia coli Rep helicase. *J. Biol. Chem.*, 267:7596–610, 1992.
388. Yarranton GT, Gefter ML. Enzyme-catalyzed DNA unwinding: studies on Escherichia coli rep protein. *Proc. Nat. Acad. Sci. USA*, 76:1658, 1979.

389. Velankar SS, Soultanas P, Dillingham MS, Subramanya HS, Wigley DB. Crystal structures of complexes of PcrA DNA helicase with a DNA substrate indicate an inchworm mechanism. *Cell*, 97:75, 1999.
390. Korolev S, Hsieh J, Gauss GH, Lohman TM, Waksman G. Major domain swiveling revealed by the crystal structures of complexes of E. coli Rep helicase bound to single-stranded DNA and ADP. *Cell*, 90:635, 1997.
391. Ha TJ, Rasnik I, Cheng W, Babcock HP, Gauss GH, *et al.* Initiation and re-initiation of DNA unwinding by the *Escherichia coli* Rep helicase. *Nature*, 419:638–641, 2002.
392. Rasnik I, Myong S, Cheng W, Lohman TM, Ha T. DNA-binding orientation and domain conformation of the E. coli rep helicase monomer bound to a partial duplex junction: single-molecule studies of fluorescently labeled enzymes. *J. Mol. Biol.*, 336: 395–408, 2004.
393. Wang JC. Interaction between DNA and an *Escherichia coli* protein ω. *J. Mol. Biol.*, 55:523–33, 1971.
394. Liu LF, Liu CC, Alberts BM. Type II DNA topoisomerases: enzymes that can unknot a topologically knotted DNA molecule via a reversible double-stranded break. *Cell*, 19:697–707, 1980.
395. Hsieh T. Knotting of the circular duplex DNA by type II DNA topoisomerase from Drosophila melanogaster. *J. Biol. Chem.*, 258:8413–20, 1983.
396. Roca J, Wang JC. The capture of a DNA double helix by an ATP-dependent protein clamp: a key step in DNA transport by type II DNA topoisomerase. *Cell*, 71: 833–40, 1992.
397. Roca J, Berger JM, Harrison SC, Wang JC. DNA transport by a type II topoisomerase: direct evidence for a two-gate mechanism. *Proc. Nat. Acad. Sci. USA*, 93:4057–62, 1996.
398. Rybenkov VV, Ullsperger C, Vologodskii AV, Cozzarelli NR. Simplification of DNA topology below equilibrium values by type II topoisomerases. *Science*, 277:690–93, 1997.
399. Drolet M, Bi X, Liu LF. Hypernegative supercoiling of the DNA template during transcription elongation in vitro. *J. Biol. Chem.*, 269:2068–74, 1994.
400. Massé E. Relaxation of transcription-driven negative supercoils is an essential function of *E. coli* topoisomerase I. *J. Biol. Chem.*, 274:16654–58, 1999.
401. Massé E. Escherichia coli DNA topoisomerase I inhibits R-loop formation by relaxing transcription-induced negative supercoiling. *J. Biol. Chem.*, 274:16659–64, 1999.
402. Roca J, Wang JC. DNA transport by a type II DNA topoisomerase: evidence in favor of a two-gate model. *Cell*, 77:609–16, 1994.
403. Berger JM, Gamblin SJ, Harrison SC, Wang JC. Structure and mechanism of DNA topoisomerase II. *Nature*, 379:225–32, 1996.
404. Wang JC. Moving one DNA double helix through another by type II DNA topoisomerase: the story of a simple molecular machine. *Quart. Rev. Biophys.*, 31:107–44, 1998.
405. Lindsley JE, Wang JC. On the coupling between ATP usage and DNA transport by yeast DNA topoisomerase II. *J. Biol. Chem*, 268:8096–104, 1993.
406. Harkins TT, Lindsley JE. Pre-steady-state analysis of ATP hydrolysis by *Saccharomyces cerevisiae* DNA topoisomerase II. 1. A DNA-dependent burst in ATP hydrolysis. *Biochemistry*, 37:7292–98, 1998.
407. Harkins TT, Lewis TJ, Lindsley JE. Pre-steady-state analysis of ATP hydrolysis by *Saccharomyces cerevisiae* DNA topoisomerase II. 2. Kinetic mechanism for the sequential hydrolysis of two ATP. *Biochemistry*, 37:7299–312, 1998.

408. Zechiedrich EL, Khodursky AB, Bachellier S, Schneider R, Chen D, Lilley DM, Cozzarelli NR. Roles of topoisomerases in maintaining steady-state DNA supercoiling in Escherichia coli. *J. Biol. Chem.*, 275(11):8103–13, 2000.
409. Stivers JT. *Vaccinia* DNA Topoisomerase I: evidence supporting a free rotation mechanism for DNA supercoil relaxation. *Biochemistry*, 36:5212–22, 1997.
410. Levine C, Hiasa H, Marians KJ. DNA gyrase and topoisomerase IV: Biochemical activities, physiological roles during chromosome replication, and drug sensitivities. *Biochim. Biophys. Acta*, 1400:29–43, 1998.
411. Espeli O, Levine C, Hassing H, Marians KJ. Temporal regulation of topoisomerase IV activity in *E. coli. Mol. Cell*, 11(1):189–201, 2003.
412. Kirkegaard K, Wang JC. Bacterial DNA topoisomerase I can relax positively supercoiled DNA containing a single-stranded loop. *J. Mol. Biol.*, 185:625–37, 1985.
413. Dekker NH, Rybenkov VV, Duguet M, Crisona NJ, Cozzarelli NR, Bensimon D, Croquette V. The mechanism of type IA topoisomerases. *Proc. Nat. Acad. Sci. USA*, 99(19):12126–31, 2002.
414. Chan MK, Fadzil NA, Chew AL, Khoo BY. New molecular biologist perspective and insight: DNA topoisomerases production by recombinant DNA technology for medical laboratory application and pharmaceutical industry, *Electronic Journal of Biotechnology*, 16(6), 2013.
415. Koster DA, Croquette V, Dekker C, Shuman S, Dekker NH. Friction and torque govern the relaxation of DNA supercoils by eukaryotic topoisomerase IB. *Nature*, 434(7033):671–74, 2005.
416. Koster DA, Palle K, Bot ESM, Bjornsti M-A, Dekker NH. Antitumour drugs impede DNA uncoiling by topoisomerase I. *Nature*, 448(7150):213–17, 2007.
417. Charvin G, Strick TR, Bensimon D, Croquette V. Topoisomerase IV bends and overtwists DNA upon binding. *Biophys. J.*, 89(1):384–92, 2005.
418. Charvin G, Strick TR, Bensimon D, Croquette V. Tracking topoisomerase activity at the single-molecule level. *Ann. Rev. Biophys. Biomol. Struct.*, 34:201–19, 2005.
419. Peng H, Marians KJ. Escherichia coli Topoisomerase IV. Purification, characterization, sunit structure and subunit interactions. *J. Biol. Chem.*, 268(32):24481–90, 1993.
420. Hardin AH, Sarkar SK, Seol Y, Liou GF, Osheroff N, Neuman KC. Direct measurement of DNA bending by type IIA topoisomerases: implications for non-equilibrium topology simplification. *Nucl. Acid. Res.*, 39(13):5729–43, 2011.
421. Strick TR, Croquette V, Bensimon D. Single-molecule analysis of DNA uncoiling by a type II topoisomerase. *Nature*, 404:901–04, 2000.
422. Crisona N, Strick TR, Bensimon D, Croquette V, Cozzarelli N. Preferential relaxation of positively supercoiled DNA by *E. coli* topoisomerase VI in single-molecule and ensemble measurements. *Genes Dev.*, 14:2881–92, 2000.
423. Charvin G, Bensimon D, Croquette V. Single-molecule study of DNA unlinking by eukaryotic and prokaryotic type-II topoisomerases. *Proc. Nat. Acad. Sci. USA*, 100(17):9820–25, 2003.
424. Adams DE, Shekhtman EM, Zechiedrich EL, Schmid MB, Cozzarelli NR. The role of topoisomerase IV in partitioning bacterial replicons and the structure of catenated intermediates in DNA replication. *Cell*, 71:277–88, 1992.
425. Zechiedrich EL, Cozzarelli NR. Roles of topoisomeraes IV and DNA gyrase in DNA unlinking during replication in *Escherichia coli. Genes Dev.*, 9:2859–69, 1995.
426. Hiasa H, Marians KJ. Two distinct modes of strand unlinking during θ-type DNA replication. *J. Biol. Chem.*, 271:21529–35, 1996.

427. Postow L, Crisona N, Peter BJ, Hardy CD, Cozzarelli NR. Topological challenges to DNA replication: conformations at the fork. *Proc. Nat. Acad. Sci. USA*, 98:8219–26, 2001.
428. Stone MD, Bryant Z, Crisona NJ, Smith SB, Vologodskii A, Bustamante C, Cozzarelli NR. Chirality sensing by Escherichia coli topoisomerase IV and the mechanism of type II topoisomerases. *Proc. Nat. Acad. Sci. USA*, 100(15):8654–59, 2003.
429. Neuman KC, Charvin G, Bensimon D, Croquette V. Mechanisms of chiral discrimination by topoisomerase IV. *Proc. Nat. Acad. Sci. USA*, 106(17):6986–91, 2009.
430. Elowitz MB, Surette MG, Wolf P-E, Stock JB, Leibler S. Protein mobility in the cytoplasm of Escherichia coli. *J. Bacteriol.*, 181:197–203, 1999.

Index